VOLUME NINETY THREE

ADVANCES IN
VIRUS RESEARCH

VOLUME NINETY THREE

ADVANCES IN
VIRUS RESEARCH

Edited by

MARGARET KIELIAN
Albert Einstein College of Medicine,
Bronx, New York, USA

KARL MARAMOROSCH
Rutgers University, New Brunswick,
New Jersey, USA

THOMAS C. METTENLEITER
Friedrich-Loeffler-Institut,
Federal Research Institute for Animal Health,
Greifswald – Insel Riems, Germany

AMSTERDAM • BOSTON • HEIDELBERG • LONDON
NEW YORK • OXFORD • PARIS • SAN DIEGO
SAN FRANCISCO • SINGAPORE • SYDNEY • TOKYO
Academic Press is an imprint of Elsevier

ELSEVIER

Academic Press is an imprint of Elsevier
225 Wyman Street, Waltham, MA 02451, USA
525 B Street, Suite 1800, San Diego, CA 92101-4495, USA
125 London Wall, London, EC2Y 5AS, UK
The Boulevard, Langford Lane, Kidlington, Oxford OX5 1GB, UK

First edition 2015

Notices
Knowledge and best practice in this field are constantly changing. As new research and experience broaden our understanding, changes in research methods, professional practices, or medical treatment may become necessary.

Practitioners and researchers must always rely on their own experience and knowledge in evaluating and using any information, methods, compounds, or experiments described herein. In using such information or methods they should be mindful of their own safety and the safety of others, including parties for whom they have a professional responsibility.

To the fullest extent of the law, neither the Publisher nor the authors, contributors, or editors, assume any liability for any injury and/or damage to persons or property as a matter of products liability, negligence or otherwise, or from any use or operation of any methods, products, instructions, or ideas contained in the material herein.

ISBN: 978-0-12-802179-8
ISSN: 0065-3527

For information on all Academic Press publications
visit our website at store.elsevier.com

Working together
to grow libraries in
developing countries

www.elsevier.com • www.bookaid.org

CONTENTS

CONTRIBUTORS

Benjamas Aiamkitsumrit
Department of Microbiology and Immunology, and Center for Molecular Virology and Translational Neuroscience, Institute for Molecular Medicine and Infectious Disease, Drexel University College of Medicine, Philadelphia, Pennsylvania, USA

Maxime Boutier
Immunology-Vaccinology (B43b), Department of Infectious and Parasitic Diseases, Fundamental and Applied Research for Animals & Health (FARAH), Faculty of Veterinary Medicine, University of Liège, Liège, Belgium

John P. Carr
Department of Plant Sciences, University of Cambridge, Cambridge, United Kingdom

Andrew J. Davison
MRC-University of Glasgow Centre for Virus Research, Glasgow, United Kingdom

Jagger J.W. Harvey
Biosciences eastern and central Africa-International Livestock Research Institute (BecA-ILRI) Hub, Nairobi, Kenya, and Queensland Alliance for Agriculture and Food Innovation, The University of Queensland, Brisbane, Queensland, Australia

Joanna Jazowiecka-Rakus
Immunology-Vaccinology (B43b), Department of Infectious and Parasitic Diseases, Fundamental and Applied Research for Animals & Health (FARAH), Faculty of Veterinary Medicine, University of Liège, Liège, Belgium

Gregor Meyers
Institut für Immunologie, Friedrich-Loeffler-Institut, Federal Research Institute for Animal Health, Greifswald-Insel Riems, Germany

Neena Mitter
Queensland Alliance for Agriculture and Food Innovation, The University of Queensland, Brisbane, Queensland, Australia

Léa Morvan
Immunology-Vaccinology (B43b), Department of Infectious and Parasitic Diseases, Fundamental and Applied Research for Animals & Health (FARAH), Faculty of Veterinary Medicine, University of Liège, Liège, Belgium

Gerardine Mukeshimana
Biosciences eastern and central Africa-International Livestock Research Institute (BecA-ILRI) Hub, Nairobi, Kenya, and Ministry of Agriculture and Animal Resources, Kigali, Rwanda

Michael R. Nonnemacher
Department of Microbiology and Immunology, and Center for Molecular Virology and Translational Neuroscience, Institute for Molecular Medicine and Infectious Disease, Drexel University College of Medicine, Philadelphia, Pennsylvania, USA

Ma. Michelle D. Peñaranda
Immunology-Vaccinology (B43b), Department of Infectious and Parasitic Diseases, Fundamental and Applied Research for Animals & Health (FARAH), Faculty of Veterinary Medicine, University of Liège, Liège, Belgium

Vanessa Pirrone
Department of Microbiology and Immunology, and Center for Molecular Virology and Translational Neuroscience, Institute for Molecular Medicine and Infectious Disease, Drexel University College of Medicine, Philadelphia, Pennsylvania, USA

Krzysztof Rakus
Immunology-Vaccinology (B43b), Department of Infectious and Parasitic Diseases, Fundamental and Applied Research for Animals & Health (FARAH), Faculty of Veterinary Medicine, University of Liège, Liège, Belgium

Maygane Ronsmans
Immunology-Vaccinology (B43b), Department of Infectious and Parasitic Diseases, Fundamental and Applied Research for Animals & Health (FARAH), Faculty of Veterinary Medicine, University of Liège, Liège, Belgium

David M. Stone
The Centre for Environment, Fisheries and Aquaculture Science, Weymouth Laboratory, Weymouth, Dorset, United Kingdom

Neil T. Sullivan
Department of Microbiology and Immunology, and Center for Molecular Virology and Translational Neuroscience, Institute for Molecular Medicine and Infectious Disease, Drexel University College of Medicine, Philadelphia, Pennsylvania, USA

Norbert Tautz
Institute for Virology and Cell Biology, University of Lübeck, Lübeck, Germany

Birke Andrea Tews
Institut für Immunologie, Friedrich-Loeffler-Institut, Federal Research Institute for Animal Health, Greifswald-Insel Riems, Germany

Steven J. van Beurden
Department of Pathobiology, Faculty of Veterinary Medicine, Utrecht University, Utrecht, The Netherlands

Catherine Vancsok
Immunology-Vaccinology (B43b), Department of Infectious and Parasitic Diseases, Fundamental and Applied Research for Animals & Health (FARAH), Faculty of Veterinary Medicine, University of Liège, Liège, Belgium

Alain Vanderplasschen
Immunology-Vaccinology (B43b), Department of Infectious and Parasitic Diseases, Fundamental and Applied Research for Animals & Health (FARAH), Faculty of Veterinary Medicine, University of Liège, Liège, Belgium

Francis O. Wamonje
Department of Plant Sciences, University of Cambridge, Cambridge, United Kingdom

Keith Way
The Centre for Environment, Fisheries and Aquaculture Science, Weymouth Laboratory, Weymouth, Dorset, United Kingdom

Brian Wigdahl
Department of Microbiology and Immunology, and Center for Molecular Virology and Translational Neuroscience, Institute for Molecular Medicine and Infectious Disease, Drexel University College of Medicine, Philadelphia, Pennsylvania, USA

Elizabeth A. Worrall
Queensland Alliance for Agriculture and Food Innovation, The University of Queensland, Brisbane, Queensland, Australia

CHAPTER ONE

Bean Common Mosaic Virus and Bean Common Mosaic Necrosis Virus: Relationships, Biology, and Prospects for Control

Elizabeth A. Worrall*, Francis O. Wamonje†,
Gerardine Mukeshimana‡,§, Jagger J.W. Harvey‡,*,
John P. Carr†,1, Neena Mitter*,1

*Queensland Alliance for Agriculture and Food Innovation, The University of Queensland, Brisbane, Queensland, Australia
†Department of Plant Sciences, University of Cambridge, Cambridge, United Kingdom
‡Biosciences eastern and central Africa-International Livestock Research Institute (BecA-ILRI) Hub, Nairobi, Kenya
§Ministry of Agriculture and Animal Resources, Kigali, Rwanda
1Corresponding authors: e-mail address: jpc1005@hermes.cam.ac.uk; n.mitter@uq.edu.au

Contents

Advances in Virus Research, Volume 93
ISSN 0065-3527
http://dx.doi.org/10.1016/bs.aivir.2015.04.002

Abstract

The closely related potyviruses *Bean common mosaic virus* (BCMV) and *Bean common mosaic necrosis virus* (BCMNV) are major constraints on common bean (*Phaseolus vulgaris*) production. Crop losses caused by BCMV and BCMNV impact severely not only on commercial scale cultivation of this high-value crop but also on production by smallholder farmers in the developing world, where bean serves as a key source of dietary protein and mineral nutrition. In many parts of the world, progress has been made in combating BCMV through breeding bean varieties possessing the *I* gene, a dominant gene conferring resistance to most BCMV strains. However, in Africa, and in particular in Central and East Africa, BCMNV is endemic and this presents a serious problem for deployment of the *I* gene because this virus triggers systemic necrosis (black root disease) in plants possessing this resistance gene. Information on these two important viruses is scattered throughout the literature from 1917 onward, and although reviews on resistance to BCMV and BCMNV exist, there is currently no comprehensive review on the biology and taxonomy of BCMV and BCMNV. In this chapter, we discuss the current state of our knowledge of these two potyviruses including fundamental aspects of classification and phylogeny, molecular biology, host interactions, transmission through seed and by aphid vectors, geographic distribution, as well as current and future prospects for the control of these important viruses.

1. INTRODUCTION

The genus *Potyvirus* is the largest of the eight genera currently assigned to the Family *Potyviridae* by the International Committee on Taxonomy of Viruses (ICTV, 2013). *Potyvirus* is the largest currently recognized genus of plant-infecting positive-strand RNA viruses and comprises 146 virus species (ICTV, 2013; Ivanov et al., 2014). The taxonomy of potyviruses is constantly changing due to frequent descriptions of new candidates, ambiguous serological data between viruses and isolates, variability in viral host range and differences in symptomatology between strains of the same species (Ali et al., 2006). Gibbs, Trueman, and Gibbs (2008) and Gibbs, Ohshima, et al. (2008) suggested that the *Potyviridae* family underwent a major speciation event approximately 6600 years ago, which coincides with the dawn of plant domestication. Indeed, these authors argue convincingly that it was the invention of agriculture, the subsequent migrations of humans into new arable areas, and the establishment of trade routes that spread these viruses and created the conditions for the evolutionary radiation of this family of viruses, and in particular those of the *Potyvirus* genus, into so many species (Gibbs, Ohshima, et al., 2008). This relatively recent diversification of species may explain the "star-burst" appearance of phylogenetic trees constructed for "potyvirids" with more than 100 *Potyvirus* species forming a compact cluster with a short branch to the *Rymovirus* cluster and longer branches to other potyvirid genera (Gibbs, Ohshima, et al., 2008).

Potyviruses can be separated into five basal lineages: the *Sugarcane mosaic virus* group; the *Bean yellow mosaic virus* (BYMV) group; the *Onion yellow dwarf virus* group, and the *Pea seedborne mosaic virus* group. In addition, there are two supergroups: the *Potato virus Y* (PVY) supergroup and the *Bean common mosaic virus* (BCMV) supergroup (Gibbs & Ohshima, 2010). The phylogeographic pattern of the species within the BCMV lineage suggests that this group originated in South and East Asia (Gibbs, Ohshima, et al., 2008; Gibbs, Trueman, et al., 2008). It is estimated that 3600 years ago, the BCMV lineage first emerged and now includes 19 virus species, including BCMV itself and the closely related *Bean common mosaic necrosis virus* (BCMNV) (A.J. Gibbs, personal communication; Gibbs, Trueman, et al., 2008).

BCMV and BCMNV are the most common and most destructive viruses that infect common beans (*Phaseolus vulgaris* L.) as well as a range of other cultivated and wild legumes (reviewed by Morales, 2006). Yield losses due to BCMV and BCMNV can be as high as 100% (Damayanti et al., 2008; Li et al., 2014; Saqib et al., 2010; Singh & Schwartz, 2010; Verma & Gupta, 2010). In this chapter, we aim to clarify the current taxonomy of BCMV and BCMNV by including a comprehensive list of currently known strains as well as summarizing the symptoms, geographical distribution, hosts and transmission properties of BCMV and BCMNV. This chapter also presents an up-to-date phylogenetic tree using all current BCMV and BCMNV fully sequenced genomes (GenBank, accessed September 2014). We will also summarize the current understanding of BCMV and BCMNV genome organization, gene expression, and the known or suspected functions of the viral proteins in infection and pathogenesis. Finally, we will discuss future control measures for BCMV and BCMNV with particular emphasis on the challenges of controlling these viruses in developing countries.

2. TAXONOMY

The taxonomy of BCMV and BCMNV has undergone major changes and this section aims to clarify its current status. Prior to 1934, BCMV was variously named Bean virus 1, Bean mosaic virus, and *Phaseolus* virus 1 (Morales & Bos, 1988). The originally identified BCMV isolate was lost, and all reports on the virus from its discovery in 1917 until 1943 presumed

Table 1 Reactions of Bean Cultivars for the Seven Pathotypes of *Bean Common Mosaic Virus* (BCMV) and *Bean Common Mosaic Necrosis Virus* (BCMNV)

Differential Cultivar	Pathotype						
	I	II	III	IV	V	VI	VII
Dubbele Witte	S	S	S	S	S	S	S
Sutter Pink	S	S	S	S	S	S	S
Redland's Greenleaf "C"	R	S/T	R	S	S/T	S/T	S/T
Puregold Wax	R	S/T	R	S	S/T	S/T	S
Redland's Greenleaf "B"	R	R	R	S	R	S	S
Great Northern 123	R	R	R	S	R	T	S
Sanilac	R	R	S	R	S	S	R
Red Mexican 34	R	R	S	R	S	S	R
Monroe	R	R	R	R	R	R	S
Great Northern 31	R	R	R	R	R	R	S

Table altered from Drijfhout et al. (1978), cultivars presumed to carry dominant or recessive alleles of inhibitor gene *I* grown under greenhouse conditions: 16-h daylight, 23–26 °C mean day temperature (range 20–30 °C). Key: S, susceptible: moderate-to-severe systemic mosaic, virus recoverable by assay from tip growth. T, tolerant: systemic symptoms may be mild, atypical, delayed, or absent. Virus recoverable by assay from tip growth. R, resistant: no systemic symptoms. Virus not recoverable by assay from tip growth.

that all BCMV strains were pathogenically identical (Drijfhout et al., 1978). But following the discovery of distinct BCMV pathogenic groups in 1943, differential symptoms in 10 bean cultivars were used to classify strains into the seven pathogenic groups (I–VII) (Drijfhout et al., 1978) (Table 1). Furthermore, data based on coat protein (CP, see Section 3) serology as well as analyses of proteolytic digests of CPs and the differential responses of various bean cultivars to infection led to the categorization of BCMV strains into two serotypes: A and B (Drijfhout et al., 1978; McKern et al., 1992; Vetten et al., 1992). The A and B serotypes were subsequently reclassified as distinct viral species named BCMNV and BCMV, respectively (Berger et al., 1997).

BCMNV, the former BCMV serotype A, has only five identified strains (TN-1, NL-3, NL-3K, NL-5, and NL-8) (Table. 2) (McKern et al., 1992; McKern et al., 1994; Mink & Silbernagel, 1992). BCMV, the former serotype B, contains many strains including some that were previously considered to be distinct viruses, such as peanut stripe virus (PStV), blackeye cowpea mosaic virus, *Dendrobium* mosaic virus (DeMV),

Table 2 *Bean Common Mosaic Necrosis Virus* (BCMNV) Strain Information

Strain	Pathotype	Origin	Complete Sequence (Accession #)	References
NL-3 (NL-3 (D) or Michelite)	VI	Netherlands	NC_004047 U19287 AY282577 AY138897	Fang et al. (1995) McKern et al. (1992)
NL-3K	VI	Idaho, USA	AY864314	Larsen et al. (2005) Miklas et al. (2000)
NL-5 (Jolanda)	VI	Netherlands	HQ229993	Larsen et al. (2011) McKern et al. (1992)
NL-8	III	Netherlands	HQ229994	Drijfhout et al. (1978) Larsen et al. (2011)
TN-1 (Tanzania)	VI	Tanzania	HQ229995	Silbernagel et al. (1986) Larsen et al. (2011)

Full genomic sequences were collected as of September 2014 from GenBank. Complete sequence available but no strain identification: HG792063. Key: D, indicates that the common NL-3 strain originates from Drijfhout et al. (1978); K, acknowledges the identification at Kimberly, ID, USA; NL, Netherlands; TN, Tanzania.

and Azuki bean mosaic virus (AzMV) (Table 3) (Hu et al., 1995; Khan et al., 1993; McKern et al., 1992; Mink & Silbernagel, 1992; Tsuchizaki & Omura, 1987). Currently, there are 22 complete BCMV genome sequences and 9 complete BCMNV genome sequences available through NCBI. Sequence data for CP genes and 3'-untranslated regions have been used extensively to place BCMV isolates taxonomically (Fang et al., 1995; Sharma et al., 2011). As a result, most partial or complete sequences of BCMV or BCMNV are for the CP gene. Figure 1 shows the phylogenetic tree of the full genomes of available BCMV and BCMNV strains as of September 2014 in GenBank. In the phylogenetic tree, 22 BCMV sequences show more diversity when compared to the 9 tightly grouped BCMNV sequences (with PVY as the outgroup).

Mixed potyvirus infections can result in interspecific recombination that can facilitate the genesis of new virus strains. It was demonstrated under experimental conditions that by doubly inoculating bean plants that were resistant to one but susceptible to the other virus, recombinants of BCMV US-5 and BCMNV NL-8 could be generated (Silbernagel et al., 2001). Subsequently, an isolate of BCMNV with enhanced pathogenicity,

Table 3 *Bean Common Mosaic Virus* (BCMV) Strain Information

Strain	Pathotype	Origin	Complete Sequence (Accession #)	References
AzM	0	Japan	–	Collmer et al. (1996) Matsumoto (1922) McKern et al. (1992) Mink et al. (1999) Tsuchizaki and Omura (1987)
BlCM	I	Florida, USA	AJ312437 AJ312438 AY575773 NC_003397	Lana et al. (1988) Lima et al. (1979) Tsuchizaki and Omura (1987) Wang and Fang (2004), Zheng et al. (2002)
CH-1 (Chile-2)	IV	Chile	–	Mink et al. (1999)
CH-2 (Chile-2)	IV	Chile	–	McKern et al. (1992)
DeM	0	Hawaii, USA	–	Hu et al. (1995) Mink et al. (1999)
NL-1 (Westlandia)	I	Netherlands	AY112735	McKern et al. (1992)
NL-1n	I[a]	India	–	Kapil et al. (2011)
NL-2 (RM)	V	Netherlands	–	McKern et al. (1992)
NL-4 (Great Northern)	VII	Netherlands	DQ666332	Bravo et al. (2008) McKern et al. (1992)
NL-6 (Colana)	IV	Netherlands	–	McKern et al. (1992)
NL-7	II	Peru	–	Drijfhout and Bos (1977)
NL-7n	II[a]	India	–	Kapil et al. (2011)
NVRS	IV	England	–	Walkey and Innes (1978)
PR-1 (PR9M)	I	Puerto Rico	–	McKern et al. (1992)

Table 3 *Bean Common Mosaic Virus* (BCMV) Strain Information—cont'd

Strain	Pathotype	Origin	Complete Sequence (Accession #)	References
PSt (Stripe)	0	China	AY968604 KF439722 U05771 U34972	Demski et al. (1984) Flasinski et al. (1996) Gunasinghe et al. (1994) Mink et al. (1999) Wang et al. (2005)
RU-1 (Russian)	VII	USDA-P1	AY863025 GQ219793 KF919297 KF919298 KF919300	Feng et al. (2014) Larsen et al. (2005) McKern et al. (1992) Naderpour et al. (2010)
US-1 (type)	I	Washington, USA	–	McKern et al. (1992)
US-2 (Z) (NY-15 (Z))	V	New York, USA	–	McKern et al. (1992)
US-2 (D) (NY-15 (D))	V	Idaho, USA	–	Dean and Hungerford (1946) McKern et al. (1992)
US-2 (P) (NY-15 (P))	V	New York, USA	–	Kyle and Provvidenti (1987) McKern et al. (1992)
US-3 (Idaho 123)	IV	Idaho, USA	–	McKern et al. (1992)
US-4 (Western)	IV	Washington, USA	–	McKern et al. (1992)
US-5 (Florida)	IV	Florida, USA	–	McKern et al. (1992)
US-6 (Mexican)	VII	Mexico	–	McKern et al. (1992)
US-7 (R220)	II	Washington, USA	–	Drijfhout et al. (1978)
US-10 (NW-63)	VII	Washington, USA	KF919299	Feng et al. (2014) McKern et al. (1992)

[a]Identical to pathotype except for their necrotic reaction on cultivar Jubila at high temperature. Full genomic sequences were collected as of September 2014 from GenBank. Complete sequence available but no strain identification: EU761198, KC832501, KC832502, KC478389, KJ508092, and HG792064. AzM, Azuki mosaic; BICM, Blackeye cowpea mosaic; CH, Chile; DeM, *Dendrobium* mosaic; NL, Netherlands; PR, Puerto Rico; PSt, Peanut stripe; RU, Russia; US, USA; NY, New York.

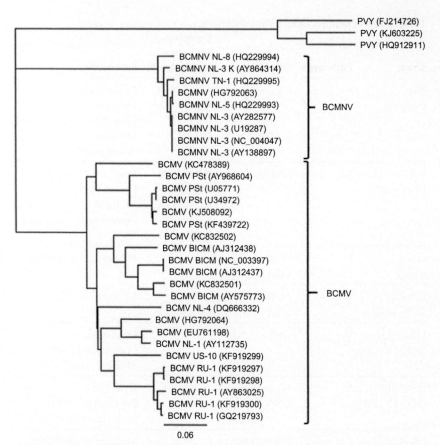

Figure 1 Phylogenetic tree of *Bean common mosaic virus* (BCMV) and *Bean common mosaic necrosis virus* (BCMNV). Nine full-length genomic sequences were found for BCMNV and 22 for BCMV. BCMNV forms a tight cluster without much variation while BCMV shows higher diversity. A neighbor-joining phylogenetic tree was generated using Geneious tree builder (Tamura-Nei default settings). Scale bar: 0.06 nucleotide substitutions per site. Analysis based on full genomes available through GenBank (September 2014) and outgroup *Potato virus Y* (PVY).

NL-3K, was shown to be a naturally occurring recombinant virus that was derived from the NL-3 D strain of BCMNV but that possessed sequences from the BCMV RU-1 strain (Larsen et al., 2005). The recombination event that created BCMNV NL-3 D incorporated an RNA sequence encoding part of the BCMV RU-1 P1 protein. The chimeric P1 protein encoded by the NL-3K recombinant BCMNV comprises 415 amino acids, compared with 317 residues for BCMNV NL-3 D. The sequence of the N-terminal 114 residues of the BCMNV NL-3K P1 protein was 98%

identical to the corresponding sequence of BCMV RU-1 P1 but the remainder of the chimeric P1 sequence was 98% identical to the NL-3 D BCMNV P1 protein (Larsen et al., 2005).

A recombinant strain of BCMV that induces temperature-independent systemic necrosis in plants possessing the dominant *I* resistance gene and a recessive resistance gene (*bc-1*) (see Section 4) was recently described (Feng et al., 2014). This BCMV strain, named RU-1M, is a recombinant of BCMV RU-1OR and an unknown potyvirus in which an insertion of 770 nucleotides into the BCMV P1 coding region has occurred between nucleotides 513 and 1287 (Feng et al., 2014). This study and that of Larsen et al. (2005) show that interspecies recombination is likely to be a rich source of novel BCMV and BCMNV strains and, potentially, the generation of new potyviral species. Both studies highlight the importance of P1 and the P1-HC-Pro precursor protein in pathogenesis and the breakage of genetic resistance (see Section 3).

3. GENOME AND STRUCTURE

Like all potyviruses, BCMV and BCMNV are monopartite, single-stranded positive-sense RNA viruses that form flexuous rod-shaped virions (Ivanov et al., 2014 and see Profile 63, Family: *Potyviridae* in Hull, 2014). The virions are 750 nm in length with a diameter of 11–13 nm and contain a genomic RNA molecule that is approximately 10 kb long that possesses a $3'$-terminal polyA tail (El-Sawy et al., 2013; Fang et al., 1995). Virions are principally constructed from CP molecules that are arranged helically around the genomic RNA. An additional protein, the VPg (viral protein genome-linked), which is covalently linked to the $5'$-terminus of the RNA, projects out at the tip of the virus particle and serves a purpose similar to the cap structure found on most cellular mRNAs, as well as other functions (see below).

The potyviral genomic RNA has a long open reading frame (ORF) that can be directly translated by host ribosomes to yield a polyprotein that self-cleaves to yield 10 proteins. Additionally, frame-shifting allows an additional protein to be produced (Fig. 2). Self-processing of the polyprotein is catalyzed by the proteolytic activities of P1, helper component-proteinase (HC-Pro), and nuclear inclusion protein-a proteinase (NIa-Pro). Polyprotein processing results in the production of the mature forms of these proteins plus formation of other mature viral proteins including P3, cylindrical inclusion (CI), VPg, nuclear inclusion-b (NIb: the viral RNA-dependent RNA polymerase) (reviewed in Hull, 2014; Ivanov et al.,

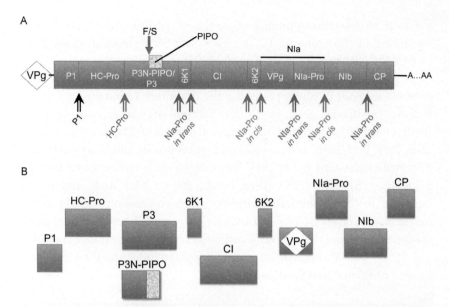

Figure 2 Potyvirus gene expression (A). The positive-sense single-stranded RNA genomes of BCMV and BCMNV (approximate length 10 kb) and those of other potyviruses encode 11 proteins. A molecule of VPg (virus-protein-genome linked) is covalently attached to the 5'-end of the genomic RNA, which has a 3' polyA tail. Translation by host cell ribosomes of a single long ORF encoded by the genomic RNA yields a polyprotein; the precursor polypeptide for the mature viral proteins (B). The polyprotein contains protein sequences that possess proteinase activity (P1, HC-Pro, and the NIa-Pro). HC-Pro cleaves *in cis* (gray arrow) to release a P1/HC-Pro fusion protein (not shown) that is process *in cis* by P1 to yield the mature proteins P1 and HC-Pro. The NIa-Pro cleaves both *in cis* and *in trans*. *Cis*-cleavage by NIa-Pro activity releases the 6K2 protein and the VPg/NIa-Pro fusion protein, NIa (cleavage sites indicated by blue (gray in the print version) arrows). Cleavage *in trans* by NIa-Pro releases mature VPg, coat protein (CP), the NIb (RNA-directed RNA polymerase) protein, the P3 protein, and the 6K1 protein (cleavage sites indicated with red (gray in the print version) arrows). Frame-shifting at a site (F/S) within the P3 cistron can allow low level expression of an embedded protein coding sequence (PIPO: pretty interesting *Potyviridae* ORF, indicated by a stippled box) (A). The resulting product is released by HC-Pro proteinase activity (gray arrow) to yield P3N-PIPO, a protein possessing the N-terminal sequence of P3 fused to the PIPO amino acid sequence (B). P3N-PIPO is produced in far smaller amounts than the other 10 products of potyviral RNA translation.

2014; Oana et al., 2009; Riechmann et al., 1992). Two 6 kDa polypeptides called 6K1 and 6K2 are also released during polyprotein processing of which only 6K2 has a known function, which is as a membrane anchor for the viral replication complexes (reviewed by Ivanov et al., 2014). Cleavage of seven out of the nine polyprotein cleavages are performed by NIa-Pro, which cuts both in *cis* and *trans*, whereas both P1 and HC-Pro cut only once in *cis*

(Adams et al., 2005) (Fig. 2). P1 (a serine protease) cuts only at the P1/HC-Pro border. The amino acid sequence of P1 proteins of BCMV subgroup viruses differs from the sequences of P1 proteases encoded by other potyviruses in that the catalytic triad sequence is His-(X_{7-11})-Glu-(X_{30-36})-Ser rather than the canonical His-(X_{7-11})-Asp-(X_{30-36})-Ser (Adams et al., 2005). Although other functions of the P1 protease have remained somewhat mysterious, a recent paper on the *Tobacco etch virus* (TEV) ortholog suggests that it may enhance viral protein synthesis through interaction with both ribosomal subunits, a function that is apparently antagonized by HC-Pro (Martínez & Daròs, 2014).

Until recently, it was thought that the polyprotein gave rise to all the mature potyviral proteins. However, the application of gene-finding algorithms to the large number of available potyviral RNA sequences (including those of BCMV and BCMNV) revealed the existence of a highly conserved short ORF known as PIPO (pretty interesting *Potyviridae* ORF), which is embedded within the P3 cistron of all potyviruses (Chung et al., 2008). P3N-PIPO is required for potyviral intercellular movement (see Section 4). P3N-PIPO expression in BCMV, BCMNV, and TuMV was recently shown to depend on polymerase slippage, which gives rise to a viral RNA containing a frameshift within the P3 coding sequence (Olspert et al., 2015), and a similar phenomenon is thought to occur for *Plum pox virus* (PPV) (Rodamilans et al., 2015).

The HC-Pro factors encoded by BCMV and BCMNV have not been studied as extensively as those encoded by a number of other potyviruses including TEV, *Turnip mosaic virus* (TuMV), and PVY. However, it seems likely that the HC-Pro factors encoded by BCMV and BCMNV play similar, multiple biological roles as those of other potyviruses. The potyviral HC-Pro is not only required for polyprotein maturation as a cysteine protease but also plays vital roles in aphid-mediated virus transmission (see Section 7) and in disruption of RNA silencing pathways leading to suppression of silencing-based antiviral defense (allowing increased viral RNA amplification) and symptom induction (see Section 4). It is likely that HC-Pro has additional, but currently less well elucidated, effects on host cell biology (reviewed by Ivanov et al., 2014).

4. INFECTION AND SYMPTOMS

In plants of susceptible bean lines, BCMV and BCMNV can have very similar symptoms that include dwarfing, mosaic, leaf curling, and chlorosis (Flores-Estévez et al., 2003) (Fig. 3). However, even "symptomless" infection of hosts in which these effects are not manifested can decrease crop yield to

Figure 3 Typical symptoms induced on bean plants systemically infected with BCMV and BCMNV. The common bean (*Phaseolus vulgaris*) variety Dubbele Witte is compatible with both BCMV and BCMNV (see Table 1 and Drijfhout et al., 1978). Growth of plants systemically infected with BCMNV (A) or BCMV (B) is inhibited (scale bar = 4 cm). Leaf development on infected Dubbele Witte bean plants is disturbed and leaves of plants systemically infected with BCMNV (C) and BCMV (D) display a distinct curling pheno-type. (E) In the Kenyan common bean variety Wairemu, BCMNV infection but not BCMV infection (not illustrated) also leads to systemic necrosis, which is visible here at an early stage in the veins and leaf tips. (F) BCMV and BCMNV can be transmitted in a nonper-sistent manner by a variety of aphid species including the black bean aphid (*Aphis fabae*), a common insect pest of bean. The wingless adult shown here has been attached to a fine gold wire using silver-containing paint to permit electrophysiological monitor-ing of its feeding behavior on bean tissue (EPG: see Section 7.2).

the extent of 50% (Morales, 2006). The most striking difference in sympto-mology between BCMV and BCMNV relates to the hypersensitive necrotic reaction (black root disease) that occurs in bean plants containing the domi-nant *I* resistance gene (Ogliari & Castao, 1992). Early experiments showed that this gene conferred strong hypersensitive resistance to BCMV (Ali, 1950). The hypersensitive reaction is a form of resistance that involves local-ization of a pathogen to the initially infected cells or to a zone of cells in the immediate vicinity of the primary infection site and that is often associated with programmed death of cells in this zone (see Loebenstein, 2009).

The resistance to BCMV conferred by the *I* gene was discovered in the early 1930s by Ralph Corbett (Pierce, 1934). The variety "Corbett Refugee" was derived from a remaining healthy plant observed in a field of the susceptible cultivar "Refugee Green" following BCMV infection. Subsequently, the resistance trait was introgressed into many other bean cultivars. Unfortunately, in some cases this *I*-gene-mediated restriction of virus spread is incomplete and the virus escapes from the primary infection zone, causing a trailing necrosis due to resistance-associated cell death. This systemic necrosis resulting from incomplete resistance results in the so-called "black root" disease of bean (Drijfhout et al., 1978; Silbernagel et al., 2001). Black root disease was first noted in beans bred from "Corbett Refugee" in the late 1930s (Jenkins, 1940). The plants showed severe wilting and chlorosis of the lower leaves and necrotic streaks running along the stem above and below the cotyledonary node (Jenkins, 1941).

BCMV is not usually associated with black root disease but two of its strains (NL-2 and NL-6) are able to trigger this reaction in *I*-gene-possessing plants at higher temperatures (above 30 °C) and hence this form of black root is termed "temperature-dependent." Importantly, *all* BCMNV strains trigger black root disease in bean plants harboring the *I* gene, independently of temperature (Silbernagel et al., 2001). Indeed, the endemic presence of BCMNV in certain regions may make deployment of common bean cultivars possessing the *I* gene problematic (see Section 5). But in some circumstances, this distinction between BCMV and BCMNV may not be absolute; for example, in the case of a recombinant isolate belonging to the BCMV RU-1 strain group that induces temperature-independent necrosis in the *I* gene containing "Jubila" bean variety (Feng et al., 2014) (see Section 2). Table 1 lists the symptoms observed with 10 bean cultivars that were used to distinguish the seven pathotypes of BCMV and BCMNV. However, modern techniques for strain differentiation no longer solely rely on differential symptoms but instead utilize multiple properties including *CP* or other gene and amino acid sequences (Flores-Estévez et al., 2003; Xu & Hampton, 1996) (see Section 2).

Although BCMV and BCMNV replication and movement through bean plants have not been subjects of intense study at the molecular level, it is likely that lessons from other potyvirus–plant interactions can be applied. Efficient translation of potyviral genomic RNA is facilitated by interactions between the VPg molecule covalently attached to the 5′-end of the viral RNA and specific host translation factors of the eIF4E or eIF4(iso)E type (reviewed by Truniger & Aranda, 2009). Certain variant alleles encoding versions of these factors that do not interact with potyviral VPg proteins impair the ability of these viruses to exploit the host's translational machinery. Indeed, this provides the mode of action for many recessive resistance genes that condition resistance to potyviruses and at least one recessive gene conditioning resistance to potyviruses in bean, *bc-3*, has been proved to be associated with *eIF4E* sequence variants at the *Bc-3* locus (Hart & Griffiths, 2013) (see Section 8).

It is unlikely that BCMV and BCMNV replication differ significantly from that of TuMV, TEV, or PPV, with synthesis of nascent viral RNA dependent upon the activities of a core set of viral proteins: the NIb RNA-dependent RNA polymerase; the CI (which has a helicase activity); the NIa-VPg protein, and the 6K2 membrane-remodeling factor (outlined on pages 936–937 and 953–954 in Hull, 2014). Potyvirus infection is associated with dramatic changes in host cell membrane organization (Grangeon et al., 2012) and virus replication occurs within host membrane-derived vesicles (virus factories) formed through the 6K2 protein activity, and virus factory formation is dependent upon the cell secretory system (Agbeci et al., 2013). During infection, virus factories move to the periphery of infected cells toward the vicinity of plasmodesmata utilizing the host cell cytoskeletal microfilament network (Agbeci et al., 2013).

HC–Pro is a multifunctional protein possessing diverse functions ranging from suppression of RNA silencing to facilitation of aphid-mediated virus transmission (Hasiów-Jaroszewska et al., 2014; Rojas et al., 1997; Urcuqui-Inchima et al., 2001). HC-Pro possesses three modules: the N-terminal Domain I; the central Domain II, and C-terminal Domain III. Domains I and II are essential for transmission (Hasiów-Jaroszewska et al., 2014) (see Section 7). Domain II is involved in genome amplification predominantly through suppression of RNA silencing. Domain II contains the IGN motif and the FRNK box, which are crucial for RNA-silencing suppression and efficient viral RNA accumulation (Kasschau et al., 1997; Sahana et al., 2014; Shiboleth et al., 2007). Domain III is also involved in cell-to-cell movement as well as possessing proteinase activity for polyprotein processing (Hasiów-Jaroszewska et al., 2014).

For potyviruses, the accumulation of nascent viral RNA and local and systemic spread of infection is to a great extent dependent upon the inhibition of antiviral RNA silencing by HC-Pro and/or its precursor fusion protein, P1/HC-Pro. HC-Pro was one of the first two viral suppressors of RNA silencing to be identified, the other being the 2b counter-defense protein of *Cucumber mosaic virus* (CMV) (Anandalakshmi et al., 1998; Brigneti et al.,

1998; Kasschau & Carrington, 1998). The HC-Pro factors of potyviruses inhibit antiviral RNA silencing through binding of 21 nucleotide short-interfering RNAs and also acts as symptom determinants through interactions with, among other things, Argonaute factors, calmodulin-like factors, and the proteasome (Anandalakshmi et al., 2000; Nakahara et al., 2012; Sahana et al., 2014; Shiboleth et al., 2007). So far, cellular proteins interacting with the HC-Pro or P1/HC-Pro factors of BCMV and BCMNV remain unidentified. Recently, another potyviral gene product, the VPg/NIa fusion was found to interfere with host defense, most likely through interaction with the host nucleolar protein fibrillarin (Ivanov and Mäkinen, 2012; Rajamäki & Valkonen, 2009).

Work with TuMV showed that P3N–PIPO, together with the CI protein and the host plasma membrane protein PCaP1, form complexes required for potyvirus intercellular movement (Vijayapalani et al., 2012; Wei et al., 2010). The VPg, which atomic force microscopy shows to project from the tip of potyvirus particles, might be able to penetrate plasmodesmata and interact with eIF4-type factors in the next cell, which raised the possibility of a cotranslational mechanism in which viral RNA molecules are drawn through plasmodesmata and translated simultaneously by ribosomes (Torrance et al., 2006). However, another potential mechanism may be provided by the transport through plasmodesmata of virus replication factories contained in 6K2-induced membranous vesicles (Grangeon et al., 2013). Transport via membranous virus replication factories appears to be the currently favored mechanism of intercellular transport for potyviruses (see figure 3 in Ivanov and Mäkinen, 2012). It remains unknown whether intercellular movement of BCMV or BCMNV occurs through either of these mechanisms (cotranslational vs. coreplicational) in bean. In an early study using microinjection, it was found that BCMV HC-Pro and CP (expressed in and purified from bacteria) could traffic through plasmodesmata, which was suggestive of these factors having movement protein properties (Rojas et al., 1997). Conceivably, these factors could play roles in enhancing movement (be it cotranslational or coreplicational) by altering the gating properties of the plasmodesmata but this remains, especially for BCMV and BCMNV, unexplored territory. Meanwhile, information is lacking on host factors controlling systemic movement of BCMV and BCMNV in bean. For example, it is unknown if gene products analogous to the *Arabidopsis thaliana* phloem factors RTM ("restriction of TEV": Chisholm et al., 2001) control long-distance potyvirus movement in bean.

5. GEOGRAPHIC DISTRIBUTION

Believed to originate in South or East Asia (Gibbs, Ohshima, et al., 2008; Gibbs, Trueman, et al., 2008), BCMV has now spread worldwide and can be found wherever legumes are grown. We collated reports in the scientific literature from 1917 to September 2014 to generate a heat map of the global incidence and distribution of BCMV and BCMNV (Fig. 4). The largest number reports of BCMV and BCMNV were from the USA (21 scientific records), while the second highest number originated from India (6 scientific records). However, the map cannot be claimed to

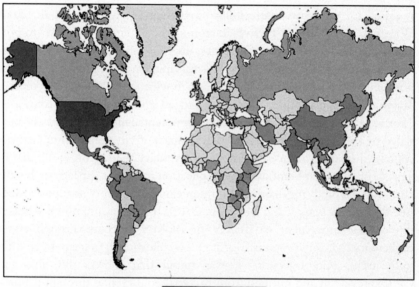

Number of scientifically reported cases   21.0   11.0   1.00

Figure 4 Heat map of cases reported in the scientific literature of incidences of *Bean common mosaic virus* (BCMV) and *Bean common mosaic necrosis virus* (BCMNV) from 1917 to September 2014. Collected from a total of 84 papers. Number of BCMV and/or BCMNV scientific recordings: Argentina: 1, Australia: 2, Belize: 1, Brazil: 2, Burundi: 2, Canada: 1, Chile: 2, China: 5, Colombia: 3, Congo: 1, Costa Rica: 2, Cuba: 1, Dominican Republic: 3, Egypt: 1, El Salvador: 2, Ethiopia: 2, Guatemala: 2, Haiti: 2, Honduras: 1, India: 6, Indonesia: 5, Iran: 4, Japan −2, Jamaica: 1, Kenya: 3, Lebanon: 1, Lesotho: 1, Malaysia: 1, Malawi: 4, Mexico: 5, Myanmar: 1, Netherlands: 5, New Zealand: 3, Nicaragua: 2, Nigeria: 1, Panama: 2, Paraguay: 2, Peru: 5, Philippines: 2, Poland: 1, Puerto Rico: 1, Russia: 1, Rwanda: 2, Serbia: 2, South Korea: 2, Spain: 4, Swaziland: 2, Taiwan: 4, Tanzania: 2, Thailand: 3, Turkey: 2, United Kingdom: 1, United States: 21, Uganda: 3, Vietnam: 2, Zambia: 2 and Zimbabwe: 2.

definitively reflect BCMV and BCMNV incidence since there has been less surveillance for these viruses in some regions, especially within African countries (see Section 9.2).

Even in regions where BCMV has been long established, there are reports of its occurrence in novel hosts and cultivation areas (Li et al., 2014). BCMNV distribution is not as extensive as that of BCMV (Deligoz & Soken, 2011; Melgarejo et al., 2007; Petrovic et al., 2010; Sáiz et al., 1995). This is probably because BCMNV evolved from BCMV more recently (Gibbs, Ohshima, et al., 2008; Gibbs, Trueman, et al., 2008; Spence & Walkey, 1995). After a survey of 13 African countries, Spence and Walkey (1995) concluded that, due to its prevalence in Central and Eastern Africa, BCMNV most likely evolved in Central or Eastern Africa. Consequently, this is where the greatest BCMNV strain diversity is found (Spence & Walkey, 1995). However, BCMNV has occasionally been transported to other parts of the world through contaminated bean seed (Beaver et al., 2003). Indeed, the high efficiency of seed transmission BCMV and BCMNV (see Section 7) has aided the long-distance dissemination of both viruses to many regions including: Europe (Drijfhout et al., 1978; Pasev et al., 2014; Sáiz et al., 1995), North America (Flores-Estévez et al., 2003; Kelly et al., 1983; Provvidenti et al., 1984; Tu, 1986), South America (Melgarejo et al., 2007), as well as throughout Africa itself (Njau & Lyimo, 2000; Sengooba et al., 1997; Silbernagel et al., 1986; Spence & Walkey, 1995).

Both viruses are major constraints on bean production and can cause serious crop losses (Morales, 2006). In Africa, and in particular in Central and East Africa, BCMNV is endemic and infects many wild legumes (Sengooba et al., 1997; Spence & Walkey, 1995) (see Section 6). The prevalence of BCMNV limits the usefulness of the genetic resistance to BCMV provided by the dominant *I* gene since in these hosts BCMNV induces black root disease (see Section 4).

Thus, the use of the *I* gene cultivars in bean cultivars in Central and Eastern Africa where BCMNV strains exist can be counterproductive leaving farmers with very few choices for controlling bean-infecting potyviruses (see Section 8). From a study of reactions of US bean cultivars to a Tanzanian BCMNV isolate (at that time classed as a BCMV isolate), Silbernagel et al. (1986) recommended that in regions where BCMNV occurs breeding efforts should focus on pyramiding of *I* with recessive resistance genes; a strategy that should enhance the durability of resistance (Kelly et al., 1995).

Potyviruses are the most economically destructive plant-infecting viruses in Australia (Coutts et al., 2011). Around half of the potyviruses found in

Australia are indigenous virus species not found elsewhere. Interestingly, these native potyviruses belong to one lineage of the BCMV group (see Section 2) but tend to be found in native and naturalized wild plants rather than in cultivated plants (Coutts et al., 2011; Gibbs, Ohshima, et al., 2008; Gibbs, Trueman, et al., 2008; Kehoe et al., 2014). However, a recent study indicated that indigenous legume-infecting potyviruses can jump from native to introduced plants, which could conceivably pose a threat to cultivated plants in the future (Kehoe et al., 2014). BCMV has been isolated from cultivated beans and from wild legume hosts in New South Wales, Queensland, Tasmania, Victoria and Western Australia but as yet not in the Northern territory or South Australia (Moghal & Francki, 1976; Saqib et al., 2005, 2010). None of the BCMV strains previously considered to be distinct potyvirus species (PStV, DeMV, and AzMV) have been reported in Australia although one, PStV, was detected as close as Papua, infecting peanut/groundnut crops and in imported material stopped by quarantine authorities (see Saqib et al., 2010, and references therein). Currently, we know of no reports of BCMNV in Australia.

6. HOST RANGE

BCMNV and BCMV were formally separated into distinct virus species in 1992, following the initial discovery boom regarding this group of viruses in the 1970s and 1980s. Thus, when examining the earlier literature, it should be remembered that before the reclassification, the host range of BCMNV was not considered separately from that of BCMV. There are six susceptible host families for BCMV, which are the: *Amaranthaceae*, *Chenopodiaceae*, *Leguminosae-Caesalpinioideae*, *Leguminosae-Papilionoideae*, *Solanaceae*, and *Tetragoniaceae* (Bos & Gibbs, 1995). However, BCMV is well known for naturally infecting wild and crop legumes and *Passiflora* species. The most notable cultivated legume hosts are *Phaseolus* species (predominately *P. vulgaris*), *Vicia faba*, *Arachis hypogaea* (peanut/groundnut), and *Vigna unguiculata* (cowpea) (Bos & Gibbs, 1995; Morales & Bos, 1988). In Australia, BCMV has been reported in *Macroptilium atropurpureum* (siratro), *Phaseolus vulgaris*, *Arachis hypogaea*, *Vigna trilobata*, and *Vigna unguiculata* (Coutts et al., 2011; Saqib et al., 2005, 2010). Meanwhile, in Africa, BCMV and BCMNV have been reported in a multitude of hosts. These include *Phaseolus vulgaris*, *Crotalaria incana*, *Rhynchosia* sp., *Macroptilium atropurpureum*, and *Cassia occidentalis* (Spence & Walkey, 1995). A survey of wild and forage legumes in Uganda in 1997 identified BCMNV in *Centrosema pubescens*,

Crotalaria incana, *Lablab purpureus*, *Phaseolus lunatus*, *Senna bicapsularis*, *Senna sophera*, and *Vigna vexillate* (Sengooba et al., 1997) (see Section 5).

Diagnostic host species that are susceptible to BCMV are: *Chenopodium quinoa*; *Macroptilium lathyroides*; *Phaseolus vulgaris*; *Pisum sativum*, and *Vicia faba* (Bos & Gibbs, 1995; Morales & Bos, 1988). While the indicator hosts *Cucumis sativus*, *Medicago sativa*, *Nicotiana tabacum*, *Nicotiana glutinosa*, and *Pisum sativum* are nonsusceptible (Bos & Gibbs, 1995; Morales & Bos, 1988).

Other hosts susceptible to BCMV and/or BCMNV include, but are not limited to: *Cajanus cajanus*, *Canavalia ensifromis*, *Cassia tora*, *Chenopodium amaranticolor*, *Cicer arietinum*, *Crotalaria juncea*, *Crotalaria spectabilis*, *Crotalaria striata*, *Cyamopsis tetragonoloba*, *Dendrobuim* spp., *Glycine max* (a confirmed host for BCMNV), *Gomphrena globosa*, *Lens culinaris*, *Lens esculenta*, *Lupinus albus*, *Lupinus angustifolius*, *Lupinus luteus*, *Macropitilium lathyroides*, *Melilotus albus*, *Nicotiana benthamiana*, *Nicotiana clevelandii*, *Phaseolus coccineus*, *Phaseolus lathyroides*, *Phaseolus sativum*, *Rhynchosia minima*, *Sesbania exaltata*, *Tetragonia tetragonioides*, *Trifolium incarnatum*, *Trifolium subterraneum*, *Trigonella foenum-graecum*, *Vicia sativa*, *Vicia villosa*, *Vigna angularis*, *Vigna radiata*, *Vigna sesquipedalis*, and *Vanilla planifolia* (Bhadramurthy & Bhat, 2009; Bos & Gibbs, 1995; Morales & Bos, 1988; Spence & Walkey, 1995; Zheng et al., 2002).

7. TRANSMISSION

All known strains of BCMV and BCMNV can be transmitted through seed or spread by aphids. Thus, an outbreak can be triggered by the use of contaminated seed stock and amplified by aphid–mediated virus transmission to generate an epidemic (Hampton, 1975). Alternatively, the viruses may infect a healthy crop population when viruliferous aphids immigrate from infected wild plants (Sengooba et al., 1997). Most published studies of BCMV and BCMNV transmission were conducted before reclassification of these viruses into separate species in 1992; thus, the literature is not clear on what, if any, differences there are between the transmission characteristics of BCMV and BCMNV.

7.1 Seed Transmission

Ekpo and Saettler (1974) determined that seed transmission of BCMV in bean occurred not through contamination of the testa but from internal seed transmission of the virus through infection of the embryo or cotyledons. This conclusion was vindicated by electron micrographic evidence showing that arrays of BCMV particles were present in virtually all cells and tissues in dormant and germinating bean seeds obtained from experimentally infected plants (Hoch & Provvidenti, 1978). Currently, there is no information on the molecular determinants of seed transmission of BCMV or BCMNV. However, a study of seed transmission of another potyvirus, *Soybean mosaic virus*, indicates that CP/HC-Pro interactions may be important for this process; an intriguing parallel with processes determining successful aphid-mediated transmission (see Section 7.3) (Jossey et al., 2013).

Remarkably, seeds of *Phaseolus vulgaris* can retain infectious BCMV for at least 30 years (Pierce & Hungerford, 1929). This stability of virus in the embryo is likely a major contributory factor for seed being the most important route for long-distance worldwide dissemination of BCMV and BCMNV through trade. For example, it is believed that the 1977 BCMV epidemics in Europe and America were most likely initiated by seed stock contamination. Previously considered to be resistant to many BCMV strains, the following cultivars succumbed to BCMV infection during epidemics triggered by seed contamination in the northwestern USA in 1977 and 1981: "Black Turtle," "Columbia Pinto," "Great Northern UI 31," "Great Northern UI 1140," "Montcalm Red Kidney," "Pinto UI 114," "Red Mexican NW 59," "Red Mexican NW 63," "Rufus" and "Viva." In some cases, seed contamination of "resistant" cultivars was probably due to segregation of resistance alleles in notionally homozygous lines (Hampton et al., 1983).

Another BCMV epidemic occurred in the United States due to contaminated seed stock in 1988/89 (Forster et al., 1991; Klein et al., 1988). In 1988, the US Department of Agriculture *Phaseolus* germplasm collection was surveyed for BCMV and strains that would later be classified as BCMNV (Klein et al., 1988). BCMNV was not observed in the tested plants, while 7% (591/8147) were positive for BCMV in the *Phaseolus vulgaris*, *Phaseolus acutifolius* (tepary bean), *Phaseolus aborgineus*, and *Phaseolus angustifolius* accessions (Klein et al., 1988). In 1989, the Idaho Crop Improvement Association tested another certified seed collection for BCMV (Forster et al., 1991). Of the stock tested, two foundation seed lots of line UI 114 produced in 1987 and 1988 and one foundation seed lot of UI 60 produced in 1982 were contaminated with the virus (Forster et al., 1991). Perhaps not surprisingly, the bean crop in south-central and south-western Idaho was heavily affected by this contamination. Consequently, 650 acres of bean crop were rejected by the Idaho Crop Improvement Association due to excessive BCMV incidence (Forster et al., 1991).

The rate of BCMV and BCMNV seed transmission depends on a range of factors including host cultivar, virus strain, stage of infection, and environment (Sastry, 2013). The percentage of infected seed varies from 0.67% to 98% (Chiumia & Msuku, 2001; Deligoz & Soken, 2011; Galvez & Morales, 1989; Hampton, 1975; Shankar et al., 2009). However, rates of seed transmission range from 10% to 30% (Galvez & Morales, 1989). This variability in infection rate is to some extent explained by the season and developmental stage of the plants at which a crop is exposed to infection. Thus, the rates of peanut/groundnut seed infection in Georgia (USA) with the BCMV strain PStV ranged from 37% in summer to 18% in winter, while it ranged from 19% to 11% in a Spanish cultivar (Demski & Warwick, 1986). Morales and Castano (1987) discovered that the incidence of seed transmission of five BCMV strains in 14 mosaic-susceptible bean cultivars is significantly affected by the inoculation date. Plants inoculated within the first 20 days of their vegetative period had a significant increase in the percentage of seed transmission. The highest incidence was observed at 10 days after sowing, while BCMV-infected seed was significantly decreased 30 days after sowing (Morales & Castano, 1987). Plants infected after their flowering stage tend to produce a lower yield of BCMV-infected seeds (Galvez & Morales, 1989). Thus, seed transmission of BCMV and BCMNV is a significant problem in bean and even certified seed stocks may contain as much as 1% infected seed (Morales, 2006). This is a pathogen infection rate that would not be tolerated in other major crops and clearly there is room for major improvements in the quality of "pathogen-free" bean seed stocks.

7.2 Aphid-Mediated Transmission

The impact of aphid-mediated transmission on the spread of BCMV and BCMNV can be significant. Studies at the International Center for Tropical Agriculture (CIAT) in Colombia during a period when aphid populations reached high levels reported that an initial BCMV seed infection rate of 2–6% was amplified to 100% plant infection due to aphid-mediated transmission (Galvez & Morales, 1989). In California in 1996, BCMNV incidence was reported at 5–10% during early field inspections but this increased to 70–90% during surveys conducted late in the growing season (Guzman et al., 1997).

Aphids transmit BCMV and BCMNV in a nonpersistent manner, which is the mode of aphid-mediated transmission utilized by potyviruses generally (Galvez & Morales, 1989; Morales, 2006). Studies using the electrical penetration graph (EPG, Fig. 3F) method show that this transmission mechanism requires a brief stylet penetration during probing of the epidermal cells of only a few seconds (3–15 s) for both virus acquisition and virus transmission (e.g., see Powell, 2004). Aphids retain potyviruses on their stylets for only a limited time if further feeding does not take place (Westwood & Stevens, 2010). Most of the literature describing aphid-mediated transmission of BCMV was published in the period 1950–1990. By and large, this work has not been followed up using more modern techniques such as EPG or revisited, following the reclassification of BCMV into two species, to take into account potential differences in the relationships of BCMV or BCMNV with their hosts or vectors.

Aphid vectors reported for BCMV and BCMNV include *Macrosiphum solanifolii*, *Macrosiphum pisi*, *Macrosiphum ambrosiae*, *Myzus persicae*, *Aphis rumicis*, *Aphis gossypii*, *Aphis medicaginis*, *Hyalopterus atriplicis*, and *Rhopalosiphum pseudobrassicae* (Zaumeyer & Meiners, 1975; Zettler & Wilkinson, 1966). In 1994, Halbert and colleagues studied the transmission of BCMV by five cereal aphids (*Diuraphis noxia*, *Metopholophium dirhodum*, *Rhopalosiphum padi*, *Schizaphis graminum*, and *Sitobion avenae*) as well as *Myzus persicae* (Halbert et al., 1994). They found that of all aphid species tested, only *D. noxia* transmitted BCMV (Halbert et al., 1994). Both BCMV and BCMNV are transmitted by *Acyrthosiphon pisum*, *Aphis craccivora*, *Aphis fabae*, and *Myzus persicae* (Silbernagel et al., 2001).

Most BCMV and BCMNV transmission experiments have used *Myzus persicae* and *Aphis fabae* as vectors (Galvez & Morales, 1989; Meiners et al., 1978; Melgarejo et al., 2007; Omunyin et al., 1995; Spence & Walkey, 1995). Typically, aphids are reared on a suitable host like pepper or cabbage, starved, and then transferred to the desired plants for viral acquisition or transmission (Galvez & Morales, 1989; Meiners et al., 1978; Melgarejo et al., 2007; Omunyin et al., 1995; Spence & Walkey, 1995). However, all aphids do not transmit all viruses with equal efficiency and aphid species-specific transmission characteristics are likely to be important factors in viral epidemiology.

In an investigation of the efficiency of various aphid species in transmitting CMV in bean, it was found that the bean specialist aphid *Aphis fabae* and a generalist aphid, *Myzus persicae*, were, in fact, poor vectors for this virus (Gildow et al., 2008). Some surveys have correlated the incidence of BCMV with the presence of *Aphis fabae* on infected plants (Omunyin et al., 1995) but it does not follow necessarily that this aphid was responsible for inoculation. Similar tests have also been done to determine the transmission efficiency of aphid species in transmitting PVY to potato. The results showed that the generalist aphid *Myzus persicae* had a transmission efficiency of 83.3%, followed by *Sitobion avenae* (a cereal specialist), transmitting PVY at a rate of 23%. Surprisingly, the transmission efficiency for the potato aphid, *Macrosiphum euphorbiae*, considered to be a specialist on this host had a transmission efficiency of only 3%, while two other aphid species in the study, *Rhopalosiphum padi* and *Aphis fabae*, did not transmit PVY (Boquel, Ameline, & Giordanengo, 2011). It is important to carry out more of these types of experiment for a wider range of crop–vector–virus pathosystems for two reasons. First, because too much may be assumed from the presence of specialist aphids on infected plants about their role in transmitting viruses to crops; generalists and nonspecialists may be more potent vectors. Second, in certain agricultural systems, for example in African smallholder farming, the intercropping of multiple plant species is common (Mucheru-Muna et al., 2010). While having other benefits, intercropping may exacerbate aphid, and therefore vector, diversity and availability.

7.3 Molecular Determinants of Aphid-Mediated Transmission and Host Effects

Like most potyviral proteins, the CP is multifunctional and required for multiple stages of the infection cycle. The CP is required for viral RNA amplification, aphid-mediated transmission, cell-to-cell and systemic movement, as well as for encapsidation of viral RNA and formation of the virions (Ivanov & Mäkinen, 2012; Rojas et al., 1997; Shukla, Frenkel, & Ward, 1991; Urcuqui-Inchima et al., 2001). But for successful aphid-mediated potyvirus transmission, the HC-Pro is essential. This was initially discovered by Govier and Kassanis (1974), who found that PVY-infected plants contained a soluble factor (the "helper component") that was loosely bound to virions and was needed for acquisition of the virus by aphids. Subsequent molecular analyses of potyviral CP and HC-Pro proteins showed that HC-Pro sequences contain the amino acid domain KITC, which allows interaction with a stylet receptor (Blanc et al., 1997, 1998), and that potyviral CPs possess the sequence DAG, which interacts with the PTK motif located in domain II of HC-Pro (Hasiów-Jaroszewska et al., 2014; Rojas et al., 1997; Urcuqui-Inchima et al., 2001). Through these interactions, dimers or higher-order oligomers of HC-Pro create a "bridge" linking the insect stylet and the CP subunits of the virion, without which the virus cannot be acquired by aphids (reviewed by Ng & Falk, 2006).

Analysis of the CP and HC-Pro sequences of a number of BCMV and BCMNV strains indicate that the "bridge" model explains the molecular interactions required for vectored transmission of these viruses. Sequencing of BCMV strains DeMV, NL-1, NL-1n, NL-4, NL-7, and NL-7n, as well as some BCMV strains collected in Australia showed that the BCMV CP possess the DAG motif (Coutts et al., 2011; Hamid et al., 2013; Hu et al., 1995; Sharma et al., 2011). Hamid et al. (2013) also identified a BCMV-specific conserved motif in the CP amino acid sequence (MVWCIDN) in strains NL-1, NL-4, and NL-7. Bhadramurthy and Bhat (2009) suggest that BCMV CP also has a conserved QMKAAA motif as well as the MVWCIDN motif. HC-Pro factors encoded by BCMV and BCMNV possess the PTK motif located in Domain II. Examination of published sequences for BCMV and BCMNV (e.g., see Bravo et al., 2008) shows that the places of the isoleucine and threonine residues of the KITC domain are taken by leucine and serine, and thus the HC-Pro proteins of BCMV and BCMNV possess a conserved KLSC motif. This substitution is seen in the HC-Pro sequences of certain other potyviruses, which group together phylogenetically with BCMV and BCMNV (Gibbs & Ohshima, 2010), such as *Zucchini yellow mosaic virus* (ZYMV) and *Soybean mosaic virus* (Domier et al., 2003; Maia et al., 1996).

Currently, the aphid receptor molecule(s) responsible for binding of potyvirus particles within the stylet have not been definitely identified. However, a *Myzus persicae* protein (MpRPS2) with homology to a ribosomal protein has been shown to bind TEV HC-Pro but not to a mutant HC-Pro with a modified KITC domain (Fernández-Calvino et al., 2010) and four potential interacting *Myzus persicae* proteins were found for the HC-Pro of ZYMV (which like the HC-Pro factors of BCMV and BCMNV has a KLSC sequence) (Dombrovsky et al., 2007).

Virus infection profoundly alters plant biochemistry (reviewed by Handford & Carr, 2006). There is growing evidence that these alterations may benefit viruses by, among other things, affecting the interactions of plants with insect vectors (Jones, 2014). This can be due to changes in the emission of volatile compounds to make plants more or less attractive to potential vectors (Mauck et al., 2010), or by modification of soluble plant metabolites to alter "taste," thereby influencing feeding behavior; phenomena which have been observed in the interactions of aphids (including *Myzus persicae* and *Aphis gossypii*) with plants infected with CMV (Carmo-Sousa et al., 2014; Mauck et al., 2010; Westwood et al., 2013; Ziebell et al., 2011). Thus, it has been hypothesized that the virus transmission mode (nonpersistent vs. persistent) may influence the coevolution of plant–virus–vector interactions to shape the biochemical reactions of infected plants to indirectly facilitate vectored transmission (Ingwell et al., 2012; Mauck et al., 2012; Westwood et al., 2013). Indeed, virus-induced changes in host biochemistry that influence the onward transmission of the virus might be considered to be "extended" viral phenotypes (discussed in Palukaitis et al., 2013).

This may also be the case for potyviruses. Boquel, Giordanengo, and Ameline (2011) observed that feeding behavior of the aphids *Macrosiphum euphorbiae* and *Myzus persicae* was affected differentially on potato plants infected with PVY, with feeding deterrence induced against the potato specialist (*Macrosiphum euphorbiae*) but resistance to feeding by *Myzus persicae* (a polyphagous aphid) was diminished. The effects of potyviruses on plant–aphid interactions are conditioned by specific viral gene products. Aphid (*Myzus persicae*) performance was enhanced on plants of *A. thaliana* and *N. benthamiana* following infection with TuMV (Casteel et al., 2014). In these TuMV-infected plants, it was found that deposition of callose (which has defensive properties against aphids) was decreased and the levels of certain amino acids were elevated in the phloem sap and that these changes were attributable to the TuMV NIa-Pro (Casteel et al., 2014). In contrast, Westwood et al. (2014) found that infection of *N. benthamiana* plants with PVY decreased the reproduction of *Myzus persicae* although the growth of the individual aphids placed on the infected plants was not inhibited. Although transgenic expression of the PVY HC-Pro in *N. benthamiana* inhibited jasmonic acid-induced plant gene expression (a potential anti-insect defense) and enhanced aphid reproduction, it was concluded that this viral factor is not the predominant determinant of virus-induced effects on aphid–plant interactions during infection (Westwood et al., 2014). Currently, it is not known if BCMV or BCMNV alter the behavior or performance (survival, growth, or reproduction) of aphids on infected bean plants.

It can be seen that the effects of virus infection on aphid–plant interactions are complex, and it has been proposed that the same virus can have different effects on different hosts in order to promote transmission between some, while enhancing survival and reproduction of vectors on other hosts (Westwood et al., 2013). This complex set of relationships among hosts, vectors, and viruses may offer a new route to control virus infection (reviewed by Bragard et al., 2013) and will be discussed further in Section 8.

8. CONTROL MEASURES

8.1 Genetic Resistance and Future Potential for Its Exploitation

BCMV and BCMNV can cause major losses in bean and several other leguminous crops. In response, common bean cultivars have been specifically bred for genetic resistance to both viruses. Cultivars with the dominant *I* gene are resistant to BCMV but susceptible to BCMNV-induced black root disease (see Section 4). Recessive resistance genes are available but they are virus strain–specific, and thus, it is difficult to breed bean cultivars that possess a broad resistance to many strains of BCMV and BCMNV based on one of these genes alone. Genetic markers linked to resistance genes for BCMV and BCMNV have been identified and could be used to underpin marker-assisted selection (Haley et al., 1994; Melotto et al., 1996; Morales & Kornegay, 1996). Therefore, some breeding efforts are focused on utilizing molecular markers to pyramid recessive genes such as *bc-u*, *bc-1*, *bc-1²*, *bc-2*, *bc-2²* and *bc-3*, with the dominant *I* gene in attempts to provide the broadest possible resistance (Kelly et al., 1995; Mukeshimana et al., 2005; Pasev et al., 2014).

Investigation of the mechanisms underlying recessive resistance have provided important insights into potyviral biology, in particular the exploitation of cellular eIF4E/eIF(iso)4E and eIF4G type translation initiation factors (Robaglia & Caranta, 2006; Truniger & Aranda, 2009). Bean genotypes carrying the *bc-3* gene for BCMV resistance were found to carry homozygous nonsilent mutations at codons 53, 65, 76, and 111 in a *PveIF4E* coding sequence and these mutations closely resembled a pattern of mutations determining potyvirus resistance in other plants (Naderpour et al., 2010). Hart and Griffiths (2013) showed that the *Bc-3* locus can have three *eIF4E* genes associated with it. This indicates a high level of variability at the locus, perhaps driven by coevolution with bean–infecting potyviruses.

Deeper understanding of the mechanistic basis of recessive resistance can provide a basis for the artificial generation of resistance (Wang & Krishnaswamy, 2012). For example, RNA silencing of an eIF4E gene (*Cm-eIF4E*) in transgenic melon (*Cucumis melo* L.) induced resistance to the potyviruses *Cucumber vein yellowing virus*, *Melon necrotic spot virus*, *Moroccan watermelon mosaic virus*, and ZYMV, indicating that the *Cm-eIF4E* gene controls melon susceptibility to these four viruses and therefore is an efficient target for the identification of new resistance alleles able to confer broad-spectrum virus resistance in melon (Rodríguez-Hernández et al., 2012).

The precision of genetic engineering permitted specific downregulation of the *Cm-eIF4E* transcript to be achieved without affecting expression of *Cm-eIF(iso)4E* (Rodríguez-Hernández et al., 2012). In a similar development, transgenic plum (*Prunus domestica*) plants expressing a silencing construct directed against *PdeIF(iso)4E* exhibited resistance to PPV, which causes the economically important Sharka disease in *Prunus* species (plums, peaches, apricots, cherries, and related ornamentals) (Wang et al., 2013). The approach exploited the interaction occurring between the PPV VPg and PdeIF(iso)4E to deprive the pathogen of a key factor needed for successful infection (Wang et al., 2013).

In the future, it may be possible to further refine the creation of artificial recessive resistance phenotypes using genome editing. This methodology, which produces mutations that are indistinguishable from those that can occur naturally, could be used to alter specific amino acid residues in eIF4-type translation factors rather than to downregulate entirely the expression of the gene encoding the factor. This would have the advantage of depriving a potyvirus of a cognate eIF4-type factor, while minimizing any negative effects on host plant fitness. Current genome editing systems are based on CRISPR (Clustered Regularly Interspaced Short Palindromic Repeats)/Cas (CRISPR-associated), TAL effector nuclease (TALEN), or zinc finger systems. CRISPR has certain advantages over methods using zinc finger and TALEN and has been used successfully to introduce mutations in genes in rice, wheat, sorghum, *Arabidopsis*, and *Nicotiana* species (Belhaj et al., 2013; Miao et al., 2013; Shan et al., 2013). As far as we are aware, no manipulation of viral host factors utilizing transformation with either conventional transformation technologies or genome editing has been attempted in bean.

The *I* gene confers resistance to BCMV but renders plants susceptible to a trailing systemic necrosis in response to infection with BCMNV (Section 4). To the best of our knowledge, the *I* gene is yet to be cloned. However, the *I* locus maps to a cluster of sequences with homology to *R* genes encoding plant immune receptors (R proteins) of the TIR–NB–LRR (Toll/interleukin-1-nucleotide binding site-leucine rich repeat) class (Vallejos et al., 2006). Although the specific member of this cluster conferring resistance to BCMV has not yet been identified, it is reasonable to suppose that it will be an R gene of the TIR–NB–LRR type. This means that once the *I* gene has been isolated there will be considerable scope for altering its properties so as to expand its ability and make it effective against BCMNV.

The rationale for suggesting this is based on work on the *Rx* gene of potato, which confers strong resistance to *Potato virus X* (PVX) but not against another potexvirus, *Poplar mosaic virus* (PopMV). Farnham and Baulcombe (2006) created a mutant *Rx* gene (*RxM1*) that, when expressed in transgenic plants, enabled them to respond to PopMV with a hypersensitive-like reaction. However, this response was not sufficient to completely restrict PopMV and virus continued to spread, leading to systemic necrosis and host death. The response of *RxM1*-transgenic plants to PopMV appears directly analogous to the response of *I*-gene-containing bean plants to BCMNV. That is, both types of plant possess an R gene product that mediates a weak recognition of a virus. In more recent work from the same group, Harris et al. (2013) found that additional mutagenesis of the RxM1 sequence to introduce changes in the amino acid sequence of the nucleotide-binding pocket of the Rx protein yielded novel immune receptors able to provide resistance against both PopMV and PVX. Thus, stepwise mutation of the *I* gene would be a logical step in expanding its ability to confer resistance not only against BCMV but also against BCMNV.

Expression of virus-derived gene sequences to in transgenic plants (pathogen-derived resistance) is a highly effective method of protection against viruses. Its use has been explored in a variety of crops, but despite its effectiveness, it has not been applied as widely as was anticipated in its early days (Gottula & Fuchs, 2009; Reddy et al., 2009). Over the past 29 years, various sequences encoding virus proteins such as CPs, replicases, and mutant movement proteins have been used to engender resistance (Gottula & Fuchs, 2009; Reddy et al., 2009). However, the most predictable and effective results appear to be attainable using transgenes expressing transcripts (double-stranded RNA (dsRNA) or artificial microRNAs) that trigger RNA interference (RNAi) (Gottula & Fuchs, 2009; Reddy et al., 2009). Indeed, this is the basis of one of the best-known uses of pathogen-derived resistance to protect against a potyvirus; the production of transgenic papaya resistant to *Papaya mosaic virus* (Fuchs & Gonsalves, 2007).

Unfortunately, the application of technologies based on plant transformation has been very slow in common bean; partly due to its relatively low efficiency for *Agrobacterium tumefaciens*-mediated transformation, which can be remedied by careful choice of *A. tumefaciens* strain (Mukeshimana et al., 2013), but most critically due to the recalcitrance of this plant to *in vitro* regeneration (Hnatuszko-Konka et al., 2014). The only successful study of which we are aware in which a transgenic approach has been used in common bean to generate resistance to a virus is for the *Begomovirus Bean golden mosaic virus* (BGMV) (Bonfim et al., 2007). Transgenic lines of a Pinto variety of common bean were produced using particle bombardment to introduce a T-DNA expressing a hairpin (RNAi-inducing) construct directed against the viral sequence encoding AC1, the "Rep" protein that facilitates replication of this DNA virus by the host DNA replication system. A BGMV-resistant line (EMBRAPA 5.1) has been field-tested under commercial conditions under which the inserted T-DNA appears to be inherited in a stable manner (Aragão & Faria, 2009; Aragão et al., 2013), and it had been reported that this line would be released for public consumption in Brazil in the near future (Tollefson, 2011). From the data published so far, it is not clear how the transgene engenders resistance; although virus sequence-specific short-interfering RNA molecules were detected in the plants of EMBRAPA 5.1 (Aragão et al., 2013; Bonfim et al., 2007), it is not clear if these are directing degradation of viral RNA transcripts encoding AC1 or methylation of the AC1 coding region of the viral genome. So far, the EMBRAPA 5.1 line appears to be the only successful instance of a transgenic bean crop with enhanced virus resistance.

A novel approach utilizing RNAi to obtain resistance without producing stably transformed plants exploits RNAi by the topical application of exogenous, sequence-specific dsRNA (Gan et al., 2010; Robinson et al., 2014; Tenllado & Diaz-Ruiz, 2001; Tenllado et al., 2003; Yin et al., 2009). Currently, there are limitations for the spray application of dsRNA for field use. However, a discussion of the technical aspects of this approach is beyond the scope of this chapter and readers are directed to Robinson et al. (2014) and Tenllado et al. (2004). Further research is being conducted by Worrall and Mitter (unpublished data) to design RNAi constructs targeting conserved regions of BCMV and potentially BCMNV (HC-Pro, PIPO, NIb, and CP). The aim is to validate these constructs for both transgenic and topical spray applications. Topical application will be aided by nanoparticle-based delivery of RNAi effector molecules in order to overcome the instability of dsRNA as well as to facilitate a controlled delivery at the leaf surface.

8.2 Interference with Transmission of BCMV and BCMNV

Inhibition of virus transmission could potentially provide resistance at the crop and population levels by inhibiting the initial spread and subsequent propagation of an epidemic (Westwood & Stevens, 2010). Conventional insecticide sprays have little impact on the transmission of nonpersistently transmitted viruses such as BCMV and BCMNV because virus acquisition and transmission occurs faster than the onset of toxicity (Westwood & Stevens, 2010). In addition, the number of insecticides available for control of aphids is diminishing due to environmental or health concerns and the increasing incidence of pesticide resistance in aphid populations (Westwood & Stevens, 2010). In any case, these chemicals are relatively expensive, rendering them impractical for use by many small-holder farmers in developing countries.

Early studies revealed diluted mineral oil to be an effective, nontoxic spray method effective against transmission by aphids of TuMV, *Beet yellows virus*, BCMV, and BYMV in the form of diluted mineral oil, which inhibited aphid transmission of these viruses (Walkey & Dance, 1979). Walkey and Dance (1979) found that the sprays diluted to 2.5% and 5% effectively controlled viral transmission by *Myzus persicae*, *Brevicoryne brassicae*, and *Aphis fabae*. Work in this area has continued. For example, recently Boquel et al. (2013) found that mineral oil was effective at inhibiting the acquisition of PVY by *Rhopalosiphum padi* from infected potato plants. However, it is unknown whether this method would be affordable or cost-effective for small-scale farmers.

Studies of interactions of plants with aphids, viruses, and with beneficial insects have suggested new directions for using natural volatile signal chemicals (often termed "semiochemicals") to inhibit or deter aphids from approaching plants. This can be done by utilizing known alarm pheromones such as (*E*)-β-farnesene, through for example, engineering crop plants to emit this compound, or other semiochemicals (reviewed by Pickett et al., 2012). Wheat plants transformed to biosynthesize (*E*)-β-farnesene are undergoing field trials in the United Kingdom (Pickett et al., 2012). Investigation of how viruses themselves alter the emission of volatiles to aid in vectoring (see Section 7.3) may reveal novel semiochemicals or blends to use to deter or trap aphids. Application of these methodologies to bean will require improvements in transformation (see Section 9). In the meantime, there is potential for utilizing plant extracts as homemade semiochemical-based aphid repellents by smallholder farmers (Pickett et al., 2012).

An additional route to inhibiting BCMV or BCMNV infection may lie with plant resistance to aphid feeding or infestation, which can be conferred by naturally occurring genetic resistance or engendered by genetic engineering (reviewed by Westwood & Stevens, 2010 and Dogimont et al., 2010). Just as NBS-LRR immune receptors can trigger hypersensitive resistance following specific recognition of a pathogen effector molecule (see Section 8), there are also plant *R* genes (known to or thought likely to encode NBS-LRRs) that mediate recognition of and resistance to specific aphid species. Examples of these include *Vat* in melon that mediates hypersensitive resistance to the aphid *Acyrthosiphon pisum* (Sarria

Villada et al., 2009) and *RAP1* in *Medicago truncatula*, which confers extreme resistance to *Acyrthosiphon pisum* (Stewart et al., 2009). Another interesting example is the tomato *Mi-1* gene, which confers resistance to aphids (*Macrosiphum euphorbiae*) in leaf tissue and to nematodes in roots. Goggin et al. (2006), however, found that when expressed transgenically in *Solanum melongena* (eggplant: also known as aubergine or brinjal) *Mi-1* retained only its antinematode property. This may show how *R* gene transfer between even relatively closely related plants (in this case, *S. lycopersicum* to *S. melongena*) may not always be straightforward.

Another genetically based resistance strategy to combat herbivorous invertebrates is generation of transgenic plants expressing RNA precursors for short-interfering RNAs or artificial microRNAs directed against targets in the pest; it is envisaged that the feeding invertebrate will take up these molecules, which will be processed (or will have been preprocessed in the plant) to a mature form that will interfere with gene expression (reviewed in Westwood & Stevens, 2010). This approach was shown by Pitino et al. (2011) to be viable against aphids. These workers showed that dsRNAs containing aphid sequence-specific sequences expressed in transgenic *A. thaliana* were processed into short-interfering RNAs, taken up during feeding by *Myzus persicae* and inhibited expression of a gut-specific and a salivary-gland-specific transcript. Small but statistically significant decreases in the reproduction of aphids placed on these plants were noted for lines expressing either of these constructs (Pitino et al., 2011). This suggests that expression of combinations of "anti-aphid" dsRNAs could be highly effective in decreasing the impact of aphid infestation. Regrettably, however, because of the rapidity with which BCMV, BCMNV, and other nonpersistently transmitted viruses are injected by aphids into a healthy host or acquired by them from an infected plant, that neither this approach nor the use of natural aphid specific *R* genes would inhibit virus spread in a crop.

9. GRAND CHALLENGES: AN AFRICAN PERSPECTIVE ON CONTROL OF BCMV AND BCMNV

9.1 BCMV and BCMNV: Threats to African Smallholder Survival and Prosperity

Common bean was introduced into Africa about 400 years ago (Greenway, 1945) and has become a major crop underpinning nutritional and economic security for smallholder farmers in Eastern, Southern, and Central

Africa, as well as a major commercial crop in these regions (Morales, 2006). Across this area, over 150 million people rely on common bean as a staple crop, and around 3 million tons are produced each year, on almost 5 million ha. (Broughton et al., 2003; Jansa et al., 2011). Beans are a growing part of the export market (Beebe et al., 2013). For example, snap (or French) beans are a major export from a number of African countries. Over 90% of the production in East Africa is exported internationally and it has been estimated that over a million people in Kenya profit from Snap Bean cultivation (Highlights CIAT Africa http://ciat-library.ciat.cgiar.org/Articulos_Ciat/highlight31.pdf).

Smallholder farmers, who usually intersperse bean with other key crops, produce a considerable proportion of this yield. As an intercrop bean supports cultivation of maize, cassava, and other starch crops by: soil enrichment through nitrogen fixation (Broughton et al., 2003; Mucheru-Muna et al., 2010), protection from pests and diseases (including in some configurations of the "push–pull" pest and disease control system: Khan et al., 2009), and by enhancing water use efficiency in mixed bean-maize crops in arid zones (Walker & Ogindo, 2003). Bean can be highly productive and in some areas within Eastern, Southern, and Central Africa as many as 3 bean crops per year are possible.

Per capita consumption of common bean in Africa is the highest globally and bean is estimated to be second and third most important source of dietary protein and calories, respectively, in Eastern and Southern Africa (Broughton et al., 2003). In Rwanda and Burundi, consumption exceeds 40 kg/year and in rural Kenya and Uganda it reaches 66 and 58 kg/year, respectively. In Rwanda, for urban and rural populations alike, bean provides over 60% of dietary protein (Jansa et al., 2011). Common bean is consumed across all of its growing stages; in the form of leaves, green pods, as well as fresh and dry grain. Importantly, common beans can be stored dry for long periods, making this crop of critical value for food security in the region (CIAT, 2009). Bean is a crop of particular economic value to women. In sub-Saharan Africa, bean is often grown and traded by women and it has been noted that in Africa increased wealth creation by female members of the community has a particularly strong positive impact on food security, nutrition, child health, and school attendance rates (Kevane, 2012).

While staples like maize, cassava, sweet potato, enset (*Ensete ventricosum*), or rice provide dietary starch for most smallholder farmers in Africa, common beans and other legume crops provide the most significant sources of dietary protein (Broughton et al., 2003). Beans also enrich the diet with several vitamins but perhaps more importantly they provide mineral nutrients, most notably iron and zinc, as well as calcium, copper, magnesium, manganese, and phosphorus (Broughton et al., 2003). The contribution of iron to the diet is particularly vital in developing countries in which nutritional anemia due to iron deficiency is widespread and, among other health impacts, can increase maternal mortality (Anon, 2014; Broughton et al., 2003). World Health Organization studies have found that the prevalence of anemia in Africa ranges from 15% to 50% for adult females, 35–72% for pregnant women, 30–60% for children under the age of 15, and 6–17% for adult males. Fortunately, in parts of Kenya and Uganda, bean consumption is high enough to exceed the minimum dietary iron requirement, and for many other countries in the region it provides well over half of the requirement (Broughton et al., 2003).

However, bean production in Eastern, Southern, and Central Africa is not currently meeting demand and regionally yields are very low. Indeed, resource-poor farmers in Africa currently harvest around 500 kg/ha. but the potential yield is an order of magnitude higher (Verdoodt et al., 2004) and BCMV and BCMNV infection contributes to these losses (Beebe et al., 2013). The situation is thought likely to be exacerbated by climate change that could increase challenges to crops from abiotic stresses such as drought or increased biotic attack due to alterations in the range or populations of aphid vectors (Canto et al., 2009).

9.2 Combating BCMV and BCMNV

The good news is that sources of resistance are available for breeding varieties of common bean resistant to BCMV and BCMNV (see Section 8). However, as discussed earlier in this chapter, deployment of the *I* gene in Africa can lead to problems when BCMNV is encountered (see Section 5). Indeed, BCMNV is found in wild legumes and weeds around and within smallholder farmer fields (Coyne et al., 2003; Mukeshimana et al., 2005; Sengooba et al., 1997). The *I* gene coupled with the recessive resistance gene *bc-3* can provide resistance to all known isolates of BCMV and BCMNV (Singh et al., 2008), and Mukeshimana et al. (2005) have identified markers that are being used to breed these two genes into farmer-preferred varieties in Africa. In the future, it is likely that other sources of resistance may be identified in *Phaseolus vulgaris*, given that <5% of the genetic diversity of this species has been incorporated into cultivated varieties, with only a portion of this diversity present in Africa (Beebe et al., 2013; Broughton et al., 2003; Wortmann et al., 1995). Bean breeders working within African National Agricultural Research Systems (NARS) have a wealth of expertise and knowledge about farmer preferences and the constraints faces by smallholders. A range of resources is available to empower African plant breeders to further their efforts to develop virus-resistant, farmer-preferred varieties. In addition to the materials already developed and in use by national breeding programs, CIAT provides breeding materials and other support for incorporation of resistance into varieties by African NARS breeders. For additional technical support, genotyping services and availability of high-end biosciences laboratories for breeding-related research, other initiatives in the region such as the BecA-ILRI Hub (http://hub.africabiosciences.org/) provide an increasingly available set of capacity building and molecular breeding support services to NARS breeding programs. For a detailed description of how African NARS breeding programs and their international partners apply these types of approaches, see Kelemu et al. (2012). National bean breeders of 29 African countries are further supported across the value chain by the Pan African Bean Research Alliance (www.pabra-africa.org).

Establishment of clean seed supply chains is also a critical focus for addressing the challenge of these seed-transmitted viruses. Contaminated bean seed can provide foci for aphid-mediated transmission. However, lack of reliable tissue culture and regeneration protocols stands in the way of both tissue culture-based production of clean planting materials and inhibit the development of virus- or vector-resistant transgenic bean varieties (see Section 7). Detailed information about the geographic distribution and genetic diversity of the viruses and their vectors in Africa, however, is lacking. But in the future this will be required to inform breeding, clean seed production strategies, and other interventions to fight these pathogens. Efforts are underway to characterize the viruses present in bean on smallholder farms in East Africa, which can be extended across the region. To underpin this, the development of appropriate diagnostic tests for use by African breeding programs, phytosanitary organizations, and NARS and African universities is underway. New, facile methods of diagnosis under development at the BecA-ILRI Hub include immunodiagnostics, loop-mediated isothermal amplification (LAMP), or other assays amenable to use with minimal laboratory equipment or in the field.

Decreasing the impact of bean-infecting potyviruses on production of beans by smallholder farmers in Africa will enhance the well-being of growers and consumers. As we have seen, many potential technical solutions exist or are in development. However, a key aspect of deployment of these in Africa must be the engagement and involvement of African scientists as well as the smallholder farmers themselves in developing the most appropriate suite of crop protection strategies.

ACKNOWLEDGMENTS

We thank Valentine Aritua, Robin Buruchara, and Mathew Abang for stimulating discussions on bean–virus interactions. We also acknowledge Andrew Firth, Allan Olspert, and Betty Chung for discussion on potyvirus gene expression mechanisms, and Alex Murphy, Trisna Tungadi, Jack Westwood, Niels Groen, Heiko Ziebell, Toby Bruce, and John Pickett for discussions of virus–aphid–plant interactions. We thank Christine Alexander for help with tracking down difficult-to-find references from the older literature. Work on bean-infecting viruses and potyviruses in the labs of J.J.W.H and J.P.C. is funded by Sustainable Crop Production Research for International Development (SCPRID), a research programme funded by the UK Biotechnology and Biological Sciences Research Council (BBSRC) with cofunding from the UK Department for International Development, the Bill & Melinda Gates Foundation, the Department of Biotechnology of India's Ministry of Science and Technology, and the Indian Council of Agricultural Research (Ref. BB/J011762/1). Other work on plant–virus interactions in the JPC lab are funded by Grants from the BBSRC (BB/J015652/1) and the Leverhulme Trust (RPG-2012-667). Elizabeth Worall's Ph.D. program with Associate Prof. Mitter is supported by a scholarship from the University of Queensland. The work on BCMV management strategies in the Mitter lab is supported by Accelerate Partnership Grant from the Queensland Government with Nufarm Australia Ltd. as industry partner.

REFERENCES

Adams, M. J., Antoniw, J. F., & Beaudoin, F. (2005). Overview and analysis of the polyprotein cleavage sites in the family *Potyviridae*. *Molecular Plant Pathology*, 6, 471–478.

Agbeci, M., Grangeon, R., Nelson, R. S., Zheng, H., & Laliberte, J. F. (2013). Contribution of host intracellular transport machineries to intercellular movement of *Turnip mosaic virus*. *PLoS Pathogens*, 9(10), e1003683. http://dx.doi.org/10.1371/journal.ppat.1003683.

Ali, M. A. (1950). Genetics of resistance to the common bean mosaic virus (bean virus 1) in the bean (*Phaseolus vulgaris* L.). *Phytopathology*, 40, 69–79.

Ali, A., Natsuaki, T., & Okuda, S. (2006). The complete nucleotide sequence of a Pakistani isolate of Watermelon mosaic virus provides further insights into the taxonomic status in the *Bean common mosaic virus* subgroup. *Virus Genes*, 32, 307–311.

Anandalakshmi, R., Marathe, R., Xin, G., Herr, J. M., Jr., Mau, C., Malory, A., et al. (2000). A calmodulin-related protein that suppresses posttranscriptional gene silencing in plants. *Science*, 290, 142–144.

Anandalakshmi, R., Pruss, G. J., Xin, G., Marathe, R., Mallory, A. C., Smith, T. H., et al. (1998). A viral suppressor of gene silencing in plants. *Proceedings of the National Academy of Sciences of the United States of America*, 95, 13079–13084.

Anon. (2014). Nutritional anemia (chapter 9). In: Mother and child nutrition in the tropics and subtropics. pp. 339–356, in Journal of Tropical Pediatrics On-Line Supplement http://www.oxfordjournals.org/our_journals/tropej/online/mcnts_chap9.pdf (accessed October 25 2014).

Aragão, F. J. L., & Faria, J. C. (2009). First transgenic geminivirus-resistant plant in the field. *Nature Biotechnology, 27*, 1086–1088.

Aragão, F. J. L., Mogueira, E. O. P. L., Tinoco, M. L. P., & Faria, J. C. (2013). Molecular characterization of the first commercial transgenic common bean immune to the *Bean golden mosaic virus. Journal of Biotechnology, 166*, 42–50.

Beaver, J. S., Rosas, J. C., Myers, J., Acosta, J., Kelly, J. D., Nchimbi-Msolla, S., et al. (2003). Contributions of the bean/cowpea CRSP to cultivar and germplasm development in common bean. *Field Crops Research, 82*, 87–102.

Beebe, S., Rao, I., Mukankusi, C., & Buruchara, R. (2013). Improving resource use efficiency and reducing risk of common bean production in Africa, Latin America and the Caribbean. In C. H. Hershey & P. Neate (Eds.), *Eco-efficiency, from vision to reality* (pp. 117–134): Cali, Colombia: CIAT. http://ciat.cgiar.org/wp-content/uploads/2013/04/eco-efficiency_book.pdf.

Belhaj, K., Chapparo-Garcia, A., Kamoun, S., & Nekrasov, V. (2013). Plant genome editing made easy: Targeted mutagenesis in model and crop plants using the CRISPR/Cas system. *Plant Methods, 9*, 39.

Berger, P., Wyatt, S., Shiel, P., Silbernagel, M., Druffel, K., & Mink, G. I. (1997). Phylogenetic analysis of the *Potyviridae* with emphasis on legume-infecting potyviruses. *Archives of Virology, 142*, 1979–1999.

Bhadramurthy, V., & Bhat, A. (2009). Biological and molecular characterization of *Bean common mosaic virus* associated with vanilla in India. *Indian Journal of Virology, 20*, 70–77.

Blanc, S., Ammar, E. D., Garcia-Lampasona, S., Dolja, V. V., Llave, C., Baker, J., et al. (1998). Mutations in the potyvirus helper component protein: Effects on interactions with virions and aphid stylets. *Virology, 79*, 3119–3122.

Blanc, S., Lopez-Moya, J. J., Wang, R. Y., Garcia-Lampasona, S., Thornbury, D. W., & Pirone, T. P. (1997). A specific interaction between coat protein and helper component correlates with aphid transmission of a potyvirus. *Virology, 231*, 141–147.

Bonfim, K., Faria, J. C., Nogueira, E. O., Mendes, É. A., & Aragão, F. J. (2007). RNAi-mediated resistance to *Bean golden mosaic virus* in genetically engineered common bean (*Phaseolus vulgaris*). *Molecular Plant-Microbe Interactions, 20*, 717–726.

Boquel, S., Ameline, A., & Giordanengo, P. (2011). Assessing aphids' potato virus Y transmission efficiency: A new approach. *Journal of Virological Methods, 178*, 63–67.

Boquel, S., Giguere, M. A., Clark, C., Nanayakkara, U., Zhang, J. A., & Pelletier, Y. (2013). Effect of mineral oil on *Potato virus Y* acquisition by *Rhopalosiphum padi*. *Entomologia Experimentalis et Applicata, 148*, 48–55.

Boquel, S., Giordanengo, P., & Ameline, A. (2011). Divergent effects of PVY-infected potato plant on aphids. *European Journal of Plant Pathology, 129*, 507–510.

Bos, L., & Gibbs, A. J. (1995). *Bean common mosaic* potyvirus. *Plant viruses online—descriptions and lists from the VIDE database.* http://sdb.im.ac.cn/vide/descr068.htm.

Bragard, C., Caciagli, P., Lemaire, O., Lopez-Moya, J. J., MacFarlane, S., Peters, D., et al. (2013). Status and prospects of plant virus control through interference with vector transmission. *Annual Review of Phytopathology, 51*, 171–201.

Bravo, E., Calvert, L. A., & Morales, F. J. (2008). The complete nucleotide sequence of the genomic RNA of *Bean common mosaic virus* strain NL4. *Revista de la Academia Colombiana de Cienccias Exactas, Fisicas y Naturales, 32*, 37–46.

Brigneti, G., Voinnet, O., Li, W. X., Ji, L. H., Ding, S. W., & Baulcombe, D. C. (1998). Viral pathogenicity determinants are suppressors of transgene silencing in *Nicotiana benthamiana. The EMBO Journal, 17*, 6739–6746.

Broughton, W. J., Hernandez, G., Blair, M., Beebe, S., Gepts, P., & Vanderleyden, J. (2003). Beans (*Phaseolus* spp.)—model food legumes. *Plant and Soil, 252*, 55–128.

Canto, T., Aranda, M. A., & Fereres, A. (2009). Climate change effects on physiology and population processes of hosts and vectors that influence the spread of hemipteran-borne plant viruses. *Global Change Biology, 15*, 1884–1894.

Carmo-Sousa, M., Moreno, A., Garzo, E., & Fereres, A. (2014). A non-persistently transmitted-virus induces a pull-push strategy in its aphid vector to optimize transmission and spread. *Virus Research, 186*, 38–46.

Casteel, C. L., Yang, C. L., Nanduri, A. C., DeJong, H. N., Whitham, S. A., & Jander, G. (2014). The NIa-Pro protein of *Turnip mosaic virus* improves growth and reproduction of the aphid vector, *Myzus persicae* (green peach aphid). *Plant Journal, 77*, 653–663.

Chisholm, S. T., Parra, M. A., Anderberg, R. J., & Carrington, J. C. (2001). Arabidopsis *RTM1* and *RTM2* genes function in phloem to restrict long-distance movement of tobacco etch virus. *Plant Physiology, 127*, 1667–1675.

Chiumia, L. M., & Msuku, W. A. B. (2001). Status of common bean mosaic virus in common beans in Malawi. In *Proceedings of bean/cowpea collaborative research support program—East Africa*, Arusha, Tanzania: Washington State University.

Chung, B. Y.-W., Miller, W. A., Atkins, J. F., & Firth, A. E. (2008). An overlapping essential gene in the Potyviridae. *Proceedings of the National Academy of Sciences of the United States of America, 105*, 5897–5902.

CIAT. (2009). *Common bean: The nearly perfect food*. Cali, Colombia: CIAT. www.ciat.org/ciatinfocus/beans.htm.

Collmer, C. W., Marston, M. F., Albert, S. M., Bajaj, S., Maville, H. A., Ruuska, S. A., et al. (1996). The nucleotide sequence of the coat protein and 3' untranslated region of Azuki bean mosaic potyvirus, a member of the bean common mosaic virus subgroup. *Molecular Plant-Microbe Interactions, 9*, 758–761.

Coutts, B. A., Kehoe, M. A., Webster, C. G., Wylie, S. J., & Jones, R. A. C. (2011). Indigenous and introduced potyviruses of legumes and *Passiflora* spp. from Australia: Biological properties and comparison of coat protein nucleotide sequences. *Archives of Virology, 156*, 1757–1774.

Coyne, D. P., Steadman, J. R., Godoy-Lutz, G., Gilbertson, R., Arnaud-Santana, E., Beaver, J. S., et al. (2003). Contribution of the Bean/Cowpea CRSP to management of bean disease. *Field Crops Research, 82*, 87–102.

Damayanti, T., Susilo, D., Nurlaelah, S., Sartiami, D., Okuno, T., & Mise, K. (2008). First report of *Bean common mosaic virus* in yam bean [*Pachyrhizus erosus* (L.) Urban] in Indonesia. *Journal of General Plant Pathology, 74*, 438–442.

Dean, L. L., & Hungerford, C. W. (1946). A new bean mosaic in Idaho. *Phytopathology, 36*, 324.

Deligoz, I., & Soken, M. A. (2011). Differentiation of Bean Common Mosaic Virus (BCMV) and Bean Common Mosaic Necrosis Virus (BCMNV) strains infecting common bean in Samsun province. *The Journal of Turkish Phytopathology, 37*, 1–14.

Demski, J. W., Reddy, D. V. R., Sowell, G., & Bays, D. (1984). Peanut stripe virus-a new seed-borne potyvirus from China infecting groundnut (*Arachis hypogaea*). *Annals of Applied Biology, 105*, 495–501.

Demski, J. W., & Warwick, D. (1986). Testing peanut seeds for peanut stripe virus 1. *Peanut Science, 13*, 38–40.

Dogimont, C., Bendahmane, A., Chovelon, V., & Boissot, N. (2010). Host plant resistance to aphids in cultivated crops: Genetic and molecular bases, and interactions with aphid populations. *Comptes Rendus Biologies, 333*, 566–573.

Dombrovsky, A., Gollop, N., Chen, S., Chejanovsky, N., & Raccah, B. (2007). *In vitro* association between the helper compoenet-proteinase of zucchini yellow mosaic virus and cuticle proteins of *Myzus persicae*. *Journal of General Virology, 88*, 1602–1610.

Domier, L. L., Lattore, I. J., Steinlage, T. A., McCoppin, N., & Hartman, G. L. (2003). Variability and transmission by *Aphis glycines* of North American and Asian *Soybean mosaic virus* isolates. *Archives of Virology, 148*, 1925–1941.

Drijfhout, E., & Bos, L. (1977). The identification of two new strains of bean common mosaic virus. *Netherlands Journal of Plant Pathology, 83*, 13–26.

Drijfhout, E., Silbernagel, M. J., & Burke, D. W. (1978). Differentiation of strains of *Bean common mosaic virus*. *Netherlands Journal of Plant Pathology, 84*, 13–26.

Ekpo, E. J. A., & Saettler, A. W. (1974). Distribution pattern of bean common mosaic virus in developing bean seed. *Phytopathology*, *64*, 269–270.

El-Sawy, M. A., Mohamed, H. A. E., & Elsharkawy, M. M. (2013). Serological and molecular characterisations of the Egyptian isolate of *Bean common mosaic virus*. *Archives of Phytopathology and Plant Protection*, *47*, 1–13.

Fang, G. W., Allison, R. F., Zambolim, E. M., Maxwell, D. P., & Gilbertson, R. L. (1995). The complete nucleotide sequence and genome organization of *Bean common mosaic virus* (NL3) strain. *Virus Research*, *39*, 13–23.

Farnham, G., & Baulcombe, D. C. (2006). Artificial evolution extends the spectrum of viruses that are targeted by a disease resistance gene from potato. *Proceedings of the National Academy of Sciences of the United States of America*, *103*, 18828–18833.

Feng, X., Poplawsky, A. R., Nikolaeva, O. V., Myers, J. R., & Karasev, A. (2014). Recombinants of bean common mosaic virus (BCMV) and genetic determinants of BCMV involved in overcoming resistance in common bean. *Phytopathology*, *104*, 786–793.

Fernández-Calvino, L., Goytia, E., López-Abilla, D., Giner, A., Urizarna, M., Vilaplana, L., et al. (2010). The helper-component protease transmission factor of tobacco etch potyvirus binds specifically to an aphid ribosomal protein homologous to the laminin receptor precursor. *Journal of General Virology*, *91*, 2862–2873.

Flasinski, S., Gunasinghe, U. B., Gonzales, R. A., & Cassidy, B. G. (1996). The cDNA sequence and infectious transcripts of peanut stripe virus. *Gene*, *171*, 299–300.

Flores-Estévez, N., Acosta-Gallegos, J., & Silva-Rosales, L. (2003). *Bean common mosaic virus* and *Bean common mosaic necrosis virus* in Mexico. *Plant Disease*, *87*, 21–25.

Forster, R. L., Meyers, J. R., & Berger, P. H. (1991). *Bean common mosaic virus*. Idaho: Pacific Northwest Extension Publication.

Fuchs, M., & Gonsalves, D. (2007). Safety of virus-resistant transgenic plants two decades after their introduction: Lessons from realistic field risk assessment studies. *Annual Review of Phytopathology*, *45*, 173–202.

Galvez, G., & Morales, F. (1989). Aphid-transmitted viruses. In H. F. Schwartz & M. A. Pastor-Corrales (Eds.), *Bean production problems in the tropics* (2nd ed., pp. 211–240). Cali, Colombia: Centro Internacional de Agricultura Tropical.

Gan, D., Zhang, J., Jiang, H., Jiang, T., Zhu, S., & Cheng, B. (2010). Bacterially expressed dsRNA protects maize against SCMV infection. *Plant Cell Reports*, *29*, 1261–1268.

Gibbs, A. J., & Ohshima, K. (2010). Potyviruses and the digital revolution. *Annual Review of Phytopathology*, *48*, 205–223.

Gibbs, A. J., Ohshima, K., Phillips, M. J., & Gibbs, M. J. (2008). The prehistory of potyviruses: Their initial radiation was during the dawn of agriculture. *PLoS One*, *3*, e2523.

Gibbs, A. J., Trueman, J., & Gibbs, M. J. (2008). The bean common mosaic virus lineage of potyviruses: Where did it arise and when? *Archives of Virology*, *153*, 2177–2187.

Gildow, F. E., Shah, D. A., Sackett, W. M., Butzler, T., Nault, B. A., & Fleischer, S. J. (2008). Transmission efficiency of *Cucumber mosaic virus* by aphids associated with virus epidemics in snap bean. *Phytopathology*, *98*, 1233–1241.

Goggin, F. L., Jia, L. L., Shah, G., Hebert, S., Williamson, V. M., & Ullman, D. E. (2006). Heterologous expression of the *Mi-1* gene form tomato confers resistance against nematodes but not aphids in eggplant. *Molecular Plant-Microbe Interactions*, *19*, 383–388.

Gottula, J., & Fuchs, M. (2009). Towards a quarter century of pathogen-derived resistance and practical approaches to plant virus disease control. *Advances in Virus Research*, *75*, 161–183.

Govier, D. A., & Kassanis, B. (1974). A virus induced component of plant sap needed when aphids acquire potato virus Y from purified preparations. *Virology*, *61*, 420–426.

Grangeon, R., Agbeci, M., Chen, J., Grondin, G., Zheng, H., & Laliberté, J. F. (2012). Impact on the endoplasmic reticulum and Golgi apparatus of turnip mosic virus infection. *Journal of Virology*, *86*, 9255–9265.

Grangeon, R., Jiang, J., Wan, J., Agbeci, M., Zheng, H., & Laliberté, J. F. (2013). $6K_2$–induced vesicles can move cell to cell during turnip mosaic virus infection. *Frontiers in Microbiology, 4*, 351.

Greenway, P. J. (1945). Origins of some East African food plants. *East African Agricultural Journal, 10*, 34–39.

Gunasinghe, U. B., Flasinski, S., Nelson, R. S., & Cassidy, B. G. (1994). Nucleotide sequence and genome organization of peanut stripe potyvirus. *The Journal of General Virology, 75*, 2519–2525.

Guzman, P., Rojas, M. R., Davis, R. M., Kimble, K., Stewart, R., Sundstrom, F. J., et al. (1997). First report of bean common mosaic necrosis potyvirus (BCMNV) infecting common bean in California. *Plant Disease, 81*, 831.

Halbert, S. E., Mink, G. I., Silbernagel, M. J., & Mowry, T. M. (1994). Transmission of *Bean common mosaic virus* by cereal aphids (*Homoptera: Aphididae*). *Plant Disease, 78*, 983–985.

Haley, S. D., Afanador, L., & Kelly, J. D. (1994). Identification and application of a random amplified polymorphic DNA marker for the I gene (potyvirus resistance) in common bean. *Phytopathology, 84*, 157–160.

Hamid, A., Ahmad, M., Padder, B., Shah, M., Saleem, S., Sofi, T., et al. (2013). Pathogenic and coat protein characterization confirming the occurrence of *Bean common mosaic virus* on common bean (*Phaseolus vulgaris*) in Kashmir, India. *Phytoparasitica, 46*, 1–6.

Hampton, R. O. (1975). The nature of bean yield reduction by bean yellow and bean common mosaic virus. *Phytopathology, 65*, 1342–1346.

Hampton, R. O., Silbernagel, M. J., & Burke, D. W. (1983). Bean common mosaic virus strains associated with bean mosaic epidemics in the northwestern United States. *Plant Disease, 67*, 658–661.

Handford, M. G., & Carr, J. P. (2006). Plant metabolism associated with resistance and susceptibility. In G. Loebenstein, & J. P. Carr (Eds.), *Natural resistance mechanisms of plants to viruses* (pp. 315–340). Netherlands: Springer.

Harris, C. J., Slootweg, E. J., Goverse, A., & Baulcombe, D. C. (2013). Stepwise artificial evolution of a plant disease resistance gene. *Proceedings of National Academy of the Sciences of the United States of America, 110*, 21189–21194.

Hart, J. P., & Griffiths, P. D. (2013). A series of *eIF4E* alleles at the *Bc-3* locus are associated with recessive resistance to *Clover yellow vein virus* in common bean. *Theoretical and Applied Genetics, 126*, 2849–2863.

Hasiów-Jaroszewska, B., Fares, M. A., & Elena, S. F. (2014). Molecular evolution of viral multifunctional proteins: The case of potyvirus HC-Pro. *Journal of Molecular Evolution, 78*, 75–86.

Hnatuszko-Konka, K., Kowalczyk, T., Gerszberg, A., Wiktorek-Smagur, A., & Kononowicz, A. K. (2014). Phaseolus vulgaris—Recalcitrant potential. *Biotechnology Advances, 32*(7), 1205–1215. http://dx.doi.org/10.1016/j.biotechadv.2014.06.001 (available on-line).

Hoch, H., & Provvidenti, R. (1978). Ultrastructural localization of bean common mosaic virus in dormant and germinating seeds of *Phaseolus vulgaris*. *Phytopathology, 68*, 327–330.

Hu, J., Ferreria, S., Wang, M., Borth, W., Mink, G. I., & Jordan, R. (1995). Purification, host range, serology, and partial sequencing of *Dendrobium* mosaic potyvirus, a new member of the *Bean common mosaic virus* subgroup. *Phytopathology, 85*, 542–546.

Hull, R. (2014). *Plant virology* (5th ed.). New York: Academic Press.

ICTV. (2013). *Virus taxonomy: 2013 release.* http://www.ictvonline.org/virusTaxonomy.asp.

Ingwell, L. L., Eigenbrode, S. D., & Bosque-Pérez, N. A. (2012). Plant viruses alter insect behavior to enhance their spread. *Scientific Reports, 2*, 578.

Ivanov, K. I., Eskelin, K., Lõhmus, A., & Mäkinen, K. (2014). Molecular and cellular mechanisms underlying potyvirus infection. *Journal of General Virology, 95*, 1415–1429.

Ivanov, K. I., & Mäkinen, K. (2012). Coat proteins, host factors and plant viral replication. *Current Opinion in Virology, 2,* 712–718.

Jansa, J., Bationo, A., Frossard, E., & Rao, I. M. (2011). Options for improving plant nutrition to increase common bean productivity in Africa. In A. Bationo et al. (Eds.), *Fighting poverty in sub-saharan Africa: The multiple roles of legumes in integrated soil fertility management.* Dordrecht Netherlands: Springer.

Jenkins, W. A. (1940). A new virus disease of snap beans. *Journal of Agricultural Research (Washington, D.C.), 60,* 279–288.

Jenkins, W. A. (1941). A histological study of snap bean tissues affected with black root. *Journal of Agricultural Research (Washington, D.C.), 62,* 683–690.

Jones, R. A. C. (2014). Plant virus ecology and epidemiology: Historical perspectives, recent progress and future prospects. *Annals of Applied Biology, 164,* 320–347.

Jossey, S., Hobbs, H. A., & Domier, L. L. (2013). Role of *Soybean mosaic virus*-encoded proteins in seed and aphid transmission in soybean. *Phytopathology, 103,* 941–948.

Kapil, R., Prachi, S., Sharma, S. K., Sharma, O. P., Sharma, O. P., Dhar, J. B., et al. (2011). Pathogenic and molecular variability in *Bean common mosaic virus* infecting common bean in India. *Archives of Phytopathology and Plant Protection, 44,* 1081–1092.

Kasschau, K. D., & Carrington, J. C. (1998). A counterdefensive strategy of plant viruses: Suppression of posttranscriptional gene silencing. *Cell, 95,* 461–470.

Kasschau, K. D., Cronin, S., & Carrington, J. C. (1997). Genome amplification and long-distance movement functions associated with the central domain of tobacco etch potyvirus helper component–proteinase. *Virology, 228,* 251–262.

Kehoe, M. A., Coutts, B. A., Buirchell, B. J., & Jones, R. A. C. (2014). *Hardenbergia* mosaic virus: Crossing the barrier between native and introduced plant species. *Virus Research, 184,* 87–92.

Kelemu, S., Gebrekidan, B., & Harvey, J. (2012). Bringing the benefits of Sorghum genomics to Africa. In A. Paterson (Ed.), *Genomics of the Saccharinae* (pp. 519–540). New York: Springer.

Kelly, J., Afanador, L., & Haley, S. (1995). Pyramiding genes for resistance to *Bean common mosaic virus. Euphytica, 82,* 207–212.

Kelly, J., Saettler, A., & Morales, M. (1983). New necrotic strain of bean common mosaic virus in Michigan. *Annual Report of the Bean Improvement Cooperative (USA), 27,* 38–39.

Kevane, M. (2012). Gendered production and consumption in rural Africa. *Proceedings of the National Academy of Sciences of the United States of America, 109,* 12350–12355.

Khan, J. A., Lohuis, D., Goldbach, R., & Dijkstra, J. (1993). Sequence data to settle the taxonomic position of bean common mosaic virus and blackeye cowpea mosaic virus isolates. *Journal of General Virology, 74,* 2243–2249.

Khan, Z. R., Midega, C. A. O., Wanyama, J. M., Amudavi, D. M., Hassanali, A., Pittchar, J., et al. (2009). Integration of edible beans (*Phaseolus vulgaris* L.) into the push-pull technology developed for stem-borer and *Striga* control in maize-based cropping systems. *Crop Protection, 28,* 997–1006.

Klein, R. E., Wyatt, S. D., & Kaiser, W. J. (1988). Incidence of bean common mosaic virus in USDA *Phaseolus* germ plasm collection. *Plant Disease, 72,* 301–302.

Kyle, M. M., & Provvidenti, R. (1987). A severe isolate of bean common mosaic virus NY 15. *Annual Report of the Bean Improvement Cooperative (USA), 30,* 87–88.

Lana, A. F., Lohuis, H., Bos, L., & Dijkstra, J. (1988). Relationships among strains of bean common mosaic virus and blackeye cowpea mosaic virus—members of the potyvirus group. *Annals of Applied Biology, 113,* 493–505.

Larsen, R. C., Druffel, K. L., & Wyatt, S. D. (2011). The complete nucleotide sequences of bean common mosaic necrosis virus strains NL-5, NL8 and TN-1. *Archives of Virology, 156,* 729–732.

Larsen, R. C., Miklas, P. N., Druffel, K. L., & Wyatt, S. D. (2005). NL-3K strain is a stable and naturally occurring interspecific recombinant derived from *Bean common mosaic necrosis virus* and *Bean common mosaic virus*. *Phytopathology, 95,* 1037–1042.

Li, Y. Q., Liu, Z. P., Yang, Y. S., Zhao, B., Fan, Z. F., & Wan, P. (2014). First report of Bean common mosaic virus infecting Azuki bean (*Vigna angularis* Ohwi & Ohashi) in China. *Plant Disease, 98,* 1017.

Lima, J. A. A., Purcifull, D. E., & Hiebert, E. (1979). Purification, partial characterization, and serology of blackeye cowpea mosaic virus. *Phytopathology, 69,* 1252–1258.

Loebenstein, G. (2009). Local lesions and induced resistance. *Advances in Virus Research, 75,* 73–117.

Maia, I. G., Haenni, A. L., & Bernardi, F. (1996). Potyviral HC-Pro: A multifunctional protein. *Journal of General Virology, 77,* 1335–1341.

Martínez, F., & Daròs, J. A. (2014). *Tobacco etch virus* P1 protein traffics to the nucleolus and associates with the host 60S ribosomal subunits during infection. *Journal of Virology, 88,* 10725–10737.

Matsumoto, T. (1922). Some experiments with Azuki-bean mosaic. *Phytopathology, 12,* 295–297.

Mauck, K. E., Bosque-Perez, N. A., Eigenbrode, S. D., De Moraes, C. M., & Mescher, M. C. (2012). Transmission mechanisms shape pathogen effects on host-vector interactions: Evidence from plant viruses. *Functional Ecology, 26,* 1162–1175.

Mauck, K. E., De Moraes, C. M., & Mescher, M. C. (2010). Deceptive chemical signals induced by a plant virus attract insect vectors to inferior hosts. *Proceedings of the National Academy of Sciences of the United States of America, 107,* 3600–3605.

McKern, N. M., Mink, G. I., Barnett, O. W., Mishra, A., Whittaker, L. A., Silbernagel, M. J., et al. (1992). Isolates of bean common mosaic virus comprising two distinct potyviruses. *Phytopathology, 82,* 923–929.

McKern, N. M., Shukla, D. D., Barnett, O. W., Vetten, H. J., Dijkstra, J., Whittaker, L. W., et al. (1992). Coat protein properties suggest that Azuki bean mosaic virus, Blackeye cowpea mosaic virus, Peanut stripe virus, and three isolates from soybean are all strains of the same Potyvirus. *Intervirology, 33,* 121–134.

McKern, N. M., Strike, P. M., Barnett, O., Dijkstra, J., Shukla, D., & Ward, C. (1994). Cowpea aphid borne mosaic virus-Morocco and South African *Passiflora* virus are strains of the same potyvirus. *Archives of Virology, 136,* 207–217.

Meiners, J., Gillaspie, A., Jr., Lawson, R., & Smith, F. (1978). Identification and partial characterization of a strain of bean common mosaic virus from *Rhynchosia minima.* *Phytopathology, 68,* 283–287.

Melgarejo, T., Lehtonen, M., Fribourg, C., Rännäli, M., & Valkonen, J. P. T. (2007). Strains of BCMV and BCMNV characterized from Lima bean plants affected by deforming mosaic disease in Peru. *Archives of Virology, 152,* 1941–1949.

Melotto, M., Afanador, L., & Kelly, J. (1996). Development of a SCAR marker linked to the I gene in common bean. *Genome, 39,* 1216–1219.

Miao, J., Guo, D., Zhang, J., Huang, Q., Qin, G., Zhang, X., et al. (2013). Targeted mutagenesis in rice using CRISPR-Cas system. *Cell Research, 23,* 1233–1236.

Miklas, P. N., Strausbaugh, C. A., Larsen, R. C., & Forster, R. L. (2000). NL-3 (K)—a more virulent strain of NL-3 and its interaction with bc-3. *Annual Report of the Bean Improvement Cooperative, 43,* 168–169.

Mink, G. I., & Silbernagel, M. J. (1992). Serological and biological relationships among viruses in the bean common mosaic virus subgroup. In O. W. Barnett (Ed.), *Potyvirus taxonomy* (pp. 397–406). New York: Springer.

Mink, G. I., Vetten, H. J., Wyatt, S. D., Berger, P. H., & Silbernagel, M. J. (1999). Three epitopes located on the coat protein amino terminus of viruses in the bean common mosaic potyvirus subgroup. *Archives of Virology, 144,* 1173–1189.

Moghal, S. M., & Francki, R. I. B. (1976). Towards a system for the identification and clas-
    sification of potyviruses: I. Serology and amino acid composition of six distinct viruses.
    Virology, *73*, 350–360.
Morales, F. J. (2006). Common beans. In G. Loebenstein & J. P. Carr (Eds.), *Natural resistance
    mechanisms of plants to viruses* (pp. 367–382). The Netherlands: Springer.
Morales, F. J., & Bos, L. (1988). *Descriptions of plant viruses: Bean common mosaic virus*. DPV337.
    Wellesbourne UK: Association of Applied Biologists. http://www.dpvweb.net/dpv/
    showdpv.php?dpvno=337.
Morales, F. J., & Castano, M. (1987). Seed transmission characteristics of selected bean
    common mosaic virus strains in differential bean cultivars. *Plant Disease*, *71*, 51–53.
Morales, F. J., & Kornegay, J. (1996). The use of plant viruses as markers to detect genes for
    resistance to *Bean common mosaic* and *Bean common mosaic necrosis viruses*. *Annual Report of
    the Bean Improvement Cooperative*, *39*, 272–273.
Mucheru-Muna, M., Pypers, P., Mugendi, D., Kung'u, J., Mugwe, J., Merckx, R., et al.
    (2010). A staggered maize-legume intercrop arrangement robustly increases crop yields
    and economic returns in the highlands of Central Kenya. *Field Crops Research*, *115*,
    132–139.
Mukeshimana, G., Ma, Y., Walworth, A. E., Song, G., & Kelly, J. D. (2013). Factors
    influencing regeneration and *Agrobacterium tumefaciens*-mediated transformation of com-
    mon bean (*Phaseolus vulgaris* L.). *Plant Biotechnology Reports*, *7*, 59–70.
Mukeshimana, G., Paneda, A., Rodriguez-Suarez, C., Ferreira, J. J., Giraldez, R., &
    Kelly, J. D. (2005). Markers linked to the *bc-3* gene conditioning resistance to bean
    common mosaic potyviruses in common bean. *Euphytica*, *144*, 291–299.
Naderpour, M., Lund, O. L. E. S., & Johansen, I. E. (2010). Sequence analysis of expres-
    sed cDNA of Bean common mosaic virus RU1 isolate. *Iranian Journal of Virology*, *3*, 39–41.
Naderpour, M., Lund, O. L. E. S., Larsen, R., & Johansen, E. (2010). Potyviral
    resistance derived from cultivars of Phaseolus vulgaris carrying *bc-3* is associated with
    the homozygotic presence of a mutated *eIF4E* allele. *Molecular Plant Pathology*, *11*,
    255–263.
Nakahara, K. S., Masuta, C., Yamada, S., Simura, H., Kashihara, Y., Wada, T. S., et al.
    (2012). Tobacco calmodulin-like protein provides secondary defense by binding to
    and directing degradation of virus RNA silencing suppressors. *Proceedings of the National
    Academy of Sciences of the United States of America*, *109*, 10113–10118.
Ng, J. C. K., & Falk, B. W. (2006). Virus-vector interactions mediating nonpersistant and
    semipersistent transmission. *Annual Review of Phytopathology*, *44*, 183–212.
Njau, P., & Lyimo, H. (2000). Incidence of *Bean common mosaic virus* and *Bean common mosaic
    necrosis virus* in bean (Phaseolus vulgaris L.) and wild legume seedlots in Tanzania. *Seed
    Science and Technology*, *28*, 85–92.
Oana, D., Ziegler, A., Torrance, L., Gasemi, S., & Danci, M. (2009). *Potyviridae* family—
    short review. *Journal of Horticulture, Forestry and Biotechnology*, *13*, 410–421.
Ogliari, J., & Castao, M. (1992). Identification of resistant germplasm to the bean common
    mosaic virus-BCMV. *Pesquisa Agropecuária Brasileira*, *27*, 1043–1047.
Olspert, A., Chung, B.-W., Atkins, J. F., Carr, J. P., & Firth, A. E. (2015). Transcriptional
    slippage in the positive-sense RNA virus family. *Potyviridae. EMBO Reports*, in press.
Omunyin, M. E., Gathuru, E., & Mukunya, D. (1995). Pathogenicity groups of *Bean common
    mosaic virus* isolates in Kenya. *Plant Disease*, *79*, 985–989.
Palukaitis, P., Groen, S. C., & Carr, J. P. (2013). The Rumsfeld paradox: Some of the things
    we know that we don't know about plant virus infection. *Current Opinion in Plant Biology*,
    16, 513–519.
Pasev, G., Kostova, D., & Sofkova, S. (2014). Identification of genes for resistance to *Bean
    common mosaic virus* and *Bean common mosaic necrosis virus* in snap bean (*Phaseolus vulgaris* L.)

breeding lines using conventional and molecular methods. *Journal of Phytopathology*, *162*, 19–25.

Petrovic, D., Ignjatov, M., Nikolic, Z., Vujakovic, M., Vasic, M., Milosevic, M., et al. (2010). Occurrence and distribution of viruses infecting the bean in Serbia. *Archives of Biological Sciences*, *62*, 595–601.

Pickett, J. A., Aradottir, G. I., Birkett, M. A., Bruce, T. J. A., Chamberlain, K., Khan, Z. R., et al. (2012). Aspects of insect chemical ecology: Exploitation of reception and detection as tools for deception of pests and beneficial insects. *Physiological Entomology*, *37*, 2–9.

Pierce, W. H. (1934). Viroses of the bean. *Phytopathology*, *24*, 87–115.

Pierce, W., & Hungerford, C. (1929). A note on the longevity of the bean mosaic virus. *Phytopathology*, *19*, 605–606.

Pitino, M., Coleman, A. D., Maffei, M. E., Ridout, C. J., & Houogenhout, S. (2011). Silencing of aphid genes by dsRNA from plants. *PLoS One*, *6*, e25709.

Powell, G. (2004). Intracellular salivation is the aphid activity associated with inoculation of non-persistently transmitted viruses. *Journal of General Virology*, *86*, 469–472.

Provvidenti, R., Silbernagel, M., & Wang, W. (1984). Local epidemic of NL-8 strain of bean common mosaic virus in bean fields of western New York. *Plant Disease*, *68*, 1092–1094.

Rajamäki, R. L., & Valkonen, J. P. T. (2009). Control of nuclear and nucleolar localization of nuclear inclusion protein a of *Picorna*-like *Potato virus A* in *Nicotiana* species. *Plant Cell*, *21*, 2485–2502.

Reddy, D. V. R., Sudarshana, M. R., Fuchs, M., Rao, N. C., & Thottappilly, G. (2009). Genetically engineered virus-resistant plants in developing countries: Current status and future prospects. *Advances in Virus Research*, *75*, 185–220.

Riechmann, J. L., Lain, S., & Garcia, J. A. (1992). Highlights and prospects of potyvirus molecular biology. *Journal of General Virology*, *73*, 1–16.

Robaglia, C., & Caranta, C. (2006). Translation initiation factors: A weak link in plant RNA virus infection. *Trends in Plant Science*, *11*, 40–45.

Robinson, K. E., Worrall, E. A., & Mitter, N. (2014). Double stranded RNA expression and its topical application for non-transgenic resistance to plant viruses. *Journal of Plant Biochemistry and Biotechnology*, *23*, 231–237.

Rodamilans, B., Vali, A., Mingot, A., San León, D., Baulcombe, D., López-Moya, J. J., et al. (2015). RNA polymerase slippage as a mechanism for the production of frameshift gene products in plant viruses of the *Potyviridae* family. *Journal of Virology*, in press.

Rodríguez-Hernández, A. N. A. M., Gosalvez, B., Sempere, R. N., Burgos, L., Aranda, M. A., & Truniger, V. (2012). Melon RNA interference (RNAi) lines silenced for *Cm-eIF4E* show broad virus resistance. *Molecular Plant Pathology*, *13*, 755–763.

Rojas, M. R., Zerbini, F. M., Allison, R. F., Gilbertson, R. L., & Lucas, W. J. (1997). Capsid protein and helper component-proteinase function as potyvirus cell-to-cell movement proteins. *Virology*, *237*, 283–295.

Sahana, N., Kaur, H., Jain, R., Palukaitis, P., Canto, T., & Praveen, S. (2014). The asparagine residue in the FRNK box of potyviral Helper-component Protease is critical for its sRNA binding and subcellular localization. *Journal of General Virology*, *95*, 1167–1177.

Sáiz, M., De Blas, C., Carazo, G., Fresno, J., Romero, J., & Castro, S. (1995). Incidence and characterization of *Bean common mosaic virus* isolates in Spanish bean fields. *Plant Disease*, *79*, 79–81.

Saqib, M., Jones, R. A. C., Cayford, B., & Jones, M. G. K. (2005). First report of *Bean common mosaic virus* in Western Australia. *Plant Pathology*, *54*, 563.

Saqib, M., Nouri, S., Cayford, B., Jones, R. A. C., & Jones, M. G. K. (2010). Genome sequences and phylogenetic placement of two isolates of *Bean common mosaic virus* from

Macroptilium atropurpureum in north-west Australia. *Australasian Plant Pathology*, *39*, 184–191.

Sarria Villada, E., González, E. G., López-Sesé, A. I., Castiel, A. ·F., & Gómez-Guillamon, M. L. (2009). Hypersensitive response to *Aphis gossypii* Glover in melon genotypes carrying the *Vat* gene. *Journal of Experimental Botany*, *60*, 3269–3277.

Sastry, K. S. (2013). Mechanism of seed transmission. In *Seed-borne plant virus diseases* (pp. 85–100). India: Springer.

Sengooba, T. N., Spence, N. J., Walkey, D. G. A., Allen, D. J., & Femi Lana, A. (1997). The occurrence of *Bean common mosaic necrosis virus* in wild and forage legumes in Uganda. *Plant Pathology*, *46*, 95–103.

Shan, Q., Wang, Y., Li, J., Zhang, Y., Chen, K., Liang, Z., et al. (2013). Targeted genome modification of crop plants using a CRISPR-Cas system. *Nature Biotechnology*, *31*, 686–688.

Shankar, A. C. U., Nayaka, C. S., Kumar, B. H., Shetty, S. H., & Prakash, H. S. (2009). Detection and identification of the blackeye cowpea mosaic strain of *Bean common mosaic virus* in seeds of cowpea from southern India. *Phytoparasitica*, *37*, 283–293.

Sharma, P., Sharma, P. N., Kapil, R., Sharma, S. K., & Sharma, O. P. (2011). Analysis of 3'-terminal region of *Bean common mosaic virus* strains infecting common bean in India. *Indian Journal of Virology*, *22*, 37–43.

Shiboleth, Y. M., Haronsky, E., Leibman, D., Arazi, T., Wassenegger, M., Whitham, S. A., et al. (2007). The conserved FRNK box in HC-Pro, a plant viral suppressor of gene silencing, is required for small RNA binding and mediates symptom development. *Journal of Virology*, *81*, 13135–13148.

Shukla, D. D., Frenkel, M., & Ward, C. W. (1991). Structure and function of the potyvirus genome with special reference to the coat protein coding region. *Canadian Journal of Plant Pathology*, *13*, 178–191.

Silbernagel, M. J., Mills, L., & Wang, W. (1986). Tanzanian strain of bean common mosaic virus. *Plant Disease*, *70*, 839–841.

Silbernagel, M. J., Mink, G. I., Zhao, R. L., & Zheng, G. Y. (2001). Phenotypic recombination between bean common mosaic and bean common mosaic necrosis potyviruses in vivo. *Archives of Virology*, *146*, 1007–1020.

Singh, S. P., & Schwartz, H. F. (2010). Breeding common bean for resistance to diseases: a review. *Crop Science*, *50*, 2199–2223.

Singh, S. P., Teran, H., Lema, M., Dennis, M. G., Hayes, R., & Robinson, C. (2008). Breeding for slow-darkening, high-yielding, broadly adapted dry bean pinto 'Kimberly' and 'Shoshone'. *Journal of Plant Registrations*, *2*, 180–186.

Spence, N. J., & Walkey, D. G. A. (1995). Variation for pathogenicity among isolates of bean common mosaic virus in Africa and a reinterpretation of the genetic relationship between cultivars of *Phaseolus vulgaris* and pathotypes of BCMV. *Plant Pathology*, *44*, 527–546.

Stewart, S. A., Hodge, S., Ismail, N., Mansfield, J. W., Feys, B. J., Prospéri, J. M., et al. (2009). The *RAP1* gene confers effective, race-specofoc resistance to the pea aphid in *Medicago truncatula* independent of the hypersensitive reaction. *Molecular Plant-Microbe Interactions*, *22*, 1645–1655.

Tenllado, F., & Diaz-Ruiz, J. R. (2001). Double-stranded RNA-mediated interference with plant virus infection. *Journal of Virology*, *75*, 12288–12297.

Tenllado, F., Llave, C., & Diaz-Ruiz, J. R. (2004). RNA interference as a new biotechnological tool for the control of virus diseases in plants. *Virus Research*, *102*, 85–96.

Tenllado, F., Martínez-García, B., Vargas, M., & Díaz-Ruíz, J. R. (2003). Crude extracts of bacterially expressed dsRNA can be used to protect plants against virus infections. *BMC Biotechnology*, *3*, 3.

Tollefson, J. (2011). Brazil cooks up transgenic bean. *Nature*, *478*, 168.

Torrance, L., Andreev, I. A., Gabrenaite-Verhovskaya, R., Cowan, G., Makinen, K., & Taliansky, M. E. (2006). An unusual structure at one end of potato potyvirus particles. *Journal of Molecular Biology, 357,* 1–8.

Truniger, V., & Aranda, M. A. (2009). Recessive resistance to plant viruses. *Advances in Virus Research, 75,* 119–159.

Tsuchizaki, T., & Omura, T. (1987). Relationships among bean common mosaic virus, blackeye cowpea mosaic virus, Azuki bean mosaic virus and soybean mosaic virus. *Annals of the Phytopathological Society of Japan, 53,* 478–488.

Tu, J. (1986). Isolation and characterization of a new necrotic strain(NL-8) of bean common mosaic virus in southwestern Ontario. *Canadian Plant Disease Survey, 66,* 13–14.

Urcuqui-Inchima, S., Haenni, A.-L., & Bernardi, F. (2001). Potyvirus proteins: A wealth of functions. *Virus Research, 74,* 157–175.

Vallejos, C. E., Astua-Monge, G., Jones, V., Plyler, T. R., Sakiyama, N. S., & Mackenzie, S. A. (2006). Genetic and molecular characterization of the *I* locus of *Phaseolus vulgaris. Genetics, 172,* 1229–1242.

Verdoodt, A., van Ranst, E., & Ye, L. M. (2004). Daily simulation of potential dry matter production of annual field crops in tropical environments. *Agronomy Journal, 96,* 1739–1753.

Verma, P., & Gupta, U. (2010). Immunological detection of *Bean common mosaic virus* in French bean (*Phaseolus vulgaris L.*) leaves. *Indian Journal of Microbiology, 50,* 263–265.

Vetten, H., Lesemann, D.-E., & Maiss, E. (1992). Serotype A and B strains of bean common mosaic virus are two distinct potyviruses. *Archives of Virology, 5,* 415–431.

Vijayapalani, P., Maeshima, M., Nagasaki-Takeuchi, N., & Miller, W. A. (2012). Interaction of the trans-frame potyvirus protein P3N-PIPO with host protein PCaP1 facilitates potyvirus movement. *PLoS Pathogens, 8,* e1002639.

Walker, S., & Ogindo, H. O. (2003). The water budget of rainfed maize and bean intercrop. *Physics and Chemistry of the Earth, 28,* 919–926.

Walkey, D. G. A., & Dance, M. (1979). The effect of oil sprays on aphid transmission of turnip mosaic, beet yellows, bean common mosaic, and bean yellow mosaic viruses. *Plant Disease Reporter, 63,* 877–881.

Walkey, D. G. A., & Innes, N. L. (1978). Resistance to bean common mosaic virus in dwarf beans (*Phaseolus vulgaris L.*). *Journal of Agricultural Science, Cambridge, 92,* 101–108.

Wang, H. L., Chang, Y. Y., & Chang, C. A. (2005). Molecular sequencing and analysis of the viral genome of *Peanut stripe virus* Ts strain. *Plant Pathology Bulletin, 14,* 211–220.

Wang, H. L., & Fang, C. C. (2004). Molecular sequencing and analysis of the viral genomic regions of Blackeye cowpea mosaic virus Taiwan strain. *Plant Pathology Bulletin, 13,* 117–126.

Wang, X., Kohalmi, S. E., Svircev, A., Wang, A., Sanfaçon, H., & Tian, L. (2013). Silencing of the host factor eIF(iso)4E gene confers plum pox virus resistance in plum. *PLoS One, 8,* e50627.

Wang, A., & Krishnaswamy, S. (2012). Eukaryotic translation initiation factor 4E-mediated recessive resistance to plant viruses and its utility in crop improvement. *Molecular Plant Pathology, 13,* 795–803.

Wei, T. Y., Zhang, C. W., Hong, J. A., Xiong, R. Y., Kasschau, K. D., Zhou, X. P., et al. (2010). Formation of complexes at plasmodesmata for potyvirus intercellular movement is mediated by the viral protein P3N-PIPO. *PLoS Pathogens, 6,* e1000962.

Westwood, J. H., Groen, S. C., Du, Z., Tungadi, T., Lewsey, M. G., Luang-In, V., et al. (2013). A trio of viral proteins tunes aphid-plant interactions in *Arabidopsis thaliana. PLoS One, 10,* 1371.

Westwood, J. H., Lewsey, M. G., Murphy, A. M., Tungadi, T., Bates, A., Gilligan, C. A., et al. (2014). Interference with jasmonic acid-regulated gene expression is a general property of viral suppressors of RNA silencing but only partly explains virus-induced changes in plant-aphid interactions. *Journal of General Virology, 95,* 733–739.

Westwood, J. H., & Stevens, M. (2010). Resistance to aphid vectors of virus disease. *Advances in Virus Research, 76,* 179–210.

Wortmann, C. S., Lunze, L., Ochwoh, V. A., & Lynch, J. P. (1995). Bean improvement for low fertility soils in Africa. *African Crop Science Journal, 3,* 469–477.

Xu, L., & Hampton, R. (1996). Molecular detection of bean common mosaic and bean common mosaic necrosis potyviruses and pathogroups. *Archives of Virology, 141,* 1961–1977.

Yin, G., Sun, Z., Liu, N., Zhang, L., Song, Y., Zhu, C., et al. (2009). Production of double-stranded RNA for interference with TMV infection utilizing a bacterial prokaryotic expression system. *Applied Microbiology and Biotechnology, 84,* 323–333.

Zaumeyer, W., & Meiners, J. (1975). Disease resistance in beans. *Annual Review of Phytopathology, 13,* 313–334.

Zettler, F. W., & Wilkinson, R. E. (1966). Effect of probing behavior and starvation of *Myzus persicae* on transmission of bean common mosaic virus. *Phytopathology, 56,* 1079–1082.

Zheng, H., Chen, J., Chen, J., Adams, M. J., & Hou, M. (2002). *Bean common mosaic virus* isolates causing different symptoms in asparagus bean in China differ greatly in the 5′-parts of their genomes. *Archives of Virology, 147,* 1257–1262.

Ziebell, H., Murphy, A. M., Groen, S. C., Tungadi, T., Westwood, J. H., Lewsey, M. G., et al. (2011). Cucumber mosaic virus and its 2b RNA silencing suppressor modify plant-aphid interactions in tobacco. *Scientific Reports, 1,* 187.

CHAPTER TWO

The Molecular Biology of Pestiviruses

Norbert Tautz*, Birke Andrea Tews[†], Gregor Meyers[†,1]
*Institute for Virology and Cell Biology, University of Lübeck, Lübeck, Germany
[†]Institut für Immunologie, Friedrich-Loeffler-Institut, Federal Research Institute for Animal Health, Greifswald-Insel Riems, Germany
[1]Corresponding author: e-mail address: gregor.meyers@fli.bund.de

Contents

Advances in Virus Research, Volume 93
ISSN 0065-3527
http://dx.doi.org/10.1016/bs.aivir.2015.03.002

47

Abstract

Pestiviruses are among the economically most important pathogens of livestock. The biology of these viruses is characterized by unique and interesting features that are both crucial for their success as pathogens and challenging from a scientific point of view. Elucidation of these features at the molecular level has made striking progress during recent years. The analyses revealed that major aspects of pestivirus biology show significant similarity to the biology of human hepatitis C virus (HCV). The detailed molecular analyses conducted for pestiviruses and HCV supported and complemented each other during the last three decades resulting in elucidation of the functions of viral proteins and RNA elements in replication and virus–host interaction. For pestiviruses, the analyses also helped to shed light on the molecular basis of persistent infection, a special strategy these viruses have evolved to be maintained within their host population. The results of these investigations are summarized in this chapter.

1. INTRODUCTION

The first description of a pestivirus-induced disease was a report on classical swine fever (CSF). It dates from 1833, long before viruses were recognized as pathogens of sub-bacterial size. Much later, reports on bovine viral diarrhea (BVD) and mucosal disease (MD) (1940s) as well as border disease of sheep (BD) (1950s) were published. The observation of serological cross-reactivity between the agents causing CSF and BVD initiated the idea of pestiviruses as a group of related viruses (Darbyshire, 1969). Nucleotide sequencing allowed consolidation of the classification resulting in the genus *Pestivirus* that nowadays comprises the four species classical swine fever virus (CSFV), bovine viral diarrhea virus types 1 and 2 (BVDV-1 and BVDV-2), and border disease virus (BDV) of sheep as well as several more exotic viruses (Simmonds et al., 2012). More recently, further molecular characterization of new genus members has led to the proposal of a classification with eight pestivirus species which is under discussion by the responsible ITCV study group.

Initially, a significant lack of data on the molecular characteristics of pestiviruses led to a wrong classification into the family *Togaviridae*. However, the determination of the first genomic sequences and the assignment of viral proteins to defined coding regions of the viral RNA revealed fundamental differences between togaviruses and pestiviruses with regard to genome organization and strategy of gene expression so that the genus *Pestivirus* was moved into the family *Flaviviridae*, whose members share basic features like formation of small enveloped virus particles, a single-stranded RNA genome of positive polarity with a similar organization of the coding regions, and the strategy of gene expression via one long open reading frame (Lindenbach, Murray, Thiel, & Rice, 2013). These basic features of the pestiviral genome together with a diagram of the virus particle and electron micrographs are shown in Fig. 1.

Apart from the above-described similarities, many features of the *Flaviviridae* family members are quite diverse and reflect their adaptation to different propagation strategies and hosts. However, the molecular biology of hepaciviruses, pegiviruses, and pestiviruses is strikingly similar so that pestiviruses have been widely used as surrogate system for human HCV, a virus that could not be propagated in tissue culture until 2005 (Lindenbach et al., 2005; Wakita et al., 2005; Zhong et al., 2005).

Progress in pestivirus analysis was quite slow during the early years soon after recognition of pestiviruses as animal pathogens. This was due to the

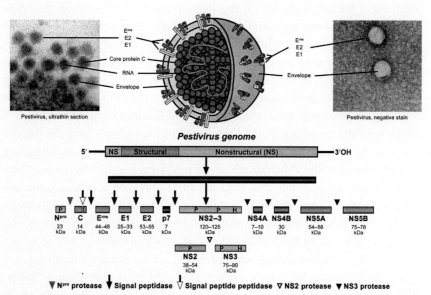

Figure 1 Basic features of pestiviruses. The figure shows a diagram of a pestiviral particle flanked by electron microscopic pictures (BVDV in ultrathin section on the left; CSFV in negative stain on the right). Below, the viral genome is shown in a schematic representation with the single long ORF encoding a polyprotein indicated below. Processing of the polyprotein by the proteases specified at the bottom leads to the shown viral proteins. P, protease domain; H, helicase. Electron microscopy: Harald Granzow and Frank Weiland, Friedrich-Loeffler-Institut; graphic design: Mandy Jörn, Friedrich-Loeffler-Institut. Details shown in the figure are addressed in the text. (See the color plate.)

absence of cell culture systems allowing high titer propagation and difficulties to establish efficient purification protocols for pestiviruses. Moreover, pestivirus particles do not exhibit a very characteristic morphology and their size of 40–60 nm is in the range of many enveloped cellular structures, rendering electron microscopic studies of pestiviruses and pestivirus-infected cells difficult. Most of the problems in pestivirus research are inherent to these viruses. Modern molecular biology has fueled detailed investigation of many aspects of pestivirus biology, but working with these viruses is still demanding.

Pestiviruses are among the economically most important pathogens of livestock (Lindenbach et al., 2013; Thiel, Plagemann, & Moennig, 1996). Stamping-out strategies, vaccination, as well as strict biosafety measures have been employed for control of these pathogens but, despite considerable efforts, pestiviruses continue to cause severe losses in livestock farming and the risk of reintroduction represents a constant threat for countries having managed to eradicate a given pestivirus. The difficulties to control pestiviruses are not only due to the high frequency of transport and trade of animals susceptible to pestivirus infection but also originate from specific survival strategies these pathogens gained during evolution. Among them, the most important factor is their ability to establish long-lasting persistent infection in their host animals (Thiel et al., 1996). Small numbers of virus carriers are sufficient to ensure continuous virus spread since they live for considerable periods of time while shedding high amounts of virus. Intensive research on the molecular biology of these interesting pathogens during the last three decades has elucidated many of their basic features providing a key to understanding the molecular basis of pestivirus survival strategies. This review summarizes recent data with the aim to provide a look into the fascinating world of pestiviruses and their interplay with their host's biology. There will be a clear focus on the molecular biology of these viruses restricting description of pathogenesis and epidemiology of pestivirus-induced diseases and the details of exotic pestivirus classification to a necessary minimum. For further information, the reader is referred to other excellent reports (Baker, 1987; Bauermann, Ridpath, Weiblen, & Flores, 2013; Brackenbury, Carr, & Charleston, 2003; Chase, 2013; Hamers et al., 2001; Lindberg, 2003; Luo, Li, Sun, & Qiu, 2014; Moennig, 2000; Moennig, Floegel-Niesmann, & Greiser-Wilke, 2003; Moennig, Houe, & Lindberg, 2005; Moennig & Plagemann, 1992; Passler & Walz, 2010; Ridpath, 2003, 2010; Simmonds et al., 2012; Tao et al., 2013; Thiel et al., 1996; Vilcek & Nettleton, 2006).

2. TAXONOMY, HOSTS, AND DISEASES

2.1 Taxonomy: Approved and Tentative Species

The genus *Pestivirus* belongs to the virus family *Flaviviridae* that also comprises the genera *Flavivirus*, *Hepacivirus*, and *Pegivirus* (Simmonds et al., 2012). In addition to the typical members of the four recognized species, CSFV, BVDV-1, BVDV-2, and BDV, a considerable number of novel pestiviruses have been isolated from different animal species (Arnal et al., 2004; Becher, Orlich, Kosmidou, et al., 1999; Becher, Schmeiser, Oguzoglu, & Postel, 2012; Becher et al., 1997, 2003; Kirkland et al., 2007; Schirrmeier, Strebelow, Depner, Hoffmann, & Beer, 2004; Stahl et al., 2007; Stalder et al., 2005; Thabti et al., 2005; Vilcek, Ridpath, Van Campen, Cavender, & Warg, 2005). These viruses have been proposed to represent additional pestivirus species but have not been approved as such yet (Fig. 2). The Ninth report of the International Committee on Taxonomy of Viruses (ICTV) lists four tentative species, namely Giraffe-1 pestivirus, Pronghorn antelope pestivirus, Atypical pestivirus, and Bungowannah virus. Among them, the Giraffe-1 pestivirus was described first. It was isolated during an outbreak of disease in giraffes in Kenya (Harasawa, Giangaspero, Ibata, & Paton, 2000). The PG-2 virus detected in bovine cells originating from Kenya was recently found, after complete genome sequencing, to represent a second member of this tentative species (Becher et al., 2014). The prototype for the second tentative pestivirus species was detected in a Pronghorn antelope in the United States (Vilcek et al., 2005). The Atypical pestiviruses, also known as HoBi-like pestiviruses or BVDV-3, represent an especially interesting new type of putative genus members. HoBi-like viruses were first identified in fetal bovine serum (FBS) from Brazil (Schirrmeier et al., 2004). Since that first description, several publications reported on genetically similar agents contaminating FBS and cell lines from various regions of the world (reviewed in Bauermann et al., 2013). Moreover, natural infection in buffalo and cattle has been described. Bungowannah virus represents the most recently described tentative pestivirus species. It is the most divergent pestivirus and was detected in pigs following an outbreak of stillbirths and neonatal death in Australia (Kirkland et al., 2007). Its origin and host reservoir are still not clear. Since publication of the ICTV report, a novel pestivirus was isolated from sheep in Turkey that according to sequence comparison studies clustered separately

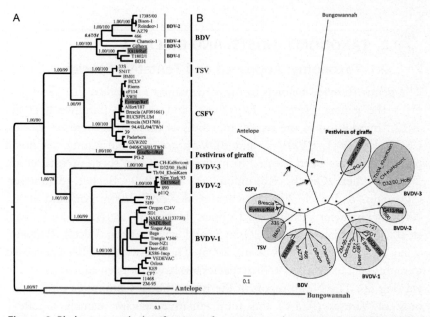

Figure 2 Phylogeny and classification of pestiviruses by Maximum likelihood and Bayesian approach. The molecular dataset contains 56 sampled pestiviruses and 2089 characters, comprising the 5'-UTR, N^{pro}, and E2 gene regions. PHYML (v2.4.4) was used for phylogeny inference according to Maximum likelihood criterion. MrBayes 3.1 was used for Bayesian analysis. This is a representative consensus tree: mid-point rooted (A) and unrooted (B). The reference sequences are highlighted. (A) All sampled pestiviruses and their relationships. The numbers at a node are posterior probability (left) and percentage of 1000 bootstrapping replicates (right). For a clear demonstration, some terminals are not displayed in (B). An * indicates strong statistical support for a node by a posterior probability value of 0.99–1.00 and by a bootstrap value of 78–100%. The scale bar represents changes per site. The arrows show the probable placements of the root for the given unrooted network. *The figure was taken from Liu, Xia, Wahlberg, Belak, and Baule (2009).*

from the approved or proposed pestivirus species (Becher et al., 2012), and it is likely that more tentative species will be detected in the future. The high number of different pestiviruses found in variant ruminant species demonstrates their broad distribution which reflects the evolutionary success of these viruses.

Within the pestivirus species, different genotypes have been proposed and their number is still increasing. This system of subdivisions may have advantages for epidemiological analyses and perhaps also predictions with regard to cross-protection. So far, many of these subdivisions are not approved and are still under debate so that we will not refer to them in this review. For more information on genotypes, we refer to corresponding publications (Jenckel et al., 2014; Luzzago et al., 2014; Pan et al., 2005; Paton et al., 2000; Postel et al., 2012; Rosamilia et al., 2014).

2.2 General Properties, Hosts, and Transmission

Pestivirus particles are enveloped and have a diameter of 40–60 nm. Four structural proteins (SPs), a basic core protein C and the envelope (E) glycoproteins E^{rns}, E1, and E2, are found in the virion. The E proteins are inserted into the viral envelope which is derived from intracellular membranes of the host cell (Lindenbach et al., 2013). Within the particle, the single-stranded RNA genome of positive polarity is located. It has a length of about 12.3 kb and contains one long open reading frame that is translated into a hypothetical polyprotein of ca. 3900 amino acids. Due to co- and posttranslational processing by viral and host proteases, 12 mature proteins are generated. Because of the lipid envelope, pestiviruses are easily inactivated by detergent treatment. The general stability of the virions is low, as reflected by a half-life of ~7 h at 37 °C and neutral pH (Depner, Bauer, & Liess, 1992). However, in contrast to other flaviviruses, pestiviruses are very resistant to low pH. The buoyant density of viral particles was determined at 1.134 g/ml in sucrose (Laude & Gelfi, 1979) and 1.090 g/ml in Iodixanol (Optiprep) (Maurer, Krey, Moennig, Thiel, & Rümenapf, 2004). Pestiviruses tend to adhere to their host cells or cellular debris which renders purification of the viruses difficult.

A special feature of all pestiviruses analyzed in detail so far is the coexistence of two different biotypes in cell culture, namely cytopathic (cp) variants leading to death of the infected cells and noncytopathic (noncp) viruses replicating in their host cells without detectable damage or considerably reduced cell viability or growth rates. Most isolates obtained from the field are noncp pestiviruses which is most likely due to the fact that only the noncp biotype can establish persistent infections (see Section 6).

Infections with pestiviruses were believed to be restricted to clovenhoofed animals. Clear evidence for the existence of pestiviruses in other animals or humans were missing, but recently, a new pestivirus was detected in rats (Firth et al., 2014). Moreover, virome analysis of bats has provided indications for novel pestiviruses in these mammals (Wu et al., 2012), but the significance of these findings still has to be supported by further analyses. A recent publication reports on a novel flavivirus with homology to pestiviruses in a soybean cyst nematode (Bekal et al., 2014).

The approved or tentative species of pestiviruses can be divided into two major groups, namely (i) pestiviruses of pigs and (ii) pestiviruses of ruminants like cattle, sheep, goats, and a large variety of wild ruminants. Transmission of pestiviruses isolated from one species of ruminants to another is frequent (Thiel et al., 1996). Moreover, ruminant pestiviruses are quite often found in pigs, whereas there is no evidence for replication of CSFV in ruminants (Tao et al., 2013; Thiel et al., 1996). However, CSFV was successfully adapted to propagation in rabbits leading to attenuation for pigs (van Oirschot, 2003). For pestiviruses of ruminants, adaptation to a small animal model has not been described.

Pestiviruses are shed in all body secretions of an infected animal, i.e., saliva, tears, nasal discharges, milk, urine, feces, and semen. Transmission can occur directly by contact between animals or indirectly via, e.g., contact with infectious secretions, contaminated food, or needles. Important for the epidemiology of pestiviruses is transmission via semen (Choi & Chae, 2002; Meyling, Houe, & Jensen, 1990).

In addition to horizontal spread between animals, in pregnant animals, all pestiviruses can also be transmitted vertically by crossing the placenta and infecting the fetus (Thiel et al., 1996).

2.3 Diseases and Other Effects Caused by Pestivirus Infection of Animals

Pestivirus infections have a major impact on livestock farming which is not only due to severe disease characterized by severe symptoms or even death but also due to side effects in clinically healthy animals. One important factor is the ability of the viruses to cross the placenta and infect the fetus. Since the placenta of pigs and ruminants does not allow transfer of antibodies from the mother to the fetus, the outcome of such an intrauterine infection is not altered by a successful immune response of the mother. The result of fetal infection is mainly dependent on the time point of gestation. It can range from resorption of the fetus to abortion, stillbirth, and a wide range of malformations to birth of weak and growth-retarded offspring (Baker, 1987; Deregt & Loewen, 1995; Thiel et al., 1996). It has to be stressed, however, that also healthy animals can be born after intrauterine infection. In part, the different results of diaplacental infection can be explained by the development of the immune system during embryogenesis. Early infection within the first ∼40 days of gestation frequently leads to death of the fetus, which is probably due to its very immature developmental stage and the absence of significant immunological defense mechanisms

(Grooms, 2004; Scherer et al., 2001). Infection between days 40 and 120 of gestation often leads to birth of animals that remain infected for life. Such persistently infected (PI) animals shed large amounts of virus and represent the main source of virus in the field, being responsible for maintaining the infectious cycle within the host population. Only noncp viruses are able to establish persistent infections in their natural host. The time point of infection is important since persistence can only be established before the immune system is able to discriminate between self- and non-self-antigens. Thus, pestiviral persistence is characterized by the absence of an adaptive immune response to the pathogen, which means that neither virus specific antibodies nor T- or B-cells can be detected in the infected animals. The timelines for developing immunological competence vary between pestivirus host species and, accordingly, the susceptible period of gestation differs but the underlying principles seem to be similar.

Experiments analyzing the outcome of an infection of the fetus with a cp virus during the period prone for establishment of persistence are not extensively described in the literature. It seems that abortion can occur as well as elimination of the virus (Deregt & Loewen, 1995). The latter is obviously the case for intrauterine infections with cp or noncp viruses late in gestation when immunological competence has already developed in a way that an effective immune response can be mounted.

Persistently infected (PI) piglets invariably develop signs of disease and die within several months after birth. In contrast, persistent infections in ruminants can last for years and pregnancy of PI animals has been reported resulting in birth of new PI animals (Thiel et al., 1996). Efficient virus replication continues during persistent infection raising the question of how some animals can appear healthy with this enormous virus load for such a long time. This finding implies that signs of disease observed upon acute infection of immunocompetent animals must, at least in part, be due to immunopathological processes. However, a clearer concept of the processes leading to disease in the latter type of infections is still missing. Below, basic data on diseases caused by pestiviruses are summarized. For more details, the reader is referred to more specific reviews (Baker, 1987; Moennig & Plagemann, 1992; Rümenapf & Thiel, 2014; Thiel et al., 1996).

2.3.1 Bovine Viral Diarrhea and Mucosal Disease

BVDV can induce very different symptoms in the host animals (Baker, 1987; Thiel et al., 1996). Diseases resulting from acute infections as well as a syndrome following persistent infection can be distinguished. In acute infections, the property of the infecting virus is of major importance for the severity of the disease. In fact, most acute postnatal infections of ruminants with BVDV are inapparent or accompanied by only mild signs of disease so that the majority of BVDV infections remains unnoticed. Symptoms associated with BVDV infection may be transient leukopenia, mild fever, diarrhea, increased nasal discharge, coughing, and other signs of abnormal respiration. Important for the outcome in the field is the immunosuppression resulting from BVDV infection which implements a significant risk of superinfections with other viruses or bacteria (Brackenbury et al., 2003; Chase, 2013; Potgieter, 1995).

First reports on severe disease observed after acute infection with BVDV were published in 1989 (Corapi, Elliott, French, & Dubovi, 1990; Corapi, French, & Dubovi, 1989; Rebhuhn et al., 1989). The affected animals showed typical signs of a hemorrhagic disease associated with high mortality. Severe thrombocytopenia was one of the characteristic parameters observed (Corapi et al., 1989). More detailed characterization of the viruses isolated from such cases showed that they represented noncp BVDV belonging to a "novel" group of viruses. In fact, these viruses were the first recognized members of the BVDV-2 species. Thorough comparison of sequence data conducted after identification of BVDV-2 revealed that members of this species had been around for long times before outbreak of the hemorrhagic disease but had not been recognized as separate species. Initially, members of the BVDV-2 species were considered more virulent than BVDV-1 strains in general and the latter were believed unable to induce hemorrhagic disease. However, more detailed analyses revealed that infection of cattle with BVDV-2 can be inapparent or accompanied by only mild symptoms and, vice versa, BVDV-1 is able to induce hemorrhagic disease (Lunardi, Headley, Lisboa, Amude, & Alfieri, 2008; Yesilbag et al., 2014). A very special case of fatal disease resulting from infection with a mixture of BVDV-2 was described recently (see Section 5) (Jenckel et al., 2014).

Analysis of tissue distribution of BVDV displaying different degrees of virulence showed dissemination into a large variety of tissues within days (Liebler-Tenorio, Greiser-Wilke, & Pohlenz, 1997; Liebler-Tenorio, Lanwehr, Greiser-Wilke, Loehr, & Pohlenz, 2000; Liebler-Tenorio, Ridpath, & Neill, 2002, 2004, 2003a, 2003b). Most of the tissues that were found virus positive came from the immune system. An obvious marker for virulence was the degree of cell depletion found in lymphoid tissues that correlated with the intensity and duration of lymphopenia and the severity of other signs of disease. Moreover, infection of bone marrow was typical for the highly virulent strains. Depletion of lymphoid tissues was also a predominant sign of BVDV infection in alpacas with a severe effect on the gastrointestinal associated lymphoid tissues (GALT) (Steffen et al., 2014). When PI calves were analyzed, virus could be detected in basically all organs (Liebler-Tenorio et al., 2004). Similar results were also found for persistently BVDV-infected white-tailed deer (Passler et al., 2012).

In addition to the above-described diseases induced by acute infection, BVDV can cause a syndrome designated as MD. MD is a sporadic disease of cattle to which most often animals of 6–24 months succumb. The most striking characteristic symptoms are bloody diarrhea accompanied by fever, anorexia, ataxia, and general weakness. Mortality is 100% with death occurring within about 2 weeks after onset of clinical signs (Baker, 1987). Necropsy reveals extensive lesions within the gastrointestinal tract affecting predominantly the mucosa, especially the GALT. Death of the diseased animals is considered to result from dehydration because of impaired resorption of water due to the destroyed intestinal mucosa.

After realizing that BVD and MD result from the same virus, researchers were puzzled for a long time by the differences between both diseases, the frequently recognized BVD with usually only mild symptoms and the sporadically occurring but lethal MD. A first clue to understanding MD came from the observation that the disease develops only in animals persistently infected after diaplacental transmission of BVDV during the first trimester of gestation (Liess, Frey, Kittsteiner, Baumann, & Neumann, 1974) resulting in immunotolerance to the persisting virus. This type of persistence is specific for pestiviruses.

A second major step toward elucidation of MD was the observation that animals that come down with MD always harbor a cp virus in addition to the persisting noncp virus. This finding provided the basis for experimental reproduction of the disease by infection with noncp and cp viruses isolated from an animal suffering from MD. Infection of pregnant heifers with the noncp virus at an appropriate time of gestation resulted in birth of persistently infected animals. Superinfection of these PI animals with the cp virus induced typical MD so that the etiology of the disease was definitely shown to be dependent on the presence of both biotypes within one animal (Bolin, McClurkin, Cutlip, & Coria, 1985; Brownlie, Clarke, & Howard, 1984). The next important finding was the detection of a very close antigenic relationship between noncp and cp viruses isolated from the same animal (Corapi, Donis, & Dubovi, 1988; Howard, 1990). This close antigenic relationship was found to be a prerequisite for the induction of MD and stands in marked contrast to the known antigenic variability of BVDV field isolates so that the hypothesis was put forward that the cp virus develops from the noncp virus. This was later proven by genome analyses which revealed an strikingly high similarity of the viral sequences from paired noncp and cp BVDV isolated from the same MD animal (Meyers, Tautz, Dubovi, & Thiel, 1991). The genome analyses also uncovered genome alterations specific for the cp isolates. These alterations were mainly due to recombination events resulting in insertion of cellular sequences, or duplications as well as deletions of viral sequences. The genetic and functional basis of cytopathogenicity of pestiviruses will be discussed in detail in Section 5.

The development of MD in consequence of a newly generated cp variant of the persisting noncp virus implies that a single cp virus resulting from a stochastic mutation/recombination event has to replicate enormously to be detected in the diseased or dead animal. Since the cp virus can easily be identified in MD animals, it has to have a significant selective advantage allowing it to catch up with the noncp viruses already circulating in the animal, so that finally both viruses are present in about equal amounts. Comparison of tissue distribution of cp viruses in MD animals with noncp BVDV in PI calves revealed a strikingly higher number of cp BVDV-infected cells in the former animals versus noncp-infected cells in PIs before outbreak of the disease supporting the conclusion of a significant advantage of the cp over the noncp virus (Greiser-Wilke et al., 1993; Liebler, Waschbusch, Pohlenz, Moennig, & Liess, 1991). The molecular basis of this selective advantage is still obscure. In acute infections, both biotypes are similar with regard to symptoms and it is impossible to differentiate between cp and noncp BVDV-induced signs of disease. In addition to the acute form of MD, a more "chronic" form of this disease has been described (Brownlie, 1991; Fritzemeier, Haas, Liebler, Moennig, & Greiser-Wilke, 1997; Moennig, Frey, Liebler, Polenz, & Liess, 1990; Moennig et al., 1993). It can develop upon superinfection of a PI animal with an exogenous cp virus. In this case, the antigenic properties of the persisting noncp and the infecting cp viruses are more divergent so that the tolerance toward the invading cp virus will not be strict. The PI animal mounts an immune response against the cp virus usually leading to its elimination. In these cases, induction of a cross-reactive immune response affecting also the noncp virus was reported. Such animals can suffer for prolonged time from clinical illness which led to the name "chronic MD."

After superinfection of PI claves with a "foreign" cp virus, so-called "late-onset MD" can follow which develops several weeks after superinfection and was, in contrast to chronic MD, reproduced experimentally (Fray, Clarke, Thomas, McCauley, & Charleston, 1998; Fritzemeier et al., 1995; Moennig et al., 1990). Genome analyses revealed that the cp viruses present in these animals represent recombinants of the persisting noncp and the superinfecting cp viruses. The recombination occurs in a way that the structural protein-encoding region of the recombinant virus is derived from the noncp virus, whereas the nonstructural protein (NSP) genes carrying the cp-specific genome alteration originate from the cp viral genome. This arrangement ensures the preservation of immunological tolerance combined with the cp phenotype and thus fulfills both prerequisites for MD development.

2.3.2 Border Disease
In general, pestivirus infections of ruminants result in similar diseases. BDV is predominantly found in sheep and BDV infections of this species resemble in many aspects the situation found in cattle infected with BVDV (Rümenapf & Thiel, 2014; Thiel et al., 1996). As for BVDV, congenital infections represent the most important issue, whereas acute infections occurring postnatally lead to only mild or even no symptoms (Shaw et al., 1969; Vantsis, Linklater, Rennie, & Barlow, 1979). As described for BVDV, BDV is able to cross the placenta when infecting pregnant ewes resulting in different malformations of the fetus as well as fetal death but also persistent infection (Barlow, 1972; Terpstra, 1981; Vantsis, Barlow, Fraser, Rennie, & Mould, 1976). Despite the continuous virus replication, inapparent PI animals may appear. In other cases, the lambs exhibit the so-called hairy shaker syndrome which in addition to hairy fleece and tremor is often characterized by general weakness and low birth weight as well as ataxia (Nettleton, Gilmour, Herring, & Sinclair, 1992). In analogy to BVDV, a fatal MD-like disease is known that is associated with the recovery of cp and noncp viruses from the animal (Nettleton et al., 1992).

Border disease was first observed in the region of the English/Welsh border but is now known to occur in most sheep-farming areas of the world. In addition to the above-described typical phenotype of the affected lambs BDV infection can also reduce conception rates within herds.

2.3.3 Classical Swine Fever
In early descriptions, CSF was characterized as a severe peracute to acute hemorrhagic disease with high mortality rates (Dunne, 1973). The time between infection and detection of severe symptoms was reported to be

short leading to early death of the infected pigs. From these early times to now, the virus has evolved into much less virulent forms. Thus, the peracute form of the disease is disappearing from the field and acute CSFV infections no longer show high mortality. This process of gradual attenuation is accompanied by development of chronic forms of the disease (Thiel et al., 1996; Wensvoort & Terpstra, 1985). The frequently observed absence of clear symptoms of CSF in adult animals may delay diagnosis and increases the chance for dissemination of the virus.

The available data support the conclusion that the variability of the clinical picture can mainly be attributed to different virulence of the virus strains. These strains are often categorized into groups of low, moderate and high virulence. So far, no clear correlation between specific sequence motifs and virulence of the different field viruses was found so that it seems reasonable to regard virulence as a result of multiple factors (Mayer, Hofmann, & Tratschin, 2004; Mayer, Thayer, Hofmann, & Tratschin, 2003; Meyers et al., 2007; Meyers, Saalmüller, & Büttner, 1999; Risatti, Borca, et al., 2005; Risatti et al., 2006; Risatti, Holinka, Fernandez, Carrillo, Lu, et al., 2007; Risatti, Holinka, et al., 2005; Sainz, Holinka, Lu, Risatti, & Borca, 2008; Tamura et al., 2012; van Gennip, Vlot, Hulst, de Smit, & Moormann, 2004). Comparison of sequences of closely related avirulent and virulent CSFV identified multiple mutations spread over the genome and thus supported this conclusion. Typical signs of severe CSF are pyrexia, significant leukopenia, respiratory and gastrointestinal symptoms, as well as hemorrhages of skin and inner organs (petechia) (Thiel et al., 1996).

3. THE GENOME

The pestivirus genome consists of a single-stranded RNA of positive polarity (Lindenbach et al., 2013). Its standard size is ca. 12.3 kb, but it can reach up to about 16.5 kb in consequence of naturally occurring recombination events (see Section 5.4) (Becher, Orlich, König, & Thiel, 1999; Meyers et al., 1992). In recombination experiments in tissue culture, even larger genomes with more than 20 kb were generated and packaged into viral particles (Gallei, Pankraz, Thiel, & Becher, 2004).

RNA of genome length represents the only viral RNA found in cells infected with standard pestiviruses. Thus, no subgenomic RNAs are transcribed during viral replication which implies that the genomic plus strand represents the only viral mRNA and codes for all viral proteins. In accordance with this finding, the genomic RNA contains one long open reading

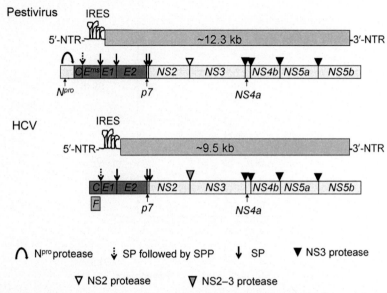

Figure 3 Comparison of the genome organization of pestiviruses and human hepatitis C virus. For both viruses, a schematic representation of the genomic RNA (upper part) and the encoded polyprotein (below) is shown. The name of the individual viral protein is given, and the proteases responsible for processing are indicated according to the scheme at the bottom. See text for further details. The F protein of HCV is expressed from a second, overlapping reading frame.

frame (ORF) coding for a hypothetical polyprotein of about 3900 amino acids (Becher, Orlich, & Thiel, 1998a; Collett, Larson, Belzer, & Retzel, 1988; Meyers, Rümenapf, & Thiel, 1989a; Moormann et al., 1990). The ORF is flanked by 5′- and 3′-nontranslated regions (NTRs) of roughly 400 and 200 nucleotides, respectively. The genome organizations of pestiviruses and HCV exhibit striking similarity (Fig. 3).

3.1 The 5′-NTR

NTRs in viral RNA can be suspected to contain *cis*-acting elements important for RNA replication and viral protein expression. In pestiviruses, the 5′-NTR starts with a sequence able to form a stable stem–loop structure (Fig. 4). This so-called Ia hairpin consists of mainly conserved stem sequences and a more variable loop region. Experiments with a BVDV replicon showed that Ia represents a bifunctional secondary structure element involved in both translation initiation and replication of the viral RNA (Grassmann, Yu, Isken, & Behrens, 2005; Yu, Isken, Grassmann, &

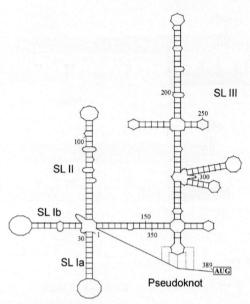

Figure 4 Secondary structure representation of the 5′-nontranslated region of the pestivirus genome. The names of the stem loops are given, and the location of the pseudoknot close to the translational start site (AUG) is indicated. The authors thank Paul Becher, Institute of Virology, University of Veterinary Medicine Hannover, Germany, for the secondary structure file from which the figure was established.

Behrens, 2000). The two functions are assigned to specific regions of Ia that only partly overlap. Formation of the stem is important for translation, whereas transcription is dependent on sequences representing part of the loop and the 5′-moiety of the stem of Ia (or the corresponding complementary sequences of the 3′-terminal region of the negative strand of the viral genome). Interestingly, only the first four residues of the conserved 5′-terminal region of the genome were essential for BVDV replication. Substitution or deletion mutants of other parts of hairpin Ia as well as of downstream hairpin Ib supported viral replication, but with lower efficiency and reduced intracellular RNA levels compared to the parental virus (Becher, Orlich, & Thiel, 2000). These mutants showed reduced replication rates in calves but were still able to induce a significant immune response (Makoschey et al., 2004).

Based on the similarity of pestiviruses and HCV, it was tested whether chimeric RNAs with 5′-NTRs composed of various segments of HCV and BVDV support RNA replication. Replacement of the complete BVDV 5′-NTR by HCV sequences impaired replication dramatically. Such chimeras improved replication efficiency during passage, and one important adaptive change was the restoration of the 5′-terminal 3–4 bases (Frolov, McBride, & Rice, 1998). This finding once again proved the importance of the utmost 5′-terminal sequence for replication of pestiviral RNA but also demonstrated a high degree of flexibility within the rest of the 5′-NTR.

Hairpin Ib of BVDV strain CP7 encompasses in the loop a stretch of a variable number of A residues as a result of spontaneous mutation. The length of the A repeat in these naturally occurring or in respective genetically engineered variants had no influence on the specific infectivity of the RNA. During tissue culture propagation of the resulting viruses, changes in the number of A residues occurred rapidly. A variant with five A residues was found to be stable (Becher et al., 2000).

The influence of hairpin Ia on translation is somewhat difficult to interpret since luciferase expression analyses in transfected cells conducted with the CSFV 5′-NTR yielded results differing considerably from those described for BVDV (Xiao et al., 2011). The CSFV experiments revealed an inhibitory effect of the 5′-proximal hairpin on translation of the viral RNA and thus contradict the data obtained for the BVDV DI9 replicon. It has to be noted that most experiments aiming at defining the role of the pestiviral 5′-NTR in translation initiation have been conducted with CSFV sequences encompassing the complete 5′-NTR including the utmost residues forming hairpin Ia. *In vitro* translation experiments with a BVDV IRES showed that hairpins Ia and Ib had no influence on translation rates (Chon, Perez, & Donis, 1998).

Viruses are completely dependent on the translation machinery of the host cell (Firth & Brierley, 2012; Lopez-Lastra et al., 2010). It is therefore essential to ensure that the viral mRNAs are accepted as substrate by these cellular components. Pestiviral RNAs lack a 5′-cap structure (Brock, Deng, & Riblet, 1992; G. Meyers, unpublished) which promotes translation initiation in most cases occurring at the next downstream AUG codon in standard cellular mRNAs (Hinnebusch, 2014). To induce efficient translation, the 5′-NTR of pestiviral RNAs contains an "internal ribosomal entry site" (IRES) that is able to directly recruit the small ribosomal subunit and position it at the translational start site. This explains why translation starts at exactly this AUG despite the presence of multiple AUG codons upstream of the actual initiation codon. Moreover, the pestiviral IRES is able to promote start of translation at the defined site even when this codon is mutated to a non AUG codon. The IRES in the pestivirus RNA was identified in the 1990s (Poole et al., 1995; Rijnbrand, van der Straaten, van Rijn,

Spaan, & Bredenbeek, 1997) and has since been studied intensively (de Breyne, Yu, Pestova, & Hellen, 2008; Fletcher, Ali, Kaminski, Digard, & Jackson, 2002; Fletcher & Jackson, 2002; Hashem et al., 2013; Hellen & de Breyne, 2007; Hellen & Pestova, 1999; Hsu et al., 2014; Jubin et al., 2000; Kolupaeva, Pestova, & Hellen, 2000b; Moes & Wirth, 2007; Pestova, de Breyne, Pisarev, Abaeva, & Hellen, 2008; Reusken, Dalebout, Eerligh, Bredenbeek, & Spaan, 2003; Thurner, Witwer, Hofacker, & Stadler, 2004). It belongs to the hepatitis C virus/pestivirus (HP) group of IRES elements encompassing the HCV and pestivirus entry sites and a certain class of picornavirus IRES. HP IRES elements are characterized by two complex stem-loop structures denoted domains II and III. The initiation site is located downstream of a pseudoknot established between sequences closely upstream of the AUG and a loop of domain III (domain IIIf). This pseudoknot is essential for IRES function (Moes & Wirth, 2007). It obviously helps positioning the AUG start codon in the P-site of the ribosome (Ji, Fraser, Yu, Leary, & Doudna, 2004; Kolupaeva, Pestova, & Hellen, 2000a; Lukavsky, Otto, Lancaster, Sarnow, & Puglisi, 2000; Pestova, Shatsky, Fletcher, Jackson, & Hellen, 1998; Sizova, Kolupaeva, Pestova, Shatsky, & Hellen, 1998) (Fig. 4).

As for other RNAs with HP IRES elements, translation of the pestivirus RNA is independent of most canonical initiation factors (Hellen & Pestova, 1999; Jackson, 2005; Ji et al., 2004; Lancaster, Jan, & Sarnow, 2006; Pisarev, Shirokikh, & Hellen, 2005). Prior to the start of protein synthesis, the small ribosomal subunit binds to IRES domain III resulting in a binary complex. This step is independent of any initiation factor and is followed by recruitment of two further components, eukaryotic initiation factor 3 (eIF3), a multiprotein complex, binding to domains IIIa/IIIb, and the so-called ternary complex composed of eIF2, GTP, and the initiator tRNA Met–tRNAi^{Met} associating with the start codon region and the 40S ribosomal subunit. This IRES/48S preinitiation complex subsequently releases eIF2 and GDP + P$_i$ and recruits the 60S ribosomal subunit. In analogy to the standard cap-dependent mode of translation initiation, this process requires initiation factors eIF5, eIF5B, and GTP (Pestova et al., 1998) but also depends on interactions of the IRES with the 40S subunit (Locker, Easton, & Lukavsky, 2007; Pestova et al., 2008).

The role of eIF3 for translation initiation at the HP IRES is not fully understood. This multiprotein complex is known to function as a scaffolding factor orchestrating the interaction with different initiation factors, mRNA and ribosome. Recent studies propose a role of eIF3 for virus propagation aside from assisting in translation initiation (Hashem et al., 2013; Khan,

Bhat, & Das, 2014). A cryoelectron microscopy reconstruction of a complex composed of 40S ribosomal subunit, eIF3, and the CSFV IRES revealed that eIF3 was completely displaced from its usual ribosomal position in the 43S complex. Instead, its ribosome-binding surface interacted with the binding site on domain III of the IRES, while the eIF3-binding surface on the 40S ribosomal subunit was occupied by part of the IRES. According to the authors, this arrangement might have two advantages for the virus: (i) prevention of competition between IRES and eIF3 for the same or over-lapping binding site(s) on the 40S subunit and (ii) reducing the formation of 43S complexes needed for translation of cellular mRNAs by sequestration of eIF3 so that translation of viral RNA is preferred. In order to evaluate this assumption, it has to be considered that during the early phase of replication, the number of viral RNA molecules will be very low compared to the num-ber of eIF3 complexes so that out-competition of binding sites on the ribo-some or reduced translation of cellular proteins will definitely not work. In later stages of the replication cycle, a shift toward preferred translation of viral over cellular RNA could occur, but a functional role is still somewhat questionable since pestiviruses are not designed for maximum replication performance (see Section 5.3).

Phosphorylation of the α subunit of eIF2 downregulates translation ini-tiation. Cells use this mechanism to restrict viral protein synthesis by activa-tion of double-stranded RNA-dependent protein kinase (PKR) and PERK (PKR-like kinase), a process playing an important role in restricting virus replication in cells (Sonenberg & Hinnebusch, 2009). eIF2 phosphorylation finally leads to a significant decrease of the ternary complex. Pestivirus and HCV translation are less sensitive to eIF2α phosphorylation than standard cap-dependent translation, due to translation initiation mediated by the HP IRES that does not require the ternary complex (Pestova et al., 2008). Met–tRNAi^{Met} binding to IRES-40S complexes is promoted by eIF5B/eIF3 resulting in translation of viral RNA in the presence of activated PKR, though at reduced levels. This might be a means to overcome this important block in the defense program of the infected cell.

The location of the exact 3′-terminal end of the pestiviral IRES has been debated (Chon et al., 1998; Frolov et al., 1998; Myers et al., 2001; Rijnbrand et al., 1997, 2001). Naturally occurring defective interfering replicons of CSFV contain deletions starting with the first nucleotide following the AUG initiation codon and ending almost 5 kb downstream (Aoki et al., 2001; Kosmidou, Ahl, Thiel, & Weiland, 1995; Kosmidou, Büttner, & Meyers, 1998; Meyers & Thiel, 1995). These replicons show efficient pro-tein expression and genome replication. Similarly, a variety of mutant genomes with deletions of the region coding for the first viral protein have been transcribed *in vitro* from infectious cDNA clones which allowed recovery of fully functional infectious virus mutants upon transfection of susceptible cells. These mutants may retain only the AUG start codon (Mayer et al., 2004; Mittelholzer, Moser, Tratschin, & Hofmann, 1997; Tratschin, Moser, Ruggli, & Hofmann, 1998). However, we and others have shown that recovery of such deletion mutants may fail or was only possible when several codons of the ORF were retained (Behrens, Grassmann, Thiel, Meyers, & Tautz, 1998; Tautz et al., 1999; G. Meyers, unpublished). These mutants often displayed considerably reduced translation rates *in vitro* and in transfected cells (G. Meyers, unpublished). *In vitro* translation analyses with a BVDV IRES demonstrated reduced translation efficiency for constructs lacking N^{pro} (Chon et al., 1998). Further analyses showed that the nucleotide sequence and not the encoded protein was critical (Tautz et al., 1999). Based on experiments with BVDV replicons, it was put forward that the absence of stable secondary structures in the RNA downstream of the IRES is important for efficient IRES-driven translation and the viability of the respective deletion mutants (Myers et al., 2001). A more structure-based analysis of the CSFV IRES led to the conclusion that the presence of single-stranded RNA in the region around and immediately downstream of the initiation codon is crucial for efficient translation (Fletcher et al., 2002).

The mode of action of the HP IRES with its ability to interact directly with the small ribosomal subunit and the low number of initiation factors needed for start of translation is reminiscent of the prokaryotic initiation mechanism. Along those lines, the IRES would represent a functional equivalent of the Shine–Dalgarno sequence in prokaryotic mRNAs (Shine & Dalgarno, 1975). Single-stranded regions around the translation initiation codon as in HP IRES-driven initiation are also found in prokaryotes (Pestova et al., 1998).

3.2 The 3′-NTR

Similar to the 5′-NTR, also the 3′-NTR of the viral RNA is expected to contain important *cis*-acting elements, characterized by specific structural and/or primary sequence features. The 3′-NTR represents most likely the first RNA element getting in contact with the viral RNA-dependent RNA polymerase (RdRP). The RdRP is the last protein encoded by the long pestiviral ORF so that it is present close to the 3′-NTR at translation termination. Since most pestiviral NSPs including the RdRP do not function when provided *in trans* (Grassmann, Isken, Tautz, & Behrens, 2001), it is likely that the replication complex assembles immediately after translation termination at the 3′-end to start production of the first minus strand of the genome. Sequence comparison of the 3′-NTR of the BVDV isolate CP7 and other pestiviruses revealed that the 192 nucleotides can be divided into a conserved and a variable part (3′C and 3′V, respectively) with the 3′C part comprising roughly 100 residues from the 3′-end (Becher et al., 1998a; Deng & Brock, 1993; Yu, Grassmann, & Behrens, 1999). The 3′-NTR is able to fold into three stem loops denoted SLI, SLII, and SLIII (Pankraz, Thiel, & Becher, 2005). SLI is located close to the 3′-end, SLII in the middle, and SLIII in the region directly following the stop codon (Fig. 5). Secondary structure probing verified stem loops SLI and SLII. SLI is located in the 3′C region and SLII is composed of the 5′-part of 3′C and part of the 3′V sequence, whereas SLIII comprises the main part of the 3′V region. SLI and SLII are separated by a single-stranded region of ca. 15 nucleotides termed SS (Yu et al., 1999). Between SLII and SLIII another single-stranded region of ca. 10 residues is located. The utmost 3′-terminal residues were also found to be single stranded as was a considerable number of nucleotides forming the loops.

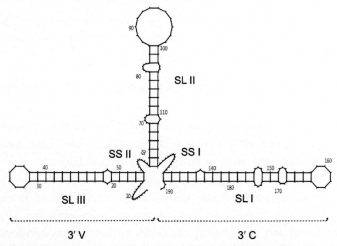

Figure 5 Secondary structure representation of the 3′-nontranslated region of the pestivirus genome. The names of the stem loops (SL) and the single-stranded regions connecting them (SS) are given. The location of the 3′ variable (3′V) and constant (3′C) regions is also shown. The authors thank Paul Becher, Institute of Virology, University of Veterinary Medicine Hannover, for the secondary structure file from which the figure was established. Numbering starts with the 5′-end of the 3′-NTR sequence.

Functional analysis proved the importance of the SLI secondary structure since viable mutants with altered primary stem sequence but conserved secondary structure were identified, whereas changes destroying SLI were deleterious (Yu et al., 1999). In addition, essential primary sequence elements are present in loop regions of SLI and in SS (Pankraz et al., 2005; Yu et al., 1999). Interestingly, mutants missing either SLII or SLIII were replication competent, although deletion of SLII reduced replication efficiency by about 4 orders of magnitude. However, passage of the virus allowed the recovery of pseudorevertants with a limited number of nucleotide exchanges close to the deletion site and growth rates increased by several orders of magnitude despite still considerably reduced intracellular RNA levels. These findings indicate that stem loop II itself is not essential for efficient virus replication. This is even more the case for SLIII since the SLIII deletion mutant replicated in the range of the wild-type virus directly after RNA transfection. In contrast, replicating mutants lacking both SLII and SLIII or SLI could not be recovered (Pankraz et al., 2005; Yu et al., 1999).

The 3′-end of the pestiviral RNA also shows important differences to canonical eukaryotic mRNAs since it does not contain a Poly(A) tail that in standard mRNAs is known to influence RNA stability. A further important function of the Poly(A) is to support translation initiation via pseudo-circularization of the RNA by Poly(A)-binding protein (PABP) interacting with both the RNA 3′-end and initiation factors assembled at the 5′-cap (Hinnebusch & Lorsch, 2012). This function seems to be taken over by NAFR proteins in pestiviruses (see Section 7).

3.3 The Polyprotein—General Organization and Processing Scheme

IRES-mediated translation of the pestiviral RNA leads to a hypothetical polyprotein of about 3900 amino acids that is processed into mature viral proteins by viral and host cellular proteases (Lindenbach et al., 2013) (Fig. 3). The general arrangement of the individual proteins in the polyprotein agrees with the scheme known from other positive strand RNA viruses containing only one ORF in the genome with the SPs in the amino-terminal part followed by the NSPs in the carboxy-terminal region of the polyprotein ending with the RdRP. In pestiviruses, one striking deviation of this scheme is found: the first protein encoded by the ORF is an NSP named N^{pro} (Stark, Meyers, Rümenapf, & Thiel, 1993). Downstream of N^{pro}, the capsid protein C and three glycosylated envelope proteins E^{rns}, E1, and E2 follow (Collett, Wiskerchen, Welniak, & Belzer, 1991; Stark, Rümenapf, Meyers, & Thiel, 1990; Thiel, Stark,

Weiland, Rümenapf, & Meyers, 1991). The presence of three surface proteins in pestiviral particles represents one unusual feature of these viruses, since related viruses including human HCV are equipped with only two envelope proteins (Lindenbach et al., 2013) (Fig. 3). Downstream of E2 seven NSPs follow which are named p7, NS2, NS3, NS4A, NS4B, NS5A, and NS5B so that the total number of so far recognized pestivirus proteins is 12 arranged in the polyprotein as NH2–N^{pro}/C/E^{rns}/E1/E2/ p7/NS2/NS3/NS4A/NS4B/NS5A/NS5B–COOH.

Processing of the polyprotein occurs cotranslationally and starts with the release of N^{pro} (Stark et al., 1993). N^{pro} is an autoprotease as its name stands for N-terminal protease. It cleaves at its own carboxy-terminus, thereby generating the amino-terminal end of the C protein (Figs. 1 and 3). The C protein is followed by a hydrophobic sequence serving as signal sequence for translocation of the E^{rns} protein into the ER. At the end of the signal peptide, a typical von Heijne motif (von Heijne, 1990) is found that is cleaved by cellular signal peptidase (SPase) (Rümenapf, Unger, Strauss, & Thiel, 1993). The resulting amino-terminal product is a C protein with a carboxy-terminal transmembrane TM sequence, a so-called anchored chore. The TM sequence is further processed by a second cellular protease, signal peptide peptidase (SPPase). It belongs to the intramembrane proteases and cleaves within TM, thereby releasing the C protein from the membrane (Heimann, Roman-Sosa, Martoglio, Thiel, & Rümenapf, 2006).

Processing at the E^{rns}/E1, E1/E2, E2/p7, and p7/NS2 borders is all done by SPase (see Section 4 for more details). The next processing site between NS2 and NS3 is special since it is involved in the regulation of pestivirus RNA replication. It is cleaved by a protease activity located within NS2.

All sites downstream of the NS2/NS3 site are processed by the NS3 protease (Elbers, Tautz, Becher, Rümenapf, & Thiel, 1996; Harada, Tautz, & Thiel, 2000; Lamp et al., 2011; Lamp, Riedel, Wentz, Tortorici, & Rümenapf, 2013; Tautz, Elbers, Stoll, Meyers, & Thiel, 1997; Wiskerchen & Collett, 1991; Xu et al., 1997). NS4A functions as a cofactor of the NS3 protease and is at least essential for processing the NS4B/NS5A and the NS5A/ NS5B sites.

4. CHARACTERISTICS AND FUNCTIONS OF PESTIVIRUS PROTEINS

4.1 N^{pro}

N^{pro} represents the first protein encoded by the pestiviral genome. It has a length of 168 amino acids. First proof for an autoproteolytic activity residing in the amino-terminal part of the polyprotein was obtained for BVDV strain NADL, but the released protein was regarded as the viral core protein (Wiskerchen, Belzer, & Collett, 1991). After cell-free expression of the amino-terminal region of the CSFV polyprotein, it was recognized that the expressed protein was further processed into products of 23 and 14 kDa. The proteolytic region was attributed to the 23-kDa amino-terminal cleavage product which was named N^{pro} for N-terminal protease, whereas the 14-kDa product was recognized as the viral core protein (Stark et al., 1993). Site-directed mutagenesis studies resulted in detection of Cys69 as the nucleophilic residue in the active center of the protease (Rümenapf, Stark, Heimann, & Thiel, 1998). Further essential residues were mapped with Glu22 and His49, but it remained unclear whether Glu22-His49-Cys69 actually build a catalytic triad in analogy to what is known for many other (viral) proteases (Rawlings, Barrett, & Bateman, 2012). These studies invalidated the earlier assumption that N^{pro} belongs to the group of papain-like cysteine proteases, since the predicted active site residue His130 was not important for protease activity (Dougherty & Semler, 1993; Rümenapf et al., 1998; Stark et al., 1993). Because of its unique sequence and novel arrangement of catalytic residues, N^{pro} has been classified into a new protease family named C53 (Rawlings et al., 2012).

The sequence of the cleavage site at the carboxy-terminus of N^{pro} is highly conserved among pestiviruses with P1 and P1' representing Cys168 and Ser169. Interestingly, the nature of the P1' residue is flexible since any of the standard amino acids except Pro could replace Ser169 without preventing cleavage (Gottipati, Acholi, Ruggli, & Choi, 2014; Rümenapf et al., 1998).

Recently, two publications reported on the crystal structure of N^{pro} (Gottipati et al., 2013; Zögg et al., 2013). Both groups used amino-terminally truncated N^{pro} expressed in bacteria for crystallization. The starting sequences were derived from a HoBi-like BVDV-3 strain or CSFV strain Alfort 187, and the N-terminal truncations encompassed residues 2–21 or 1–17 (Gottipati et al., 2013; Zögg et al., 2013). Such N-terminal deletions do not interfere with N^{pro} functions (Hilton et al., 2006; Ruggli et al., 2009; Rümenapf et al., 1998). The basic features of the N^{pro} structures deduced from these analyses were very similar. N^{pro} is different from structures found in the DALI server and represents a new fold characterized as a "clam shell" fold (Gottipati et al., 2013). It is composed of a protease domain encompassing the N-terminal ~100 amino acids plus a C-terminal peptide of ~12 residues, and a zinc-binding domain covering the C-terminal region except for the last residues. The dominating structural motif is β-sheets, which, however, are more prevalent in the C-terminal part. The amino-terminal protease domain contains mostly coils without regular secondary structure in addition to a single β-sheet composed of β1, β2, and β8 with the latter representing the utmost six C-terminal residues containing the substrate site of the N^{pro} protease (nomenclature according to Gottipati et al., 2013) (Fig. 6). The active site nucleophile Cys69 is located in a small single-turn α-helix, which for CSFV N^{pro} was detected only in a mutated version (Gottipati et al., 2013), whereas it was detectable in the wild-type protein of BVDV-3 (Zögg et al., 2013).

One important lesson learned from the structure analysis is the inclusion of the C-terminal substrate site into the inner part of the structure. The β8 region forms an essential part of the protease and remains within the substrate-binding pocket after cleavage of the Cys168/Ser169 site. This explains why N^{pro} cleaves only at its C-terminal end and is proteolytically inactive thereafter, since there is no access to the active site for additional substrate molecules. Accordingly, N^{pro} is not able to conduct *trans*-cleavage reactions and functions executed by this protein after carboxy-terminal self-cleavage will not depend on its protease activity.

More detailed analysis of the structure of the protease domain showed that it does not contain a catalytic triad as found in many other proteases. The location of Cys69 and His49 within the protein structure is in good agreement with their contribution to the protease activity. In contrast, the spatial location of the third proposed active site residue, Glu22, is too remote from the active site to be involved in catalysis. Glu22 forms a salt bridge with Arg100 and contributes to the stability of the structure which explains the deleterious effect of Glu22 mutations on N^{pro} protease activity (Rümenapf et al., 1998). Taken together, the structural data clearly support the conclusion that the proteolytic activity of N^{pro} relies on a catalytic dyad composed of His49 and Cys69, supported by further chemical groups found in the active center like the main chain amides of Gly67, Asp68, and Cys69 forming hydrogen bonds with the carboxylate of residue 168.

The substrate specificity of the protease seems to be more due to the presentation of the scissile bond to the active center than due to the actual amino-acid sequence. This finding was somewhat unexpected because of the conservation of the Cys/Ser cleavage site. As already mentioned above, Ser at P1' can be replaced by all amino acids except Pro. The P1 residue is not that flexible, but Ala or Ser can replace the conserved Cys without inhibiting cleavage (Gottipati et al., 2014; Rümenapf et al., 1998). In

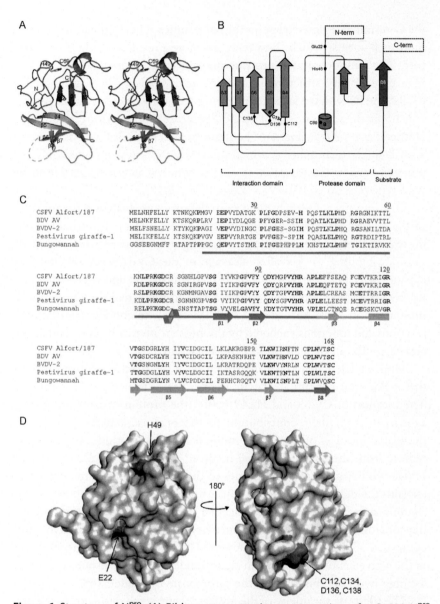

Figure 6 Structure of N^pro. (A) Ribbon representation as stereoview of a C168A N^pro mutant. The protease domain is shown in the upper part and the zinc-binding domain in the lower part. Disordered residues (residues 145–149) are indicated by a dashed line. Secondary structural elements, as well as the N- and C-termini, are labeled. (B) Schematic presentation of the arrangement of secondary structure elements of N^pro that contribute to the zinc-binding interaction domain (left), and the protease domain with the C-terminal substrate region (right). The β-sheets are shown as arrows, the α-helix with the active site residue as cylinder. The location and number of

contrast, Glu was not tolerated at position 168 which is most likely due to problems with accommodation of the bulky side chain in the substrate-binding pocket.

The conclusions with regard to the mechanism of catalysis drawn by the authors of the two different structural analysis articles are divergent. Gottipatti and coworkers propose a classical cysteine protease mechanism with formation of an imidazolium (His49)–thiolate (Cys69) in preparation of the nucleophilic attack of the scissile bond and an oxyanion hole involving the above-mentioned amides that interact with the terminal carboxylate via hydrogen bonds and stabilize the tetrahedral intermediate. In contrast, Zögg and colleagues propose a mechanism in which the catalytic Cys69 is deprotonated by a hydroxide ion found to be located close to Gly67 and His49 functioning as oxyanion hole polarizing the scissile bond. These somewhat contradicting conclusions are at least in part due to differences in the determined structures which could, however, result from technical differences. This could explain some most likely aberrant features detected in the structure analysis like the coordination of a zinc ion in the CSFV N^{pro} protease close to the active site via residues from two N^{pro} molecules located next to each other in the crystal (Gottipati et al., 2013) or the presence of a disulfide bond between Cys69 and Cys168 (Zögg et al., 2013) which should not exist in the biologically active protein in the cytoplasm of the cells. It seems highly unlikely that the N^{pro} proteins from the two pestivirus species analyzed here indeed exhibit different catalytic mechanisms so that further work is needed to refine the details underlying the N^{pro} cleavage reaction.

---

important amino acids are given. (C) Alignment of pestivirus N^{pro} sequences. The alignment includes the CSFV strain Alfort 187, Border Disease virus (BDV) strain AV, BVDV-2, Pestivirus giraffe-1, and the Bungowannah virus (GenBank Accession Numbers X87939.1, ABV54604.1, AAV69983.1, NP_777520.1, and DQ901403.1, respectively). Amino-acid numbering corresponds to the CSFV Alfort 187 sequence. The secondary structure elements are indicated below the sequence. α-Helix and β-strands are shown as coil and arrows, respectively. (D) The spatial distribution of the residues involved in N^{pro}-mediated proteasomal degradation of IRF3. The residues that are essential for the N^{pro}-mediated IRF3 degradation (dark gray) are mapped onto the surface of N^{pro}. The residues localize to two protein surfaces, one in each domain. The protease domain surface cluster (left) includes residues Glu22 and His49, and the zinc-binding domain cluster (right) is formed by the residues in the TRASH motif. *Panels (A), (C), and (D) are from Gottipati et al. (2013) with kind permission of Kyung H. Choi, Department of Biochemistry and Molecular Biology, Sealy Center for Structural Biology and Molecular Biophysics, The University of Texas Medical Branch at Galveston, Galveston, Texas, USA.* (See the color plate.)

N^{pro} is interesting not only because of its unusual protease but also because of its functions with regard to virus/host interactions. N^{pro} represents the only nonessential pestiviral gene product. Viable N^{pro} CSFV and BVDV deletion mutants were isolated (Behrens et al., 1998; Gil et al., 2006; Mayer et al., 2004; Meyers et al., 2007; Tautz et al., 1999; Tratschin et al., 1998). These viruses replicated fairly well and reduced replication efficiency in comparison with the parental wild-type viruses was attributed to less efficient translation of viral proteins due to impaired function of the IRES (see above). A major breakthrough for understanding the evolutionary conservation of the N^{pro} gene in pestiviruses despite its dispensability for virus replication was initiated by the observation that N^{pro} was able to protect cells from apoptosis via blocking the host cellular type I interferon response to virus infection and other triggers like double-stranded RNA (Bauhofer et al., 2007; Chen et al., 2007; Doceul et al., 2008; Gil et al., 2006; Hilton et al., 2006; La Rocca et al., 2005; Ruggli et al., 2003, 2005, 2009; Seago et al., 2007).

Induction of an IFN-1 response in the infected cell or in immune cells in contact with or engulfing (debris of dead) infected cells is a complex process reviewed recently in different excellent articles (Abbas, Pichlmair, Gorna, Superti-Furga, & Nagar, 2013; Brzozka, Finke, & Conzelmann, 2006; Hengel, Koszinowski, & Conzelmann, 2005; Pichlmair et al., 2012; Pichlmair & Reis e Sousa, 2007; Rieder, Finke, & Conzelmann, 2012). In a simplified view, virus infection is detected by specialized molecules like RIG-I-like receptors or Toll-like receptors recognizing so-called pathogen-associated molecular patterns (PAMPs). These PAMPs are molecular structures specific for pathogens like longer double-stranded RNA or RNA carrying 5′-triphosphate groups in addition to (short) double-stranded regions (Dixit & Kagan, 2013). Binding of such structures to the receptors initiates a signal that is transferred to the nucleus and activates transcription of IFN-1 genes. IFN-1 stimulates its own expression via an autocrine loop and activates a large number of different genes encoding proteins involved in antiviral defense. One of the key players in the transduction of signals from the PAMP receptors into the nucleus is interferon regulatory factor 3 (IRF-3). N^{pro} was shown to considerably reduce the intracellular level of IRF-3 via proteolytic degradation (Bauhofer et al., 2007; Chen et al., 2007; Doceul et al., 2008; Gil et al., 2006; Hilton et al., 2006; La Rocca et al., 2005; Ruggli et al., 2003, 2005, 2009; Seago et al., 2007). The degradation of IRF-3 is not executed by the N^{pro} protease since mutation of Cys69 had only marginal influence on N^{pro} induced reduction of IRF-3 levels. This finding is in accordance with the above-described fact that the intrinsic protease activity of N^{pro} is not designed for *trans*-cleavage reactions. Studies with proteasome inhibitors revealed that N^{pro} provokes IRF-3 degradation via the proteasomal route. This process is obviously due to interaction between N^{pro} and IRF3 which could be demonstrated by co-immunoprecipitation and eukaryotic two-hybrid experiments (Bauhofer et al., 2007; Chen et al., 2007; Ruggli et al., 2009). However, it has not been clarified so far whether this interaction requires a direct binding of the two partners or is mediated by further cellular factors.

The interaction between N^{pro} and IRF3 is dependent on the second domain of N^{pro} made up by residues ~100–160 (Fig. 6). This domain contains most of the β-sheet structure observed for N^{pro}. A total of five β-strands (β3, β4, β5, β6, and β7) form an antiparallel β-sheet (Gottipati et al., 2013; Zögg et al., 2013, nomenclature as in Gottipati et al., 2013). It was proposed that this region, though not directly involved in the catalytic activity of N^{pro}, functions as a scaffold during synthesis and folding of the protein, and shields the C-terminal β-strand (β8) which contains the cleavage site and stabilizes the N^{pro} structure. An element of major importance for N^{pro} activity in blocking the IFN-1 response is a conserved metal-binding TRASH motif that coordinates a zinc ion via Cys112, Cys134, and Cys138. This arrangement is in accordance with the TRASH motif consensus sequence C–X19–22–C–X3–C (Ettema, Huynen, de Vos, & van der Oost, 2003). The respective Cys residues are conserved among all pestiviruses including Bungowannah virus which shares the lowest homology with the other genus members (Szymanski et al., 2009). TRASH motifs are part of metal-binding domains and are found in proteins involved in heavy metal sensing and resistance, in hydrogenases, cation transporters, and transcriptional regulators (Ettema et al., 2003). Biophysical analyses revealed that N^{pro} is indeed a metalloprotein and coordinates a single zinc ion. Asp136 was identified as fourth partner in the zinc-binding site that was also found in the crystal structure of N^{pro}. Cys112 is located in a loop and the other three residues are contributed by a β-hairpin (Fig. 6). Unfortunately, neither the crystals of CSFV nor of BVDV-3 N^{pro} contained zinc. Instead, a disulfide bond connecting Csy112 and Cys134 was found in both structures. These findings do not contradict coordinative zinc binding in N^{pro} but are most likely due to oxidation during crystallization.

Binding of a zinc ion is not only important for the general stability of N^{pro} but plays a crucial role in IRF-3 degradation and IFN-1 inhibition. Mutations preventing zinc binding of N^{pro} blocked N^{pro}/IRF-3 interaction and prevented the above-described activities of N^{pro} with regard to inhibition of apoptosis and IFN-1 response to triggers like virus infection and double-stranded RNA (Szymanski et al., 2009). It was also shown that deletion of 22 residues from the N-terminus or 24 amino acids from the C-terminus of N^{pro} blocked its activity with regard to the IFN-1 system. A variety of other mutations were also found to abrogate this activity which does not necessarily mean that those residues are directly involved in the steps leading to degradation of IRF3 (Chen et al., 2007; Gil et al., 2006; Hilton et al., 2006; Ruggli et al., 2009). When looking at the N^{pro} 3D structure, these important residues are located at one side of the model but at a significant distance from each other and just on the opposite side of the TRASH motif with the zinc ion (Fig. 6D). Thus, these residues can hardly be in contact with only one cellular factor at the same time so that a destabilization effect on the protein might be an explanation for the observed prevention of IFN-1 interference.

As pointed out above, the exact mechanism underlying N^{pro}-induced degradation of IRF-3 is still obscure. However, it has been clarified that IRF-3 is polyubiquitinated and then degraded via the proteasome (Bauhofer et al., 2007; Chen et al., 2007). Polyubiquitination involves the activity of E3 ubiquitin ligases that transfer the ubiquitin to the target protein after activation and conjugation steps executed by the E1 and E2 proteins of the ubiquitination system. E3 ubiquitin ligases are known to contain zinc ions, and it therefore has been discussed whether N^{pro} could function as an E3 ubiquitin ligase. However, E3 ubiquitin ligases typically contain zinc binding so-called RING-finger motifs that are crucial for their activity. N^{pro} lacks a RING-finger motif, and it can hardly be imagined that the TRASH motif can serve as substitute. Accordingly, IRF3 is more likely polyubiquitinated by a cellular E3 ubiquitin ligase possibly recruited by N^{pro}.

Interference with antiviral activities of the host cell was recognized for both BVDV and CSFV long before clarification of the role of N^{pro} by the observation that these viruses enhanced the multiplication of Newcastle disease (ND) virus in cell culture (Inaba, Omori, & Kumagai, 1963; Kumagai, Shimizu, & Matumoto, 1958). This phenomenon was designated END (for "exaltation of ND virus") and was proposed to result from suppression of IFN production (Diderholm & Dinter, 1966; Toba & Matumoto, 1969). Shimizu and coworkers reported on an attenuated CSFV strain that did not suppress IFN production and was itself an effective IFN inducer (Shimizu, Furuuchi, Kumagai, & Sasahara, 1970). This virus interfered with ND virus replication and was designated GPE (END-negative). Later on, the GPE phenotype was found to result from mutations in N^{pro} (Tamura et al., 2014).

IRF3 is not the only cellular factor that interacts with N^{pro}. It was shown recently that in plasmacytoid dendritic cells (PDCs), the activity of IRF7, another important factor in the IFN-1 response cascade leading to a fast and strong IFN-α response, is blocked by N^{pro} (Fiebach, Guzylack-Piriou, Python, Summerfield, & Ruggli, 2011). This function of the viral protein is again dependent on the zinc-binding TRASH motif, but the observed antagonism relies on a so far unknown mechanism not involving polyubiquitination and proteasomal degradation. HAX-1, IκBα, and TRIM56, proteins involved in cell survival or the antiviral response, have also been reported to interact with N^{pro}. HAX-1 represents a pleiotropic protein with various functions including a specific antiapoptotic effect (Johns et al., 2010). IκBα was described to bind to N^{pro} of CSFV and BVDV (Doceul et al., 2008; Zahoor et al., 2010). TRIM56 is a member of the tripartite motif (TRIM) family of proteins that contains a large number of different proteins executing diverse functions. TRIM56 has E3 ubiquitin ligase activity but does not induce N^{pro} or IRF3 degradation. Nevertheless, TRIM56 interferes with BVDV replication since overexpression reduces virus production, whereas knockdown of the protein leads to increased virus yield (Wang et al., 2011). The importance of these interactions for pestiviral modulation of the innate immune response as well as the molecular mechanisms leading to these effects are unknown.

It has been a matter of debate whether N^{pro} deletion results in considerable attenuation of CSFV in piglets or adult animals. First data indicated that the deletion of N^{pro} in different CSFV isolates leads to attenuation (Mayer et al., 2004). In contrast, other results on a CSFV N^{pro} deletion mutant with a different type of N^{pro} deletion in another viral background gave no indication for considerable attenuation (G. Meyers, unpublished). The latter result is in agreement with more recent data showing that the originally described attenuating effect is not due to loss of the N^{pro}-induced repression of the IFN-1 response, but most likely results from reduced growth rates probably due to less efficient translation of the mutated genomic RNAs (Ruggli et al., 2009). This finding indicates that the N^{pro} IFN-1 repressing function is not an important virulence factor in piglets or adult animals. However, influence on pathogenesis of CSFV infection by prevention of type 1 interferon activity at local replication sites has been shown recently (Tamura et al., 2014).

4.2 The Structural Proteins

4.2.1 The Core Protein C

The highly basic core protein C is the first structural protein in the pestivirus polyprotein (Thiel et al., 1991). Its N-terminus is generated through the autocatalytic cleavage of N^{pro} (Stark et al., 1993). SPase cleaves between core and E^{rns} (Rümenapf et al., 1993), leaving the E^{rns} signal peptide at the C-terminus of C, which is further processed by SPPase (Heimann et al., 2006). C of BVDV has been shown to be an intrinsically disordered protein, which binds RNA with low affinity and no clear specificity. The minimal binding size is about 14 nucleotides per C molecule (Ivanyi-Nagy, Lavergne, Gabus, Ficheux, & Darlix, 2008; Murray, Marcotrigiano, & Rice, 2008). The core protein can be complemented *in trans* (Riedel, Lamp, Heimann, & Rümenapf, 2010). Its sequence can be modified in several regions, notably a stretch of basic amino acids can be completely deleted (Riedel et al., 2010), leading to decreased virus titers. C is flexible in length as shown by several mutations, the naturally occurring difference between different pestivirus species and the fact that virus with a core duplication was viable, although not stable in cell culture (Riedel et al., 2010). Even viruses containing fusion proteins of yellow fluorescent protein and C were viable, but with increasing size, less and less C is incorporated into infectious virus particles. Interestingly, the C gene of CSFV can be completely deleted showing that this protein is dispensable for virus propagation *in vitro*, although the resulting deletion mutants, which were dependent on a second-site mutation in the C-terminal region of NS3, yielded 30–50 times less infectious progeny and were attenuated in animals (Riedel et al., 2012). The identification of compensating mutations in NS3 in the recovered viruses with mutated or deleted C indicated that NS3 can compensate for the loss of C during virus assembly.

4.2.2 E^{rns}

E^{rns} directly follows the core protein in the pestiviral polyprotein. It is the first of the envelope glycoproteins and exhibits unusual features including its exceptional membrane anchor, the seemingly finely tuned balance between membrane bound and secreted protein, and its RNase activity. E^{rns} is unique to pestiviruses, with no homologue even in the closely related hepaci- or pegiviruses (Lindenbach et al., 2013; Stapleton, Foung, Muerhoff, Bukh, & Simmonds, 2011). E^{rns} is present on the viral particle as can be demonstrated by immunogold labeling (Fig. 7) and in the virus-free supernatant

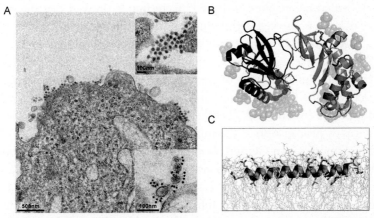

Figure 7 E^{rns} association with viral particles and structure. (A) Electron micrograph of a cell infected with BVDV. The viral particles are clearly visible on the cell surface. The inset in the upper right corner shows the viral particles at a higher magnification. The inset in the lower right corner shows CSFV particles labeled with anti-E^{rns} antibodies and gold particles which clearly demonstrates the presence of E^{rns} on the viral particle (electron microscopy: Harald Granzow and Frank Weiland, Friedrich-Loeffler Institut). (B) Structure of the E^{rns} ectodomain as a dimer as originally published (Krey et al., 2012), with each monomer in different shades. The transparent spheres show the glycosylation of E^{rns}, and disulfide bonds are shown as thin sticks. Image was rendered with PyMOL using the structure deposited in the PDB database (4DVK) (PyMOL: The PyMOL Molecular Graphics System, Version 1.7.4 Schrödinger, LLC). (C) An MD simulation of the E^{rns} membrane anchor taken from Aberle et al. (2014). (See the color plate.)

(Rümenapf et al., 1993). It is a target for (weakly) neutralizing antibodies (Weiland, Ahl, Stark, Weiland, & Thiel, 1992) and can block pestivirus infection in tissue culture when added to the culture medium (Hulst & Moormann, 1997, 2001). The protein has an apparent molecular weight of 42–48 kDa, is highly glycosylated with more than 50% carbohydrates in its mature form, and usually exists as disulfide-linked homodimers (Hulst & Moormann, 2001; Thiel et al., 1991). E^{rns} has a length of 227 amino acids and contains 8–9 conserved cysteines that form intra- and inter-molecular disulfide bonds. The most carboxy-terminal cysteine (C171) establishes a disulfide bond between the monomers in E^{rns} homodimers and is the only cysteine residue not absolutely conserved in pestivirus E^{rns} (Tews, Schürmann, & Meyers, 2009). E^{rns} is a membrane protein, but instead of a transmembrane anchor like the other two envelope proteins, it contains a long amphipathic helix at its C-terminus responsible for membrane binding (Fetzer, Tews, & Meyers, 2005; Tews & Meyers,

2007). Structural analyses of the E^{rns} C-terminus showed that the amphipathic α-helix is inserted slightly tilted into the membrane. Within this segment, a stretch of 15 amino acids shows no exchange of N-bound protons with water. These residues have either no water contact indicating complete insertion into the membrane or are fixed in a very stable helix forming upon contact with the lipid bilayer so that the N-linked protons are trapped in a rigid structure (Aberle et al., 2014) (Fig. 7). While the protein is essential for pestivirus propagation (Reimann, Semmler, & Beer, 2007; Widjojoatmodjo, van Gennip, Bouma, van Rijn, & Moormann, 2000), it is not necessary for infection with pseudotyped retroviruses carrying pestivirus glycoproteins. In this case, the envelope proteins E1 and E2 are necessary and sufficient for infection (Ronecker, Zimmer, Herrler, Greiser-Wilke, & Grummer, 2008; Wang, Nie, Wang, Ding, & Deng, 2004). It has been discussed that E^{rns} might mediate first attachment to cells through an interaction with glycosaminoglycans (Iqbal, Flick-Smith, & McCauley, 2000; Iqbal & McCauley, 2002). Increased binding to glycosaminoglycans is also a result of cell culture adaptation of pestiviruses (Hulst, van Gennip, & Moormann, 2000), which, however, can be prevented by propagation in porcine PEDSV.15 cells or in the presence of the drug DSTP 27 (Eymann-Hani, Leifer, McCullough, Summerfield, & Ruggli, 2011).

Most of the protein, even when expressed alone, is membrane bound and retained in the cell (Burrack, Aberle, Burck, Ulrich, & Meyers, 2012; Grummer, Beer, Liebler-Tenorio, & Greiser-Wilke, 2001; Tews & Meyers, 2007), but a significant percentage can also be found in the supernatant of infected cells and in the blood of infected animals (Burrack et al., 2012; Magkouras, Mätzener, Rümenapf, Peterhans, & Schweizer, 2008; Rümenapf et al., 1993; Tews & Meyers, 2007; Weiland, Weiland, Unger, Saalmüller, & Thiel, 1999; Weiland et al., 1992). Intracellular E^{rns} shows a predominantly ER-centered localization. Retention depends on the C-terminus of the protein, most notably on three bulky hydrophobic amino acids (L138, I190, L208). These amino acids are not completely conserved among all pestiviruses, but the different E^{rns} proteins also show different levels of retention in the cell (Burrack et al., 2012).

Although E^{rns} is a highly glycosylated protein, there is no detectable interaction with the ER lectin chaperones calnexin or calreticulin. Furthermore, as evaluated by determination of RNase activity, folding is independent of the glycosylation of E^{rns}. Instead, E^{rns} interacts with BiP, an ER chaperone binding to stretches of unfolded nascent proteins. Nevertheless, secretion of E^{rns} is impaired by inhibitors of α-glycosidases (Branza-Nichita, Lazar, Dwek, & Zitzmann, 2004).

One of the most interesting features of E^{rns} is the fact that it is an RNase with a preference for single-stranded RNA (Hulst, Himes, Newbigin, & Moormann, 1994; Schneider, Unger, Stark, Schneider-Scherzer, & Thiel, 1993) displaying homology to RNases of the T2 family. The RNase shows a cleavage specificity for NpU bonds (Hausmann, Roman-Sosa, Thiel, & Rümenapf, 2004). Unusual among T2 RNases, specificity is determined by the B2 rather than by the B1 site of the enzyme (Irie & Ohgi, 2001). Most T2 RNases are monomeric glycosylated proteins with no strict substrate specificity (Deshpande & Shankar, 2002). RNase activity of E^{rns} has a pH optimum of pH 6.0 and is inhibited by Zn^{2+} and Mn^{2+}, but insensitive to chelating agents (Schneider et al., 1993; Windisch et al., 1996). The crystal structure of the E^{rns} ectodomain (amino acids 1–165) has been solved (Krey et al., 2012). It contains five α helices and seven β strands (Fig. 7) with a concave and a convex face and is stabilized by four intramolecular disulfide bonds formed between residues 38/82, 68/69, 110/155, and 114/138 (numbers of BVDV sequence). The equivalent arrangement of disulfides with the unusual vicinal bond between residues 68/69 was also found in earlier biochemical assays for CSFV (Langedijk et al., 2002). Six N-glycans were also confirmed which are all located on the convex side of the molecule. A survey of the location of N-glycosylation sites in the E^{rns} sequences from different pestiviruses revealed that the majority is located on the convex side, so that it seems likely that glycosylation on the concave side would interfere with E^{rns} dimer formation or enzymatic activity. Structural analysis confirmed the T2 fold of the protein and showed that the Zn^{2+} ion bound to histidine 81 in the active center (Krey et al., 2012), which explains the inhibition of RNase activity by this ion. The highest structural similarity was observed for the plant T2 RNase MC1 that also shows a preference for NpU bonds (Krey et al., 2012). The structural data argue in favor of a two-step interaction between the RNase and its substrate. The concave side is positively charged and could bind RNA rather unspecifically to allow scanning for scissile bonds. In a second step, a highly specific interaction involving the uracil base of the residue 3′ of the cleavage site would be established. This interaction is in accordance with the structural data, showing that the B2-binding site exclusively bound uracil bases and that the binding was stabilized through a hydrogen bond between the O4 atom of the uracil ring and Asn9. In contrast, B1 could interact with any nucleotide. As mentioned above, E^{rns} forms homodimers and *in vitro* experiments with the purified enzyme have shown an approximately fourfold slower cleavage and higher affinity by the dimer compared to the monomer (Krey et al., 2012). The increased binding affinity is supposed to result from the doubled number of positively charged binding sites in the active site cavity formed when the two monomers anneal at their concave sides. As a consequence of this arrangement, the exchange of substrate and product would be impaired resulting in a slower cleavage rate.

The identification of an RNase in an RNA virus poses two important questions: (i) why is the viral genome not degraded and (ii) what is the function of this unusual activity. Viral genomic RNA was rapidly destroyed by the E^{rns} RNase proving that it was not protected from degradation (Windisch et al., 1996). Instead, analyses revealed that the formation of the active RNase requires the environment of the ER which allows proper folding assisted by glycosylation and formation of the four intramolecular disulfide bonds (G. Meyers, unpublished results). Thus, it is likely that the active enzyme is obtained only when it is already separated from the genomic RNA by a lipid membrane, and this separation is preserved throughout budding and membrane fusion. Since no active RNase is present within the cytoplasm of the infected cell, the enzymatic activity cannot have any function in the viral replication cycle. Indeed, mutation of predicted active site residues of the RNase led to inactivation of the enzyme but allowed recovery of viable virus mutants with growth characteristics similar to wild-type viruses (Hulst, Panoto, Hoekman, van Gennip, & Moormann, 1998; Meyer, Von Freyburg, Elbers, & Meyers, 2002; Meyers et al., 1999). Thus, the ability to express an active E^{rns} RNase seems to offer no significant advantage for virus replication in tissue culture. However, the RNase motifs and therefore most likely also the enzymatic activity of the protein have been conserved during evolution of pestiviruses indicating an important function of the RNase.

Animal studies pointed at a putative function of the E^{rns} RNase since viruses with inactivated RNase were attenuated in their natural hosts (Meyer et al., 2002; Meyers et al., 1999). Although animals infected with RNase-negative viruses developed initial symptoms, the disease was much less severe and the animals recovered but were protected against subsequent challenge infection. E^{rns} forms disulfide-linked homodimers through the last cysteine in its sequence (C171). Deletion or mutation of this cysteine leads to loss of the disulfide bond, and E^{rns} dimers are no longer detected, not even in native gels under mild lysis conditions (Tews et al., 2009). However, crosslinking experiments showed that E^{rns} monomers still interact at least transiently (G. Meyers, unpublished results). Viruses harboring such mutations are viable in cell culture. There is even a biologically cloned variant of the BVDV-1 prototype strain NADL that lacks C171, proving viability in the natural host (Tews et al., 2009; van Gennip, Hesselink, Moormann, & Hulst, 2005). Nevertheless, naturally occurring pestiviruses lacking C171 are very rare, indicating that E^{rns} dimerization confers a distinct biological advantage. This advantage could be seen in animal experiments where a genetically engineered CSFV without C171 proved to be attenuated in pigs (Tews et al., 2009). Furthermore, in additional animal studies with C171 negative viruses, revertants and pseudorevertants could be isolated. These viruses also partially regained virulence (G. Meyers et al., unpublished data).

Cell culture experiments showed that E^{rns} is able to suppress induction of interferon-β and the interferon-induced protein MxA after stimulation by extracellular dsRNA or ssRNA (Iqbal, Poole, Goodbourn, & McCauley, 2004; Magkouras et al., 2008; Mätzener, Magkouras, Rümenapf, Peterhans, & Schweizer, 2009). Inhibition is dose dependent and still effective at concentrations in the same order as found in the blood of a PI animal (Magkouras et al., 2008). All these facts are consistent with a role for secreted rather than virion-bound E^{rns} in the suppression of an interferon response. The special membrane anchor of E^{rns} obviously plays an important role in the finely tuned balance between secretion and retention/incorporation into the particle.

New evidence suggests that E^{rns} bound to cells is taken up via clathrin-dependent endocytosis (Zürcher, Sauter, Mathys, Wyss, & Schweizer, 2014). It has also a profound effect on PDCs. PDCs are the most important source of systemic interferon I (Reizis, Bunin, Ghosh, Lewis, & Sisirak, 2011). They produce significantly more interferon if stimulated by cells infected with CSFV than by direct stimulation through viral particles (Python, Gerber, Suter, Ruggli, & Summerfield, 2013). This is true even if infection of the contact cells occurs through assembly incompetent virus replicon particles. Different experiments showed that stimulation is even more potent in the absence of E^{rns}. In those experiments, a direct transfer of viral RNA from infected cells to PDCs could be shown. The induction of IFN could be repressed by inhibitors of TLR7. Coexpression of E^{rns} in cells infected with unrelated positive strand RNA viruses also led to decreased interferon levels. All these effects were dependent on the RNase activity of E^{rns} (Python et al., 2013).

4.2.3 E1

E1 still awaits detailed characterization. It is the only pestivirus envelope protein for which no structure or function is known. Infected animals do not develop antibodies against E1, and it has been difficult to raise antibodies against the protein after immunization with heterologously expressed protein or synthetic peptides, making detection of it difficult. Furthermore, the main focus of research has been on E2 and E^{rns} as the targets for neutralizing antibodies that bind to cellular receptor molecules and, in case of E^{rns}, have unusual enzymatic activities. E1 has often been analyzed only in context with the other two envelope proteins. E1 is a 25–33 kDa protein (depending on the virus species) with a transmembrane anchor. It forms disulfide-linked heterodimers with E2 which are present on the viral particle (Thiel et al., 1991). The amino-terminus of E1 is generated by SPase cleavage at the unusual E^{rns} membrane anchor/E1 site. The cleavage between E^{rns} and E1 is slower than between E1 and E2 (Rümenapf et al., 1993) leading to the detectable presence of E^{rns}-E1 in infected and transfected cells. SPase is also responsible for processing the E1/E2 site. In this case, a typical SPase cleavage site including both a transmembrane domain and a von Heijne consensus sequence is present (Rümenapf et al., 1993). Due to the length of the hydrophobic region at the carboxy-terminus of E1, the membrane topology of the mature E1 protein is difficult to predict and awaits experimental analysis. The transmembrane domain of HCV E1 forms a hairpin structure in the membrane before cleavage between E1 and E2, with both ectodomains situated on the lumenal side of the ER. Once cleaved, the transmembrane domain rearranges to form a single span transmembrane anchor (Cocquerel et al., 2002).

Similar to the other two envelope proteins, E1 contains several cysteine residues, most of which are exceptionally well conserved throughout the pestiviruses (Fernandez-Sainz et al., 2014). It is likely that intramolecular disulfide bonds are generated, probably stabilizing the protein. Moreover, a disulfide bridge is established with E2 leading to covalently bound heterodimers. The formation of the E1–E2 heterodimer is slow (Rümenapf et al., 1993) which might be due to the slow folding of E1 (Branza-Nichita, Durantel, Carrouee-Durantel, Dwek, & Zitzmann, 2001; Branza-Nichita, Lazar, Durantel, Dwek, & Zitzmann, 2002). E1 interacts transiently with calnexin (>30 min), and the formation of E1–E2 heterodimers starts after correct folding and release of E1 from calnexin. The interaction with calnexin is dependent on the glycosylation of E1 as shown by the use of an α-glucosidase inhibitor (Branza-Nichita et al., 2001).

Some antisera have been generated against E1 and used to elucidate E^{rns}–E1 processing (Bintintan & Meyers, 2010; Wegelt, Reimann, Zemke, & Beer, 2009). The C-terminus of E^{rns} contains a classical von Heijne cleavage motif at positions −3 and −1, but lacks the hydrophobic region that is normally required for SPase recognition, due to the special membrane anchor of E^{rns}. Mutation and inhibitor studies showed that the cleavage between E^{rns} and E1 occurs in the ER and is mediated by SPase (Bintintan & Meyers, 2010) despite the absence of the transmembrane element that in standard SPase cleavage sites is responsible for positioning the von Heijne motif in a cleavable position (von Heijne, 1990). Apparently, the amphipathic helix at the E^{rns} C-terminus can substitute for the TM element and position the cleavage site correctly. Processing is sensitive to changes in the amphipathic helix of E^{rns} and the cleavage motif and is dependent on the glycosylation of E^{rns} (Bintintan & Meyers, 2010). These results point toward delicate structural requirements that have to be fulfilled to render this peptide a substrate for SPase. Structural analysis of the E^{rns}/E1 processing site in a membrane environment is required to understand this processing in detail.

E1 and E2 are the only envelope glycoproteins necessary to mediate fusion of pseudotyped viruses (Ronecker et al., 2008; Wang et al., 2004). Furthermore, since the crystal structure of E2 did not reveal any obvious fusion peptide, it has been proposed that E1 contains the fusion peptide, with a stretch of hydrophobic amino acids (57–85) as a possible candidate. However, E1 is only about half the size of E2 and E2 is an elongated protein; thus, E1 would have to be very long and thin to bridge the distance to the target membrane after E2 has established receptor contact (El Omari, Iourin, Harlos, Grimes, & Stuart, 2013).

4.2.4 E2

E2 exhibits a size of about 53–55 kDa. It represents the receptor-binding protein of pestiviruses, is the main target for neutralizing antibodies (Deregt & Loewen, 1995; Paton, Lowings, & Barrett, 1992; Weiland et al., 1990), is responsible for the species tropism of pestiviruses (Liang et al., 2003), and, in soluble form, is able to inhibit infection. The inhibition is effective with either CSFV or BVDV on porcine and bovine cells, indicating that both virus species use the same receptor (Hulst & Moormann, 1997). In addition, an E2-derived peptide has been shown to bind to cells and inhibit infection (Li et al., 2011).

The C-terminus of E2 contains a hydrophobic sequence that acts as transmembrane anchor. The cleavage between E1 and E2, and between E2 and p7 is effected by SPase (Elbers et al., 1996; Rümenapf et al., 1993). The latter is not complete and uncleaved E2–p7 can be found after expression and in infected cells (Harada et al., 2000). The mature E2 is 373 amino acids long and contains 15 conserved cysteines, with 2 additional cysteines in BVDV strains, all of which form intra- and intermolecular disulfide bridges (El Omari et al., 2013; Li, Wang, Kanai, & Modis, 2013). Glycosylation sites are slightly less conserved with three to six potential N-glycosylation sites depending on virus species and strain.

E2 interacts with calnexin during folding, but folding is fast and dissociation of calnexin as well (~2.5 min) (Branza-Nichita et al., 2001). E2 associates with E1 to form disulfide-linked dimers. In the presence of α-glycosidase inhibitors, there is no interaction of E2 with calnexin, resulting in a certain percentage of misfolded E2. Furthermore, it seems that misfolded E2 cannot associate with E1 (Branza-Nichita et al., 2001). E2 is retained in the cell through its transmembrane anchor (Grummer et al., 2001; Köhl et al., 2004). More specifically, R1047 in the transmembrane region plays a role in the intracellular retention of E2, as mutation of this amino acid led to cell surface localization of E2 (Köhl et al., 2004). Heterodimer formation depends on the transmembrane regions of the two proteins, specifically on charged residues K671 and R674 in E1 and on R1047 in E2, the residue that also plays a role in E2 retention (Ronecker et al., 2008).

The importance of E2 as a target for neutralizing antibodies, its use in diagnosis, and the availability of a large panel of monoclonal antibodies has led to considerable efforts to characterize the protein by mapping binding sites and antigenic regions in order to get information on the structure (Paton, Lowings, & Barrett, 1990; Paton et al., 1992; van Rijn, Miedema, Wensvoort, van Gennip, & Moormann, 1994; van Rijn, van Gennip, de Meijer, & Moormann, 1993). Recently, the crystal structure for the BVDV E2 ectodomain has been solved by two different groups (El Omari et al., 2013; Li et al., 2013). Both groups found overall very similar structures, although they assigned domain names differently (Fig. 8). Omari et al. solved structures of proteins crystallized at pH 5 and 8, whereas Li et al. established crystals at pH 5.5. Glycosylation at four glycosylation sites (N117, N186, N230, and N298) was confirmed. E2 represents an elongated protein (140 Å) with several domains arranged in a linear fashion (El Omari et al., 2013). In the E2 homodimers analyzed in the crystals, the 17 cysteines form 8 intramolecular and 1 intermolecular disulfide bonds. The latter should be involved in homo- and heterodimer formation. The two most N-terminal and thus membrane distal domains show an Ig-like fold and were named DA and DB or domain I and II, respectively. In the dimer of NADL E2, DA/domain I of one monomer is shifted compared to its

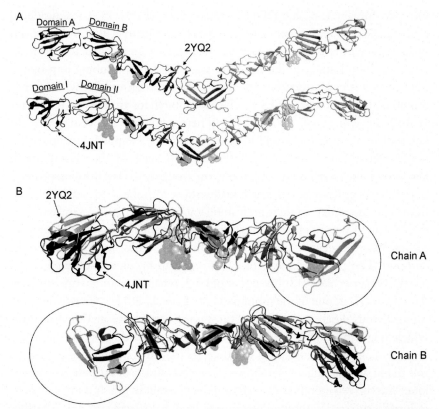

Figure 8 Structure of E2. (A) Structure of the BVDV E2 dimer at pH 8 (2YQ2 (El Omari et al., 2013)) and at pH 5.5 (4JNT (Li et al., 2013)), with both monomers in different colors. Glycans in the structures are shown as transparent spheres (the glycosylation at the base of the dimer is not present in the 2YQ2 structure as it was mutated in the crystallized protein). Indicated are domain A or I and domain B or II. (B) Monomers of the above structures superimposed. Note the domain swap in structure 2YQ2, not present in 4JNT (enclosed in the circle). Images were rendered with PyMOL using the structures deposited in the PDB database (PyMOL: The PyMOL Molecular Graphics System, Version 1.7.4 Schrödinger, LLC). (See the color plate.)

position in the other monomer, showing that the hinge between domain I and II is flexible (Li et al., 2013).

Most defined antigenic regions of neutralizing antibodies against E2 map to the two membrane distal domains, on exposed surfaces of the protein. Similarly, a host cell-binding peptide derived from CSFV E2 (Li et al., 2011) mapped to a β-hairpin exposed on domain II (DB), indicating that binding of the receptor CD46 might occur through this region. The remaining ectodomain shows no similarity to other known structures and was divided into two domains (DC and DD) by Omari et al., whereas Li et al. considered it one single domain (domain III). The data from Omari et al. show a domain swap between the two monomers in DD, whereas in the structure from Li et al., the loop leading to the structure was disordered, resulting in proposal of two possible conformations, one with a domain swap and one without. It is important to remember that E1 and E2 are required for virus entry (Ronecker et al., 2008; Wang et al., 2004) and form heterodimers (Thiel et al., 1991). In heterodimers, either E1 would have to provide the swapped domain, or E2 might take the unswapped conformation. In spite of the relation with flaviviruses, the overall fold of E2 shows no similarity to class II fusion proteins of the alpha- or flaviviruses, which furthermore form head-to-tail dimers and not tail-to-tail dimers in their prefusion (Vaney & Rey, 2011). Furthermore, the E2 structure reveals no obvious fusion peptide and no pH-dependent drastic refolding as seen in class II fusion proteins. Only small changes were detected. The most striking is that DA seems to become disordered at pH 5.0 (El Omari et al., 2013) but not at pH 5.5, where it is simply oriented differently to DB in one of the monomers (Y. Li et al., 2013). Nevertheless, the most obvious differences between the two structures are in DA/domain I, which might be simply a reaction to the pH. This domain also contains the only completely conserved histidine (H762) in E2, which might play a role in pH sensing. It has been shown before that fusion of the pestivirus envelope with the cell membrane needs more than a shift in pH. It is impossible to induce fusion from without in pestiviruses by only lowering the pH. A further trigger through the addition of reducing agents is necessary to achieve fusion, albeit rather poorly (Krey, Thiel, & Rümenapf, 2005). Thus, the structure of E2 at low pH might not represent the true post fusion state. Interestingly, HCV particles also will not fuse with the plasma membrane upon a shift in pH, suggesting similar mechanisms for both virus genera. Nevertheless, the recently determined core structure of the HCV E2 ectodomain (Khan et al., 2014; Kong et al., 2013) showed a completely different fold than pestiviral E2.

Entry of pestiviruses is probably a multistep process. CD46 could be identified as a receptor for BVDV by the use of monoclonal antibodies directed against bovine proteins that were able to inhibit infection (Schelp et al., 1995). Cells expressing $CD46_{bov}$ bind BVDV, and expression of $CD46_{bov}$ in porcine cells increases their susceptibility to BVDV infection (Maurer et al., 2004). In further studies, the viral binding site could be mapped to two short peptides in the complement control protein module 1 of $CD46_{bov}$ (Krey et al., 2006). However, CD46 is not sufficient for pestivirus entry since cells expressing E2 are resistant to BVDV infection, although they have similar levels of CD46 expression and bind BVDV to a similar extent as control cells (Tscherne, Evans, Macdonald, & Rice, 2008). CD46 is excluded from endosomes, but BVDV enters by clathrin-dependent endocytosis (Grummer, Grotha, & Greiser-Wilke, 2004; Krey et al., 2005; Maurer et al., 2004). Nonpermissive cells remain nonpermissive after CD46 expression, although they are able to replicate viral genomes (Maurer et al., 2004). CSFV E2 can inhibit infection of porcine and bovine cells by CSFV or BVDV, suggesting that they use the same receptor, but BVDV does not bind to porcine CD46 (Krey et al., 2005) and CSFV infection of bovine cells is not inhibited by antibodies against bovine CD46. Nevertheless, experiments with CSFV and antibodies against the porcine CD46 suggest that porcine CD46 plays a role for CSFV entry, but the inhibitory effect remained rather weak, suggesting involvement of additional factors (Dräger, Beer, & Blome, 2015). The related HCV was shown to use at least four different entry factors (for a review, see Ding, von Schaewen, & Ploss, 2014; Lindenbach & Rice, 2013). For pestiviruses, evidence so far suggests that a first attachment might be due to the interaction of E^{rns} with glycosaminoglycans (Hulst et al., 2000, 2001; Iqbal et al., 2000). Infection then needs CD46 (Krey et al., 2005; Maurer et al., 2004) and additional unknown factors.

As the main target for neutralizing antibodies, E2 alone is able to induce protection in animals (Bruschke, Moormann, van Oirschot, & van Rijn, 1997; Krey et al., 2005; Maurer et al., 2004). This and the fact that E2 is responsible for species tropism has been exploited in the vaccine virus cp7_E2Alf, a chimeric BVDV virus to be used for vaccination against CSFV, expressing the CSFV Alfort 187 E2 protein (Reimann, Depner, Trapp, & Beer, 2004) (see also below).

Intracellular budding has long been proposed for pestiviruses. All three envelope glycoproteins are retained in the cells (Burrack et al., 2012; Grummer et al., 2001; Köhl et al., 2004; C. Radtke & B.A. Tews, unpublished data), and biochemical data indicate that virus particles are formed intracellularly. Maturation of the carbohydrate chains of the envelope proteins occurs during transport of the particles along the secretory pathway to the cell surface (Burrack et al., 2012; Grummer et al., 2001; Jordan, Nikolaeva, et al., 2002; Köhl et al., 2004; Macovei, Zitzmann, Lazar, Dwek, & Branza-Nichita, 2006). Recent ultrastructural analyses on the pestivirus Giraffe-1 revealed the presence of particles containing the core protein or the E2 protein in the ER. Viral particles could also be detected in

Golgi stacks and outside the infected cells. These data confirm the ER as the pestiviral budding site at which the core–RNA complex is packaged in envelopes that carry the envelope proteins (Schmeiser, Mast, Thiel, & König, 2014). Viral particles are then exported from the cell via the secretory pathway.

4.3 The Nonstructural Proteins

The replication of pestiviruses takes place in contact with intracellular membranes. In contrast to other members of the *Flaviviridae*, an elaborate rearrangement of intracellular membranes was not found in pestivirus-infected cells (Schmeiser et al., 2014). Nevertheless, replication occurs in close association with membranes of the ER. In accordance with this replication strategy, most viral proteins including the NSPs, with the exception of N^{pro}, are thought to be associated with intracellular membranes either directly or indirectly via viral membrane-bound interaction partners.

The minimal set of viral proteins required for pestiviral RNA replication was found to be NS3–NS5B (Behrens et al., 1998). Remarkably, these proteins are also involved in virion morphogenesis, as described for other members of the *Flaviviridae* (Murray, Jones, & Rice, 2008).

Detailed studies on the interaction of the viral replication complex with membranes, the membrane topologies of individual proteins, as well as the roles of individual proteins in RNA replication and virion morphogenesis are still to be performed.

4.3.1 p7

p7, which is encoded downstream of E2, has not been detected in virions and thus is classified as an NSP (Elbers et al., 1996). It is released from the polyprotein by cellular SPase (Elbers et al., 1996). In pestivirus-infected cells, p7 has been detected as individual protein of about 7 kDa and as part of an E2–p7 precursor, similar to HCV (Elbers et al., 1996; Harada et al., 2000; Lin, Lindenbach, Pragal, McCourt, & Rice, 1994). This finding is remarkable since SPase cleavages are generally highly efficient. Thus, it is of interest to clarify how and why cleavage at this site is regulated in the context of the viral polyprotein. E2–p7 is essential neither for viral RNA replication nor for the generation of infectious viral progeny, while free p7 is selectively required for the latter (Harada et al., 2000). Due to its characteristics as a hydrophobic, small integral membrane protein with a charged central region, it was suggested that p7 might be a viroporin (Harada et al., 2000; Largo, Gladue, Huarte, Borca, & Nieva, 2014) functioning as ion channel (Gonzalez & Carrasco, 2003), and it was proposed that p7 of pestiviruses and HCV shares a class IIA viroporin architecture (Nieva, Madan, & Carrasco, 2012). Studies on HCV p7 demonstrated formation of intramembrane hexameric rings with Ca^{2+} channel activity triggered by low pH (Foster et al., 2011; Griffin et al., 2003). p7 of HCV and BVDV substitutes for the ion channel activity of influenza virus M2, a prototype of viroporins (Griffin et al., 2004). In analogy to HCV, pestivirus p7 may also exhibit an additional, independent role in virion morphogenesis probably via protein/protein interactions (Wozniak et al., 2010). A role of the ion channel activity of p7 in the induction/release of proinflammatory cytokines from infected macrophages was reported, indicating that, besides supporting virion formation, the disturbance of the ionic homeostasis influences also basic cellular functions (Lin et al., 2014). A small-molecule inhibitor of the p7 ion channel displayed antiviral activity against BVDV (Luscombe et al., 2010). For p7 of CSFV, a scanning mutagenesis approach identified regions which were critical for viral fitness and virulence in *vivo*; furthermore, this study verified the viroporin function of p7 by a peptide-based *in vitro* reconstitution approach and indicated differences to p7 of HCV (Gladue et al., 2012). More recently, fine mapping defined p7 amino acids 33–67 as the minimal pH-triggered pore forming peptide in liposomes (Largo et al., 2014).

The picture emerging from the work on p7 of HCV and pestiviruses points to an important role of p7 in reducing intracellular vesicle acidification required for a late stage in the assembly of infectious progeny virus (Jones, Murray, Eastman, Tasello, & Rice, 2007; Steinmann et al., 2007; Wozniak et al., 2010).

4.3.2 NS2

NS2 consists of about 450 amino acids. The highly hydrophobic N-terminal half might contain up to seven transmembrane segments followed by a cytoplasmic domain. The latter encompasses a putative Zn^{2+}-binding site, consisting of four cysteine residues, which is an integral part of an autoprotease domain residing within NS2 (de Moerlooze et al., 1990; Desport, Collins, & Brownlie, 1998; Lackner et al., 2004). Mutagenesis studies suggest that the active site of the NS2 protease is constituted by a catalytic triad of His1447, Glu1462, and Cys1512 (Lackner et al., 2004). The data obtained so far suggest that the NS2 protease is restricted to NS2-3 cleavage *in cis*. Only upon C-terminal truncation of NS2 by at least four amino acids, *trans*-cleavage of a substrate encompassing the NS2/NS3 cleavage site could be observed

(Lackner, Thiel, & Tautz, 2006). The latter experiments indicate that the C-terminus of NS2 stays in the active site of the protease after *cis*-cleavage, thereby acting as an inhibitor preventing further NS2 cleavages *in trans*. In the crystal structures of the capsid protease of alphaviruses as well as the NS2 protease of HCV an analogous, inactivating positioning of the C-termini was observed (Choi et al., 1991; Lorenz, Marcotrigiano, Dentzer, & Rice, 2006). Similarly, the N^{pro} protease is rendered inactive upon autoproteolytic cleavage by blocking the active site by the carboxy-terminal substrate strand (see above).

Interestingly, NS2 of noncp pestiviruses requires a stable interaction with a cellular chaperone of the Hsp40 family for its activity (Lackner, Müller, König, Thiel, & Tautz, 2005; Moulin et al., 2007). The role of this cofactor protein termed Jiv (J-domain protein interacting with viral protein) or DNAJ-C14 for control of RNA replication and cytopathogenicity will be described in Section 5.

4.3.3 NS2–3
Until the autoprotease activity of NS2 was identified, it was assumed that noncp pestiviruses use NS2–3 instead of NS3 for their replication. However, the minimal viral RNA replicase consists independent of the viral biotype of NS3, NS4A, NS4B, NS5A, and NS5B since NS2–3 cannot functionally replace NS3 (Behrens et al., 1998; Lackner et al., 2004; Moulin et al., 2007). Importantly, uncleaved NS2–3 has also an essential, but not well-characterized, function in virion morphogenesis. This became apparent when BVDV or CSFV strains encoding an ubiquitin monomer upstream of NS3 or containing an encephalomyocarditis virus (EMCV) IRES insertion between the NS2 and NS3 genes were generated. Since the ubiquitin moiety leads to complete cleavage by cellular ubiquitin hydrolases, replication is expected to occur in the absence of uncleaved NS2–3 (Tautz, Meyers, & Thiel, 1993). In both settings, the absence of uncleaved NS2–3 correlated with a defect in the production of infectious progeny virus, despite efficient viral RNA replication. The packaging function of NS2–3 was restored by NS2–3–4A *in trans* (Agapov et al., 1998; Moulin et al., 2007). The activity of the NS2 cysteine protease was not required for packaging (Moulin et al., 2007). Additional experiments also excluded essential functions for the helicase and NTPase activities of NS2–3 in packaging (Moulin et al., 2007). Inactivation of the NS3 serine protease, which is responsible for the release of NS4A in *cis*, interfered with *trans*-complementation by NS2–3–4A (Agapov et al., 1998; Moulin et al.,

2007). However, when NS2–3 with an accurate C–terminus together with authentic NS4A was supplied *in trans*, inactivation of the NS3 serine protease did no longer abrogate virion morphogenesis. Another complementation study also concluded that NS4A is required for complementation *in trans* and that C–terminal truncations interfere with this function (Liang et al., 2009).

Together, these data show that none of the known enzymatic activities of NS3 are required for virion morphogenesis *per se*. The fact that NS4A has to be supplied *in trans* together with NS2–3 for efficient packaging, although the virus to be rescued encodes NS4A in its polyprotein, shows that this NS4A protein is obviously not available to support the packaging function of NS2–3 expressed *in trans*. This fact may be related to cotranslational association of the polyprotein-derived NS3/NS4A and/or to a higher affinity of NS4A to NS3 when compared to NS2–3 as described in the HCV system (Welbourn et al., 2005).

Surprisingly, the need for uncleaved NS2–3 in virion morphogenesis is not conserved between pestiviruses and HCV. Insertion of an EMCV IRES between the NS2 and NS3 genes of HCV only slightly reduced viral titer (Jirasko et al., 2008; Jones et al., 2007). This unexpected finding raised the intriguing question whether some pestiviruses with ubiquitin insertions between NS2 and NS3 might be capable of virion formation. The BVDV-1 strain Osloss, which encodes one monomer of ubiquitin, was isolated from a heifer with MD and passaged in cell culture (Meyers, Rümenapf, & Thiel, 1989b; Renard, Dino, & Martial, 1987). Since fusion proteins containing ubiquitin in the N-terminal position are known to be processed very efficiently by cellular proteases, complete cleavage of the BVDV Osloss NS2/ubiquitin/NS3 protein can be expected. It remained, however, uncertain whether the virus isolate still contains a noncp BVDV strain which supports as a helper virus the replication of the ubiquitin-encoding cp BVDV strain Osloss by providing uncleaved NS2–3. To address these questions, chimeric viruses containing Osloss genome fragments were established based on the cDNA clone of noncp BVDV-1 strain NCP7. These studies revealed that the NS2-ubi-3–4A–4B* region of strain Osloss in the noncp BVDV-1 genome allowed the generation of low titers of infectious progeny in the absence of detectable amounts of uncleaved NS2–3. By passaging this chimera, a virus could be selected that was capable of efficient virion production. To finally rule out a functional role of uncleaved NS2–3, the ubiquitin-coding sequence was replaced in this chimeric virus by an EMCV IRES. The respective chimera was also shown to produce infectious progeny, finally proving that pestiviruses can be adapted to assembly in the absence of uncleaved NS2–3 (Lattwein, Klemens, Schwindt, Becher, & Tautz, 2012). Recent work identified two amino acids, one in NS2 and one in NS3, which are critical for this gain of function (N. Tautz et al., unpublished). It will be interesting to determine the mechanistic differences underlying the packaging processes with or without uncleaved NS2–3, respectively. This would help to clarify which features allow an NSP like NS3 that is designed for assisting in RNA replication to gain an additional function in virion morphogenesis.

These findings also raise the intriguing question why pestiviruses, in contrast to HCV, maintain NS2–3-dependent packaging. *In utero* infection followed by long-term persistence in the host represents the main strategy of pestiviruses for their spread in the field. Thus, any mutation which prolongs the persistence period will be selected over time. Persistence ends by the generation of a cp virus which induces lethal MD. Cp mutants are in most cases generated by RNA recombination (see Fig. 9), like insertions of substrates for cellular proteases, e.g., ubiquitin, upstream of NS3 (Becher & Tautz, 2011). However, due to the fact that pestiviruses strictly depend on uncleaved NS2–3 for virion morphogenesis, mutants with ubiquitin insertions between NS2 and NS3 will not be capable of autonomous replication and thus might not, or not efficiently, induce MD. In favor of this hypothesis, the vast majority of cp pestivirus mutants with insertions coding for substrates of cellular proteases upstream of NS3 contain, in addition, large genomic duplications encoding NS2–3–4A (Becher & Tautz, 2011). If pestiviruses would, similar to HCV, not depend on the expression of uncleaved NS2–3, a much broader set of cp mutants would be capable of autonomous replication and induction of MD. This in turn would statistically shorten the average duration of the persistent phase of such a virus in the host. Thus, rendering replication dependent on uncleaved NS2–3 represents an evolutionary advantage for the pestivirus strategy.

4.3.4 NS3

NS3 is a multifunctional protein with a size of about 80 kDa. In the N-terminal domain, a chymotrypsin-like serine protease resides (Bazan & Fletterick, 1988; Gorbalenya, Donchenko, Koonin, & Blinov, 1989), while the C-terminal part encompasses a helicase and NTPase domain (Gorbalenya, Koonin, Donchenko, & Blinov, 1989). According to mutagenesis studies, the catalytic triad of the protease is formed by residues His1658, Asp1686, and Ser1752 (Tautz, Kaiser, & Thiel, 2000; Wiskerchen & Collett, 1991).

Remarkably, in analogy to the alphavirus capsid protease, a Ser to Thr mutant still retained proteolytic activity (Tautz et al., 2000). To obtain its full activity, the NS3 serine protease requires the stable association with the central domain of its cofactor NS4A (Tautz et al., 2000; Xu et al., 1997). The NS3/4A protease is responsible for *cis*-cleavage generating the C-terminus of NS3 and all downstream cleavages which may occur *in trans*, thereby releasing NS4B, NS5A, and NS5B from the polyprotein (Wiskerchen & Collett, 1991). While in previous studies cp BVDV had been used, a recent study compared noncp and cp CSF viruses as well as cp CSFV replicons for a detailed analysis of polyprotein processing (Lamp et al., 2011). Especially upon noncp CSFV infection novel cleavage intermediates like NS4B–5A were identified besides NS4A–B and NS5A–B which had already been described (Collett, Moennig, & Horzinek, 1989; Lamp et al., 2011). This study also demonstrated that NS2–3/4A is capable of cleaving at all sites in the polyprotein, but with slower kinetics (Lamp et al., 2011). Thus, temporal restriction of NS2–3 cleavage also affects the downstream polyprotein processing events.

NS3/4A cleavage sites are characterized by Leu in the P1′ position and Ser, Ala, or Asn in P1 (Becher et al., 1998a; Schechter & Berger, 1967; Tautz et al., 1997; Xu et al., 1997). For the recently described two internal cleavages in NS3, the P1′ residues are either Leu or Ile, and the P1 positions are occupied by Met or Lys with the Leu/Lys site being processed at very low efficiency only (Lamp et al., 2013). Mapping of the C-terminal border of the protease domain of CSFV revealed that Leu1781 had to be retained for proteolytic activity. According to this and earlier studies, the length of the pestiviral NS3 protease domain is in good accordance with NS3 proteases of other flaviviruses (Lamp et al., 2013).

The recently described rather efficient internal cleavage in NS3 downstream of Leu/Ile1781 by the NS3/4A protease releases, besides a free helicase domain, a fully active minimal NS3 protease which may, depending on the activity of the NS2 protease, be preceded by the NS2 moiety. Mutations at the cleavage site abrogate this cleavage in the polyprotein which interferes with viral RNA replication. However, since these mutations also ablated the activity of the NS3/4A protease, further work is needed to understand the importance of these interesting cleavage events and the role of their processing products in the pestiviral life cycle (Lamp et al., 2013).

While the protease function of NS3/4A tolerated N-terminal fusions like GST or a deletion of two amino acids, an elongation by only one NS2-derived amino acid was sufficient to inhibit viral replication (Tautz et al., 2000; Tautz & Thiel, 2003). Along these lines, uncleaved NS2–3 was also shown to possess serine protease activity, though with reduced activity, but could not functionally substitute for NS3 in viral RNA replication (Lackner et al., 2004; Lamp et al., 2011). Together, these findings suggest a critical role for the authentic N-terminus of NS3 in RNA replication/replicase assembly, which is most likely unrelated to its serine protease activity.

In contrast to NS2–3–4A expression, ectopic expression of NS3–4A in MDBK cells induced caspase activation and apoptosis. Inactivation of the protease function ablated apoptosis induction leading to the hypothesis that NS3–4A protease is actively involved in this process (Gamlen et al., 2010).

The downstream domain of NS3, which is connected to the protease domain by a linker region, encompasses a superfamily 2 type DEXH helicase and an NTPase both of which are essential for viral RNA replication (Grassmann, Isken, & Behrens, 1999; Tamura, Warrener, & Collett, 1993; Warrener & Collett, 1995). RNA-binding activity of NS3 was observed which, however, was not specific for viral RNA (Grassmann et al., 1999).

Recently, a new and unexpected role was described for the C-terminal part of NS3 of CSFV. In studies addressing the essential properties of the viral core protein C, a massively enlarged C or even an almost complete deletion of C was compensated by single second-site mutations in NS3 between amino acids 2160 and 2256, which represents helicase subdomain 3 (Riedel et al., 2010, 2012). These findings are paralleled by observations made for other members of the *Flaviviridae*, like yellow fever virus and HCV, where important roles of the helicase domains, independent of their enzymatic functions, in virion morphogenesis were identified (Ma, Yates, Liang, Lemon, & Yi, 2008; Patkar & Kuhn, 2008). The interesting findings in the CSFV system challenge the assumed essential role of the C protein for pestiviruses. The fact that a single amino-acid substitution allows NS3 to functionally replace C in virion morphogenesis raises the question whether NS3 had a packing function in an ancestral prototype virus in the absence of a capsid gene which was acquired by a later evolutionary step and due to different requirements possibly in a novel host.

4.3.5 NS4A

NS4A has a size of about 10 kDa and contains a hydrophobic, membrane spanning N-terminal region and a C-terminal cytosolic domain (Liang et al., 2009). It represents a cofactor for the NS3 protease (Xu et al., 1997). The C-terminal domain is sufficient for protease cofactor activity which tolerates a C-terminal truncation by 7 but not by 17 amino acids. An N-terminal deletion of 18 amino acids still allowed for a residual cofactor activity (Tautz et al., 2000). For stable interaction with NS4A, the N-terminus of NS3 tolerated a truncation of up to two amino acids but not of six or more residues.

In addition to free NS4A, also a NS4A–B precursor is observed in pestivirus-infected cells (Collett et al., 1991; Lamp et al., 2011; Meyers et al., 1992). While a functional role for this precursor has yet to be established, mutant cp BVDV genomes resulting from RNA recombination events which efficiently replicate but do not encode authentic NS4A–B have been described. In these virus mutants, insertions of up to 171 nucleotides between NS4A and NS4B have been identified. Since the NS3/4A-dependent cleavage sites downstream of NS4A and upstream of NS4B were preserved, it is to be expected that these proteins are generated in an authentic way but not authentic NS4A–B (Gallei, Orlich, Thiel, & Becher, 2005).

Besides this protease cofactor function, *trans*-complementation studies also revealed a critical role of NS4A in virion morphogenesis (Agapov et al., 1998; Liang et al., 2009; Moulin et al., 2007). For both functions, the known intimate interaction between NS4A and NS3 is most likely of critical importance (Tautz et al., 2000). It will be interesting to learn how the two functions of NS4A are coordinated and whether function-specific sequence requirements can be established.

4.3.6 NS4B

NS4B is a hydrophobic protein of about 35 kDa with unknown membrane topology that exerts an essential but not well-characterized function in pestiviral RNA replication (Collett et al., 1991; Grassmann et al., 2001). According to its physicochemical properties, NS4B is predicted as an integral membrane protein localized at intracellular membranes (Weiskircher, Aligo, Ning, & Konan, 2009). In NS4B, sequences with similarity to nucleotide-binding motifs, termed Walker A and B, were identified which parallels findings for HCV although the location of the motifs in NS4B differs (Einav, Elazar, Danieli, & Glenn, 2004; Gladue et al., 2011; Walker, Saraste, Runswick, & Gay, 1982). Bacterially expressed CSFV NS4B hydrolyzed both ATP and GTP and this activity was reduced when mutations were introduced into the Walker A and/or B motifs. However, when introduced into the viral genome, no clear correlation between the effect of the mutations on NTPase activity and viral replication or even virulence could be established. Therefore, further work is needed to clarify function and relevance of these conserved sequence motifs in NS4B.

4.3.7 NS5A

NS5A has a size of about 58 kDa and was shown to be phosphorylated by cellular kinases similar to NS5A of HCV or NS5 of members of the genus *Flavivirus* (Reed, Gorbalenya, & Rice, 1998). However, in contrast to HCV, where two phosphorylated isoforms of NS5A can be discriminated by SDS-PAGE analysis, pestiviral NS5A does not show heterogeneity in electrophoretic mobility. At its N-terminus, an amphipathic α-helix serves as a membrane anchor by binding in plane to intracellular membranes and thus functionally and structurally resembles NS5A of HCV and GBV (Brass et al., 2007). For the remainder of NS5A, a three-domain structure with intervening low complexity sequences (LCSs) was suggested. Domain I was shown to bind Zn^{2+} via a zinc-coordination motif which is critical for viral replication (Tellinghuisen, Paulson, & Rice, 2006). A detailed mutagenesis study introducing individual 10 amino-acid deletions into NS5A downstream of domain I revealed that only LCS I, the N-terminal part of domain II, and, to a lesser degree, domain III tolerated deletions (Isken et al., 2014). Based on this knowledge, replicons with an m-cherry insertion replacing an N-terminal fragment of NS5A domain II were constructed and shown to replicate. In contrast, to antibody staining, detection of fluorescent NS5A in living cells allows omitting the detergent treatment of the samples and thus preserves the highly critical cellular membrane structures. Using those fluorescently labeled NS5A variants, it was shown by live cell imaging that NS5A of BVDV localizes to the surface of lipid droplets, another analogy to HCV. While NS3 did not localize to lipid droplets, it was found together with NS5A in cytoplasmic foci.

This system can be further exploited to study biology of pestivirus-infected cells by live cell imaging. According to those studies, NS5A of BVDV also tolerates some deletions and thus is suited for harboring tags for different analytical purposes. However, when compared to its HCV counterpart, NS5A of BVDV is significantly less tolerant to deletions, since even the best-tolerated deletion identified for HCV resulted in a reduction of viral RNA replication by almost a factor of 10 in BVDV (Isken et al., 2014).

Interestingly, deleterious mutations introduced into NS5A can be complemented *in trans*, which was not achieved for the other essential viral replicase constituents (Grassmann et al., 2001). This could indicate that

NS5A does not need to associate with other components of the replicase cotranslationally which allows also molecules provided *in trans* to act as substitutes.

CSFV NS5A has recently been shown to regulate viral RNA replication by either binding to the 3′-UTR of the viral RNA genome via its RNA-binding activity or the modulation of NS5B RdRp activity by direct protein–protein interactions (Chen et al., 2012; Grassmann et al., 2001; Sheng, Chen, et al., 2012; Sheng, Wang, et al., 2012; Xiao et al., 2009). Xiao et al. described a negative regulatory effect of NS5A on CSFV IRES-mediated translation of a reporter mRNA. However, other studies did not reveal obvious effects of an NS5A deletion on polyprotein translation and or processing (Isken et al., 2014). The observed discrepancies may be attributed to differences in the experimental setups and need further investigation.

Chen and coworkers identified three regions within NS5A that are required for NS5B binding, the first two of which are critical for regulation of viral replication. The authors identified amino acids W143, V145, P227, T246, and P257 as critical for RNA replication in cell culture and residues K399, T401, E406, and L413 as important for IRES-mediated translation and NS5B binding. These residues are conserved between pestiviruses and HCV, which is remarkable, due to the very low overall sequence identity between NS5A proteins of these two viruses.

NS5A of HCV is known to interact with a plethora of host factors. For NS5A of BVDV, an interaction with the alpha subunit of translation elongation factor-1 has been described (Johnson, Perez, French, Merrick, & Donis, 2001). Further studies are required to understand the emerging multiple roles of NS5A in pestiviral replication.

4.3.8 NS5B
The last protein in the pestiviral polyprotein is NS5B with a size of about 77 kDa. NS5B has sequences characteristic for an RNA-dependent RNA polymerase (RdRp) and shows polymerase activity *in vitro*. The polymerase can be stimulated by GTP, a property shared with NS5B of HCV but not with the RdRp of poliovirus, and can *in vitro* either produce copy-back products or catalyze *de novo* initiation of RNA synthesis (Collett, Anderson, & Retzel, 1988; Kao, Del Vecchio, & Zong, 1999; Lai et al., 1999; Lohmann, Overton, & Bartenschlager, 1999; Meyers et al., 1989a; Steffens, Thiel, & Behrens, 1999; Zhong, Gutshall, & Del Vecchio, 1998). However, in *vivo*, replication of the viral genome requires the concerted action of NS3 to NS5B for viral RNA synthesis.

The crystal structure of NS5B was determined for two strains of BVDV (Choi, Gallei, Becher, & Rossmann, 2006; Choi et al., 2004). It resembles a right hand with fingers, palm, and thumb domains and thus exhibits the typical general fold of RdRps and other polymerases. Compared to HCV NS5B, pestiviral NS5B has a unique N-terminal domain constituted by residues 1–133 (Choi et al., 2004, 2006). A mutagenesis study characterizing the N-terminal 300 amino acids by substituting basic residues indicated a high sensitivity especially for the unique N-terminal domain (Xiao et al., 2006). The obtained structural data suggest that some of the inhibitors developed against NS5B interfere with an interaction between the finger domain and the N-terminal domain, indicating an essential role of this interaction for enzymatic activity (Baginski et al., 2000; Choi et al., 2006).

In the context of replicons based on the moderately virulent CSFV strain Paderborn, it was shown that the NS5B coding sequence could be functionally replaced by the NS5B gene of the highly virulent CSFV strain Koslov and, less successfully, of CSFV vaccine strain Riems (Risager, Fahnoe, Gullberg, Rasmussen, & Belsham, 2013). Similar exchanges of the NS2-3 coding region did not yield functional replicons which may either be caused by incompatibilities in the protein sequences or long distance RNA interactions. Ignoring the latter possibility, these experiments implicate less stringent strain specificity for NS5B when compared to NS2–3, a finding which may have implications for the architecture of the viral replicase. Moreover, the observation that NS5B of a highly virulent CSFV strain is capable of enhancing the replication of a replicon derived from a moderately virulent CSFV strain is intriguing and might indicate a correlation between virulence and NS5B activity.

In addition to its enzymatic function, NS5B has also been shown to take part in pestiviral virion morphogenesis (Ansari et al., 2004). This unexpected function emerged when a BVD virus with NS5B elongated by a C-terminal peptide tag of 22 amino acids was engineered and found to replicate its RNA but exhibit deficiency in virion morphogenesis. This defect could be restored by a mutation in the tag sequence (Ansari et al., 2004). Thus, the C-terminus of NS5B is critical for virion morphogenesis and displays a remarkable degree of flexibility.

5. THE MOLECULAR BASIS OF CYTOPATHOGENICITY IN PESTIVIRUSES

Pestiviruses exist in two different biotypes as noncp and cp viruses. This phenomenon is of major importance for understanding pestiviral persistence and development of MD. The molecular basis for the different biotypes has long been an enigma. In this chapter, the gain of knowledge about the pestiviral biotypes will be presented in detail as it developed over time in order to illustrate how results from different approaches contributed to this subject.

Cells infected by pestiviruses of the cp biotype round up, detach from the monolayer, and die, most likely, due to the induction of apoptosis (Hoff & Donis, 1997; Jordan, Wang, Graczyk, Block, & Romano, 2002; Lambot et al., 1998; Perler, Schweizer, Jungi, & Peterhans, 2000; Schweizer & Peterhans, 1999; Vassilev & Donis, 2000; Yamane, Kato, Tohya, & Akashi, 2006), whereas replication of noncp viruses does not lead to micro-scopically detectable alterations in cells.

At protein level, a marker for cp BVDV-infected cells could be established before detailed nucleotide sequence data were available for those viruses: while in noncp and cp BVDV-infected cells, NS2–3 was observed, NS3 was only detected upon cp BVDV infection (Donis & Dubovi, 1987; Pocock, Howard, Clarke, & Brownlie, 1987).

The reason for the high prevalence of the noncp biotype of ruminant pestiviruses in the field is its ability to establish persistence upon fetal infection leading to constant virus shedding by PI animals. In contrast, viral persistence was not observed after infection of pregnant cows with cp BVDV (Brownlie, Clarke, & Howard, 1989). Thus, infections with cp BVDV represent acute infections that are cleared by the immune system in a few weeks, thereby limiting the spread of the cp viruses. Based on these aspects, cp BVDV variants are in general only isolated from PI animals after onset of MD, in addition to the persisting noncp strain (Brownlie et al., 1984; McClurkin, Coria, & Bolin, 1985). The cp virus is the trigger for the switch from the long-lasting persistence to lethal MD which was finally established by the superinfection of PI animals with cp BVDV (Bolin et al., 1985; Brownlie et al., 1984; Moennig et al., 1990). Paired cp and noncp viruses from one MD animal were observed to share a high degree of antigenic similarity, which is in sharp contrast to the high variability of BVDV (Howard, Brownlie, & Clarke, 1987; Radostits & Littlejohns, 1988; Shimizu et al., 1989). This finding led to the hypothesis that the cp virus in the MD animal may evolve from the persisting noncp virus by mutation.

5.1 Cell-Derived Insertions in Pestiviral Genomes

Initial hints on the nature of the biotype-specific genome alterations were gained when the first complete genomic sequence of CSFV strain Alfort

Tübingen was established. It was about 200 nucleotides shorter than the previously published sequences of the cp BVDV strains Osloss and NADL (National Animal Disease Laboratories; Collett et al., 1988; Meyers et al., 1989a; Renard et al., 1987). Alignments of the three genomes revealed that both BVDV RNAs contain strain-specific insertions which were not present in the CSFV sequence. The specific insertion of strain NADL was not present in the gene data bases at that time, but a Northern blot analysis using polyA RNA from bovine cells identified three bovine RNA species with homology to the insertion which was therefore termed cINS (cellular insertion; now Jiv, or DNAJ-C14) (see Fig. 9). The insertion in the BVDV strain Osloss genome turned out to encode one monomer of ubiquitin with only two amino-acid differences from the canonical mammalian ubiquitin sequence (Meyers et al., 1989b; Meyers, Rümenapf, & Thiel, 1990). Further analyses demonstrated that both insertions share a very high degree of nucleotide identity with their cellular counterparts suggesting that their incorporation, in terms of evolution, had occurred rather recently.

Interestingly, insertions were only present in the genomes of the cp pestiviruses BVDV Osloss and NADL but not in the noncp CSFV strain Alfort Tübingen. Furthermore, the insertions were located in the region of NS2–3 where the cleavage for the release of NS3, the marker protein of the cp biotype, was expected to occur. These observations led to the hypothesis that the insertions might be of importance for generation of NS3 detected in cells infected by cp pestiviruses and thus more general for the cp biotype.

5.2 Virus Pair #1: Host Cell-Derived Sequences Are cp Specific

The hypotheses that (I) cp BVDV strains evolve from persisting noncp viruses in PI animals and (II) cellular insertions cause the cp biotype were challenged by the sequence analysis of the first BVDV-1 pair consisting of strains CP1 and NCP1 (Meyers et al., 1991). This study revealed that besides a very high degree of nucleotide identity of 99.6% between both viruses, major genomic changes occurred in the cp strain. The CP1 genome contained a duplication of about 2.4 kb of viral sequences together with an insertion corresponding to a fragment of a cellular polyubiquitin mRNA (see Fig. 9; Meyers et al., 1991). The duplicated fragment encoded NS3–4A and part of NS4B, was preceded by the ubiquitin-coding insertion and inserted into the NS4B coding region. The cell-derived sequence encoded an N-terminally truncated ubiquitin-fragment followed by one complete ubiquitin monomer.

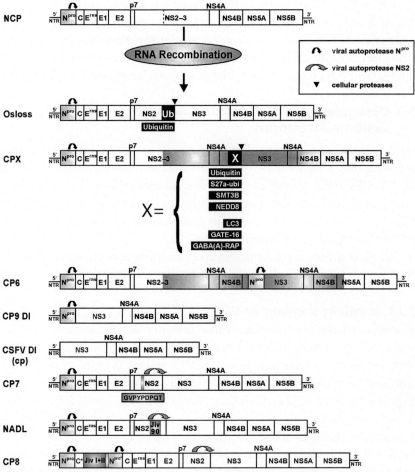

Figure 9 Genome organizations of cp pestiviruses. Schematic representation of genome organizations of autonomously replicating cp pestiviruses or subgenomic pestiviral RNAs. Those genomes evolved from noncp pestiviruses (NCP; top) by RNA recombination. The ORF including the encoded proteins as well as the flanking 5′- and 3′-NTRs are indicated. The N-terminal autoprotease N^pro is highlighted. The genome of cp BVDV Osloss contains an insertion of an ubiquitin (Ub)-coding sequence at the NS2/NS3 gene border. CPX represents cp BVDV genomes which contain a large duplication of viral sequences encoding at least NS2–4A (highlighted) together with an insertion of host-derived mRNA sequences (X) right upstream of NS3. X represents one of the sequences listed in the bars below the CPX genome: polyubiquitin, ubiquitin fusion protein S27a-ubi, ubiquitin-like proteins (SMT3B, NEDD8), or proteins with a ubiquitin-like fold (LC3, GATE-16, GABA(A)-RAP). These cell-derived proteins are substrates to cellular proteases which generate the N-terminus of NS3 (processing site indicated by black arrowhead). Analogous to CPX, the genome of CP6 encompasses a large duplication of viral sequences (highlighted). Instead of cell-derived sequences, a duplication of N^pro resides upstream of NS3; N^pro cleavage is indicated by a black arrow. In analogy,

(Continued)

The high level of sequence identity between the two viruses observed in this study strongly suggested that strain CP1 is a product of RNA recombination based on the persisting strain NCP1. Furthermore, the study revealed a correlation of the cp biotype with the insertion of a cell-derived sequence upstream of NS3 together with a large genomic duplication in strain CP1.

5.3 Deregulation of NS3 Release Induced by cp-Specific Genome Alterations

By the analysis of a large number of virus pairs or cp pestiviruses, it became apparent that a major part of cp genomes contain duplications of the NS3–4A region in combination with insertions of cellular or viral sequence upstream of NS3 and thus resemble CP1. A second group of cp pestiviruses turned out to contain insertions or an accumulation of individual mutations in NS2 (see Section 5 for details).

All these mutations identified in cp pestiviruses result in deregulation of NS3 release from the polyprotein, but by different mechanisms.

5.3.1 Insertions Upstream of NS3

The first cell-derived insertion whose function in polyprotein processing was investigated in cp BVDV-1 isolate CP14. CP14 was isolated together with its noncp counterpart from an MD animal (Corapi et al., 1988). The analysis of the CP14 genome identified an insertion of a polyubiquitin fragment between NS2 and NS3, i.e., between amino acids 1589 and 1590. *In vitro* translation studies demonstrated that neither NS2 nor NS3 is required for the removal of the ubiquitin moiety from the N-terminus of NS3. Thus, this cleavage had to be catalyzed by a cellular protease. Since at least one complete monomer of ubiquitin was required for processing at the C-terminus of ubiquitin, and since a proline residue downstream

Figure 9—Cont'd N^pro also generates the N-terminus of NS3 in the polyprotein of DI9. CSFV DI, a cp subgenome, encodes only the start methionine followed by the genomic sequence encoding NS3 to NS5B. BVDV CP7 encompasses a 27-base insertion (highlighted) central in NS2. This insertion is critical for high efficient cleavage at the NS2/3 site by the NS2 cysteine protease (indicated by a large gray arrow). Alternatively, the NS2 protease can be activated by various fragments of the cellular Jiv-mRNA (highlighted) located in the NS2 gene of BVDV NADL or within the structural protein-coding region of BVDV CP8. While the insertion in the NADL genome represents a minimal Jiv fragment, the insertion in the CP8 genome is more complex and contains besides two Jiv fragments also additional sequences as well as a duplications of core and N^pro which facilitate the generation of an authentic core protein. (See the color plate.)

of the ubiquitin inhibited the cleavage, ubiquitin C-terminal hydrolases (UCHs) (Rose & Warms, 1983) were considered responsible for this cleavage (Tautz et al., 1993).

Further studies identified other cellular insertions upstream of glycine 1590, the conserved N-terminus of NS3. They include sequences coding for a ribosomal protein S27a-ubiquitin fusion protein (Becher, Orlich, & Thiel, 1998b, 2001), ubiquitin-like proteins SMT3B and NEDD8 (Baroth, Orlich, Thiel, & Becher, 2000; Qi, Ridpath, & Berry, 1998), and a group of proteins with ubiquitin-like fold like LC3 (Fricke, Gunn, & Meyers, 2001; Fricke, Voss, Thumm, & Meyers, 2004), GABA(A)-RAP, and GATE-16 (Becher, Thiel, Collins, Brownlie, & Orlich, 2002).

All these insertions are (I) substrates for cellular proteases residing in the cytoplasm of the cell (Becher et al., 2002; Fricke et al., 2004; Hemelaar, Lelyveld, Kessler, & Ploegh, 2003; Nishida, Kaneko, Kitagawa, & Yasuda, 2001; Wu et al., 2003) and (II) are located right upstream of NS3. Furthermore, these cellular proteases, in contrast to *trans*-acting viral proteases, recognize and cleave their substrate virtually independent of downstream sequences. This allows highly efficient cleavage also in the artificial context of the viral polyprotein and thus the generation of the NS3 N-terminus.

Remarkably, this class of insertions was identified between NS2 and NS3 in only a few cases, as described for cp BVDV strains Osloss (Renard et al., 1987) and CP14 (Tautz et al., 1993). In the vast majority of cases, the insertions occurred downstream of NS4A and in combination with duplications of the NS3 to 4B or even the 5A region (Becher & Tautz, 2011) (CPX in Fig. 9). The reason for this prevalence became apparent when either an ubiquitin-coding sequence or an EMCV IRES was artificially inserted between the NS2 and NS3 genes of BVDV-1 or CSFV (Agapov et al., 1998; Moulin et al., 2007). Both insertions led to the absence of uncleaved NS2–3 and allowed for efficient RNA replication but abrogated the generation of infectious viral progeny, demonstrating that uncleaved NS2–3 plays an essential role for in virion morphogenesis (see also Section 4.3.3).

Translated in the context of the polyproteins of cp pestiviruses, insertions inducing a complete NS2–3 cleavage, like an ubiquitin insertion between NS2 and NS3, are expected to interfere with virion morphogenesis and will generate viruses which are helper dependent with respect to genome packaging. Preliminary experiments suggested also for cp BVDV CP14 a helper dependency (Tautz et al., 1993). Recombinant viruses containing parts of the BVDV Osloss genome indicate an alternative packaging pathway for this virus (Lattwein et al., 2012).

In contrast, when those insertions occur in combination with duplication of the NS3–4A region, as described for CP1, those viral genomes encode an intact NS2–3–4A region upstream of the insertion and thus are packaging competent (Fig. 9). In cp BVDV genomes, the duplicated regions enlarge the genome by about 30% (Meyers et al., 1991), but in cell culture, viruses with duplications of up to 60% of the wild-type genome have been documented. While these genomes were found to produce infectious particles, it was observed that they were truncated during passaging (Gallei et al., 2004).

Accordingly, the coupling of insertions of cellular protease substrates with large genome duplications often observed for cp pestiviruses isolated from MD animals is due to the functional constraints of the biology of pestiviruses described above.

5.3.2 Insertions of Viral Sequences Upstream of NS3

Analysis of virus pairs #6 and Pe515CP/NCP showed an insertion upstream of NS3 combined with a large genome duplication encompassing NS3–4A–4B and part of NS5B or NS3–4A and part of 4B, respectively (Meyers et al., 1992). The insertion upstream of NS3, however, is of viral origin and encompasses a duplication of the viral autoprotease N^{pro}. This protease is capable of generating the authentic N-terminus of NS3 although it was truncated N-terminally by 14 amino acids and fused to the C-terminally truncated NS4B fragment in the Pe515CP genome (Fig. 9). The tolerance of N^{pro} with respect to N-terminal truncations is in agreement with studies on the N^{pro} protease (Rümenapf et al., 1998). Thus, the duplicated viral autoprotease N^{pro} is responsible for the deregulated NS3 release.

5.3.3 Insertions of cINS/Jiv/DNAJ-C14 in NS2

Already in one of the first pestiviral genomes analyzed, namely cp BVDV-1 strain NADL, an insertion of a cellular sequence was identified which was, however, not located at the insertion site conserved for ubiquitin or ubiquitin-like proteins right upstream of Gly1590 (Meyers et al., 1991). Instead, this cell-derived insertion (cINS; now Jiv or DNAJ-C14), which is unrelated to ubiquitin, is located in NS2, 53 amino acid upstream of NS3. Accordingly, it was not to be assumed that a cellular protease will be directed by this insertion to a cleavage upstream of Gly1590, which is required to generate a functional replicase (Tautz & Thiel, 2003).

The cINS insertion present in cp BVDV NADL codes for a 90 amino acids fragment (Jiv90) of the above-mentioned cellular chaperone DNAJ-C14, that is also termed Jiv. Over time, a large number of additional cp pestivirus strains with insertions derived from the Jiv/DNAJ-C14 gene were identified, including BVDV-1 and -2, border disease virus, Hobi-like pestivirus, and an isolate from giraffe (Avalos-Ramirez, Orlich, Thiel, & Becher, 2001; Balint, Baule, Kecskemeti, Kiss, & Belak, 2005; Becher, Meyers, Shannon, & Thiel, 1996; Darweesh et al., 2015; Decaro et al., 2012; Meyers et al., 1990; Müller, Rinck, Thiel, & Tautz, 2003; Nagai et al., 2003; Ridpath & Neill, 2000; Rinck et al., 2001; Vilcek, Greiser-Wilke, Nettleton, & Paton, 2000). Most of these insertions were located in the C-terminal part of NS2 between amino acids 1528 and 1544. All insertions encompass the Jiv90 sequence identified in strain NADL under-lining the critical role of this domain. A deletion of the insertion from the NADL genome led to a viable virus which, however, exposed a noncp bio-type and a drastic reduction of NS2-3 cleavage as well as a reduction in the viral RNA levels in the infected cells but not a reduction of the infectious progeny virus titer (Mendez, Ruggli, Collett, & Rice, 1998). Thus, the insertion clearly correlated with the induction of NS2-3 cleavage in the NADL polyprotein. For understanding the principles underlying the effect of the Jiv insertions on NS2–3 cleavage, characterization of virus pairs #8 and KS86-1cp/KS86-1ncp was instrumental (Müller et al., 2003). The CP8 and KS86-1cp viruses encoded an insertion containing Jiv sequences combined with a duplication of viral sequences, however, not in the NS2–3 region but in the core gene (Müller et al., 2003; Nagai et al., 2003). Due to duplicated N^{pro} and core-coding sequences, these viruses release from their polyprotein an additional fusion protein encompassing Jiv-derived sequences including the Jiv90 domain. Upon coexpression, these fusion peptides induced NS2–3 cleavage in *trans* (Müller et al., 2003). Fur-thermore, the Jiv90 domain was identified as sufficient for NS2–3 cleavage induction *in trans*.

Studies of the cellular protein DNAJ-C14/Jiv revealed a stable interac-tion with NS2 upon ectopic expression. Therefore, DNAJ-C14 was termed J-domain protein interacting with viral protein, Jiv. Expression of Jiv fur-thermore was capable of inducing NS2–3 cleavage in *trans*. Moreover, this study demonstrated that the intracellular level of the cellular Jiv protein has the capacity to determine the viral biotype: in Jiv-overexpressing cells, noncp BVDV-1 showed efficient NS2–3 cleavage and induced a CPE (Rinck et al., 2001).

Taken together, cellular Jiv as well as its fragment Jiv90 has the capacity to induce NS2–3 cleavage also in noncp BVDV-infected cells and to regu-late RNA replication and viral biotype via the NS2–3 cleavage. However, the mechanism underlying NS2–3 cleavage and its regulation by Jiv remained unknown.

5.3.4 Small Insertions in NS2 Led to the Identification of the NS2 Protease

Virus pair #7 revealed that the cp strain CP7 contains neither a large duplication nor a cell-derived insertion. Instead, a 27-nucleotide insertion was identified within the NS2 gene. This insert is actually a duplication of a sequence located about 300 nucleotides upstream of the insertion site. However, since the insertion occurred in another reading frame, the two copies encode different peptides in the two locations. NS2–3 of CP7 was processed efficiently but not to completion when expressed transiently in BHK-21 cells. A deletion of the insertion abolished cleavage, while the 9-amino-acid insertion induced NS2–3 cleavage in the context of the NCP7 polyprotein. These findings proved that the insertion is critical for efficient NS2–3 cleavage. Mutational inactivation of the NS3 serine protease did not interfere with NS2–3 processing, excluding a functional role in this processing event (Tautz, Meyers, Stark, Dubovi, & Thiel, 1996).

These results led to the working hypothesis that pestiviral NS2 has an intrinsic protease activity which is massively enhanced by the CP7-specific 9-amino-acid insertion. Along these lines, truncations in the context of a NS2–3–GST fusion protein, expressed in BHK-21 cells, revealed that two amino acids of NS3 are sufficient to allow the removal of NS2, while N-terminal truncations of NS2 severely interfered with cleavage at the NS2/3 site. Thus, the integrity of NS2 but not of NS3 is required for cleavage at the NS2/NS3 site (Lackner et al., 2004).

The idea of a protease in pestiviral NS2 was further reinforced by the finding that the C-terminal domain of NS2 of the related HCV displays an intrinsic cysteine protease activity (Grakoui, McCourt, Wychowski, Feinstone, & Rice, 1993; Hijikata et al., 1993) which is essential for viral replication (Kolykhalov, Mihalik, Feinstone, & Rice, 2000). An alignment of the NS2 protease domain of HCV with the C-terminal part of pestiviral NS2 indicated a very limited homology between the proteins but allowed to identify candidate amino acids which could constitute the catalytic triad of a putative cysteine protease. Mutagenesis identified histidine1447, glutamic acid1462, and cysteine1532 as essential for NS2–3 cleavage and as most likely members of a catalytic triad of the NS2 protease.

Introduction of active site mutants into an infectious cDNA clone established for CP7 (Meyers, Tautz, Becher, Thiel, & Kümmerer, 1996) proved the importance of the NS2 protease activity for viral replication (Lackner et al., 2004). This led to the observation that uncleaved NS2–3 cannot functionally substitute for NS3 in viral RNA replication. Along these lines, mutational inactivation also abolished RNA replication in the context of the NCP7 RNA which in turn led to the conclusion that NS2–3 cleavage seems to be critical also for the replication of the viral noncp biotype. Finally, a time course experiment proved that free NS3 is indeed generated also in noncp BVDV-infected cells, which, however, is restricted to a short time after infection. Thus, NS2–3 cleavage in noncp-infected cells was only detected up to about 8 h postinfection, while at later time points, selectively the production of uncleaved NS2–3 was observed (Lackner et al., 2004).

In CP7-infected cells, even at 24 h postinfection, efficient NS2–3 processing did occur, indicating that the insertion renders the protease less prone to the mechanism responsible for the observed downregulation of NS2–3 cleavage.

Further studies identified insertions of similar size located at the same or almost identical positions in NS2 in other cp BVDV isolates (Balint, Baule, Palfi, et al., 2005; Balint, Palfi, Belak, & Baule, 2005). For one of these insertions, analogous effects on NS2–3 cleavage and viral cytopathogenicity have been demonstrated (Balint, Baule, Palfi, et al., 2005).

The type of cp BVDV with the least obvious genomic alterations is represented by the strain Oregon. NS2 of cpBVDV Oregon does not harbor any insertions. Instead, efficient NS2–3 cleavage depends on the presence of a set of amino acids which deviate from the pestiviral consensus sequence and are dispersed over the protease domain of NS2 (Kümmerer & Meyers, 2000; Kümmerer, Stoll, & Meyers, 1998). Based on the available data, it seems likely that also in the Oregon type of cp viruses, cleavage is catalyzed by the autoprotease in NS2. To finally resolve the molecular basis for its high activity, data on the protein structure will be required.

Structural data might also help to understand the temperature sensitive phenotype of a CP7 NS2 mutant (Y1338H at position 193 of NS2) which was cytopathogenic at 33 °C but lost this phenotype at 39.5 °C (Pankraz, Preis, Thiel, Gallei, & Becher, 2009).

Thus, several unrelated peptide insertions and most likely also a combined set of individual amino-acid exchanges can render the NS2 protease less or even unsusceptible to the temporal regulation observed for the enzyme of the noncp biotype.

5.3.5 Jiv Is a Cellular Cofactor of the Pestiviral NS2 Protease

The experiments described above made the cellular Jiv protein a prime candidate for the observed temporal regulation of NS2–3 cleavage in noncp BVDV-infected cells since Jiv binds stably to NS2 and its intracellular level correlated with NS2–3 cleavage efficiency as well as with viral cytopathogenicity (Rinck et al., 2001).

To test this hypothesis, overexpression of Jiv or knockdown in shRNA-expressing cell lines was tested for its effects after cp and noncp BVDV infection. While Jiv overexpression led to an elongated phase of NS2–3 cleavage, a significant shortening of this period was observed in the Jiv-knockdown cell lines. Furthermore, the intracellular level of viral RNA correlated with the Jiv level. The reduction of the viral RNA level in the Jiv-knockdown cells also resulted in a significant decrease in virus production upon noncp BVDV infection, indicating that Jiv might be an essential host factor for viruses of the noncp biotype (Lackner et al., 2005).

Taken together, these experiments led to the following model: Individual Jiv proteins bind stably to NS2 in the context of newly translated viral polyproteins and thereby activate the autoprotease in NS2 leading to NS2–3 cleavage *in cis*. Jiv is available only in small amounts in bovine or swine cells, at least in cell cultures. Thus, ongoing translation of the viral polyprotein leads to a situation where at around 8 h postinfection newly translated NS2–3 proteins will not have access to free Jiv/DNAJ-C14 since this cellular factor is obviously not recycled after promotion of NS2–3 cleavage. Since free Jiv/DNAJ-C14 is required as cofactor for NS2 protease activity, NS2–3 cleavage will not occur in those polyproteins resulting in the production of uncleaved NS2–3 (Lackner et al., 2005). Because NS3 is an essential constituent of the viral RNA replicase which cannot be functionally replaced by NS2–3, the shift in polyprotein processing from NS3 to NS2–3 production correlates with a massive downregulation of viral RNA replication (Lackner et al., 2005). This downregulation is a prerequisite for the noncp biotype of pestiviruses which in turn is an essential property for the establishment of persistent infections *in vivo* (see Section 6 for more detail). In cells with ectopically overexpressed Jiv, also noncp pestiviruses show a prolonged phase of NS2–3 cleavage and the induction of a CPE (Lackner et al., 2005; Rinck et al., 2001).

In the context of cp pestiviruses, Jiv insertions cancel the above-described regulatory mechanism since the translated viral polyproteins encompass the cofactor of the NS2 protease rendering the virus independent of the cellular Jiv pool.

Mechanistic studies furthermore defined an aromatic amino acid at position 39 in the Jiv90 domain as essential for its protease cofactor function but not for NS2 binding (Lackner et al., 2005). Thus, binding of Jiv to NS2 is not sufficient for protease activation. In the NS2 protease domain, two independent Jiv-binding sites were identified and Jiv can bridge these two parts of NS2 as shown by coimmunopreciptation. These findings allowed to establish the current model of protease activation which involves conformational changes induced by binding of Jiv to these two sites (Lackner et al., 2006). Also here, structural data will be required to understand the events at the atomic level.

Besides the structure and stoichiometry of this complex, it also remains to be proven which pestivirus species require Jiv as an essential host factor for their replication. A further interesting aspect for future work is how different Jiv levels in individual tissues or cell types do influence the replication kinetics of pestiviruses *in vivo*. Studies using tissue samples from cattle persistently infected with noncp BVDV revealed elevated levels of NS2–3 cleavage especially in lymphocytes (Kameyama et al., 2008). A complex gradient of DNAJ-C14 levels in different cell types or cells differing in their differentiation status, like immune cells, could be of high relevance for the remarkably successful long-term persistence of ruminant pestiviruses in their hosts.

Taken together, the unique regulatory mechanism of the NS2 autoprotease is highly critical especially for the biology of ruminant pestiviruses with their high prevalence for persistence.

5.3.6 Pestivirus DIs Are Autonomous RNA Replicons

During analysis of BVDV pairs, genomes with large deletions were identified by Northern blotting in cells infected with cp viruses. The first of these genomes, termed DI9, carried a deletion of the entire structural gene region as well as the p7- and NS2-coding sequences. Thus, in this genome, N^{pro} is located right upstream of NS3 and the autoprotease activity of N^{pro} generates the N-terminus of NS3 (Fig. 9) (Tautz, Thiel, Dubovi, & Meyers, 1994). Replication of this subgenomic RNA reduced the replication efficiency of coinfecting noncp BVDV which classified the subgenome as defective interfering RNA (DI) (Barrett & Dimmock, 1986). The presence of a noncp BVD virus allowed packaging of the DI RNA and its transfer into naive cells. Transfection experiments with RNA derived from those cells suggested that the DI RNA causes the observed CPE (Tautz et al., 1994).

With the development of a reverse genetics system for pestiviruses and the generation of a CP7/DI9 chimera, this could be finally proven (Meyers et al., 1996). Surprisingly, this recombinant form of DI9 turned out to be competent of autonomous RNA replication (Behrens et al., 1998). Based on this construct, the minimal set of viral proteins required for autonomous RNA replication has been defined as NS3 to NS5B (Behrens et al., 1998; Tautz et al., 1999). For those DIs, it was however necessary to maintain part of the N^{pro}-coding sequence possibly due to a role in IRES function or due to other constraints caused by the RNA structure (Myers et al., 2001; Tautz et al., 1999).

With respect to induction of MD, the biological relevance of this DI remains to be clarified since preliminary attempts to induce MD by the use of DI-containing material were not successful (Becher, personal communication; Stokstad et al., 2004). Thus, it is currently not clear whether the DIs isolated from MD animals are only by-products and MD was caused in these animals by autonomously replicating cp BVDV which were lost during biological cloning.

DI13 represents another BVDV subgenome which, however, maintained only the first 13 codons of the N^{pro} gene followed by 10 E1-derived codons upstream of an NS3 gene carrying a deletion of the first 4 codons. NS3 of this isolate represents an N-terminally modified fusion protein and deviates from the canonical N-terminus of NS3 (Kupfermann, Thiel, Dubovi, & Meyers, 1996). However, also in this subgenome, the 5′-part of the N^{pro} gene was maintained which is in line with the assumed critical role of this sequence in translation initiation.

Surprisingly, CSFV subgenomes isolated from cell cultures which induced a CPE after the infection with noncp CSFV Alfort Tübingen displayed a complete deletion of the coding region upstream of NS3 with the exception of the start codon (Meyers & Thiel, 1995). Analogous genome structures were also reported for other CSFV subgenomes (Aoki et al., 2001; Kosmidou et al., 1995; Kosmidou et al., 1998). In one study, no significant relevance of the DIs for virulence in the animal could be obtained (Aoki, Ishikawa, Sekiguchi, Suzuki, & Fukusho, 2003), whereas in another experiment severer symptoms were detected in addition to a shortened time between infection and onset of viremia (Kosmidou et al., 1998). However, the differences observed between the genomic structures of the DIs suggest that the requirements for translation initiation seem to differ between BVDV and CSFV.

The generation of cp BVDV recombinants within a persistently infected animal is a rare event. PI calves can live long despite high levels of viral replication. On the average, MD is observed about 2 years after birth. This apparently low frequency of establishment of a "fitting" cp virus is a consequence of the complex process of recombination (see Section 5.4). It has prompted researchers to investigate whether proof for ongoing recombination and especially precursors of cp viruses can be identified in animals before outbreak of MD or whether only one type of genome with cp typical alterations can be found. Using a RT-PCR-based strategy for examination of viral genomic RNA isolated from tissue samples of persistently infected cattle, it was shown that recombination of viral RNA was indeed detectable before occurrence of MD. Analysis of two different regions of the viral genome revealed that recombination was not restricted to particular sequences (Desport et al., 1998). In a further approach, samples derived from PI calves that were kept under isolated conditions until they came down with MD were analyzed. These experiments demonstrated that indeed viral RNA containing the same cellular insertion as present in the cp virus which was isolated after outbreak of MD was already present in the animal at least 1 year before outbreak of the disease (Fricke et al., 2001). Moreover, the presence of various closely related cp viral RNAs in the form of sets of different subgenomic viral RNAs obtained from different MD animals from one herd was shown. Interestingly, the N-terminal part of the N^{pro} gene was retained in all cases (Becher, Orlich, König, et al., 1999; Becher et al., 2001). Taken together, these results suggest that random RNA recombination processes are constantly occurring within PI animals, sometimes leading to a primary recombinant with an RNA displaying essential characteristics of a cp viral genome. In a trimming process, a set of secondary virus recombinants are generated from this hypothetical primary recombined cp RNA. These secondary recombinants display genome structures that represent variations of the basic scheme already present in the primary recombinant. This trimming process can be expected to continue until finally a variant is generated that leads to MD.

5.3.7 Designed cp CSFV Display an Attenuated Phenotype

Besides the cp DIs, no cp CSFV mutants have been described. It can be speculated that the lack of long-term persistence in pigs lowers the chance of the generation of such virus mutants. With the reverse genetics system of CSFV, the generation of an autonomously replication competent cp CSFV has been achieved. This virus encompassed the cytopathogenic module of CP8, a core/Jiv/N^{pro} fusion protein (Gallei et al., 2008). Interestingly, it was found to be attenuated in its natural host. Attenuation correlated with the induction of interferon-induced protein MX in cell culture. These results demonstrate that autonomously replicating cp CSFV mutants are viable but possess an attenuated phenotype which is expected to correlate with a counter selection *in vivo*.

5.3.8 Duplications of Genomic Regions in Noncp BVDV-2

Duplications of viral sequences have also been reported for noncp BVDV isolates. In the genome of the highly virulent strain BVDV-2 890, a duplication covering about 250 bases of the p7/NS2 gene junction was identified. At that time, it was concluded that this duplication is not connected to the highly virulent phenotype of the virus since high virulence was observed also for strains lacking this feature (Ridpath & Bolin, 1995). Deep sequencing analysis of highly virulent BVDV-2 isolates in Germany revealed similar genomic rearrangements. However, the bovine samples also contained a variant of the virus without the insertion. Using a reverse genetics system of a BVDV-2a strain with this duplication, spontaneous deletions of this sequence were observed. A model described by the authors proposes an equilibrium between the two variants in infected animals (Jenckel et al., 2014). It remains to be determined whether the viral variant without the duplication retains the highly virulent phenotype.

Another unique feature of several of these viruses is the insertion of a 16-amino-acid peptide at a site within NS2 at which Jiv insertions have been identified in cp pestiviruses (Fan & Bird, 2010; Meyer et al., 2002). Whether this insert is critical for virulence has not been finally established. In the context of noncp BVDV-1, this peptide did, however, not induce NS2–3 cleavage or cytopathogenicity (Fan & Bird, 2010).

5.4 Mechanisms of RNA Recombination and Biological Relevance of Cell-Derived Insertions in cp Pestiviruses

Insertions of cellular sequences have been described also for other RNA viruses, oncogenic retroviruses being the most prominent example with exceptional importance for providing knowledge on the basics of cell cycle regulation and cancer (Martin, 1970; Stehelin, Varmus, Bishop, & Vogt, 1976). However, in the latter, a DNA copy of the viral genome generated by reverse transcription is integrated into the host genome, which is instrumental for the recombination process. In contrast, classical RNA viruses do not undergo reverse transcription and thus have no DNA stage in their life cycle. However, also for this group of viruses, insertions of host cell sequences occur. The integration of 28S ribosomal RNA fragments was documented for influenza virus (Khatchikian, Orlich, & Rott, 1989) and poliovirus (Charini, Todd, Gutman, & Semler, 1994; McClure & Perrault, 1985), while in the genomes of Sindbis virus mutants, tRNA sequences were detected (Monroe & Schlesinger, 1983; Tsiang, Monroe, & Schlesinger, 1985). The integration of cellular protein-coding sequences is to the best of our knowledge for RNA viruses unique to pestiviruses. The findings obtained in this system demonstrate that RNA viruses have access to the entire transcribed gene pool of their host cell and obviously integrate on a random basis mRNA fragments into their genomes. The generation of cp pestivirus mutants represents, however, not an evolutionary step since the noncp biotype with its capacity to establish persistence due to neat regulatory mechanisms is more successful with regard to reproduction in the individual host as well as in the field. However, in principle, the integration of novel genes into viral genomes could be connected with a gain of fitness, thus driving viral evolution.

On the molecular level, homologous RNA recombination has been intensely studied, e.g., in poliovirus leading to the template switching or copy choice model (Kirkegaard & Baltimore, 1986). According to this model, the viral RNA polymerase starts, e.g., negative-strand synthesis using one genome as template. During synthesis, the polymerase switches from this first template to another on which RNA synthesis is completed. The resulting RNA molecule thus represents a copy originating from two different templates (Agol, 1997; King, McCahon, Slade, & Newman, 1982; Nagy & Bujarski, 1997). The integration of a cellular mRNA fragment would imply a polymerase switch from a viral template to an mRNA used as a second template and back to a viral template (Meyers & Thiel, 1996). The position at which restart of synthesis on a viral template occurs determines whether duplication or deletion of viral sequences will be found in the recombinant genome or whether an insertion is located precisely between formerly neighboring nucleotides.

An alternative mechanism to create viral genomes containing cell-derived sequences is the so-called breakage and ligation process in which viral and cellular RNAs are first cleaved by RNases and then ligated to replicating genomes. A first strong support for this mechanism was gained in poliovirus (Gmyl et al., 1999). An even more stringent approach using pestiviral genome fragments which individually do not encode an active RdRp finally proved this mechanism (Gallei et al., 2004) implying that RNA recombination based on genome fragments can lead to the generation of fully functional viral RNA genomes in the absence of a functional viral

RdRp (Austermann-Busch & Becher, 2012). Besides attractive perspectives for basic science, these findings have also critical implications for the safety of inactivated virus preparations or blood products. Further details on pestiviral RNA recombination can be found in a recent review (Becher & Tautz, 2011).

6. MOLECULAR ASPECTS OF PESTIVIRUS PERSISTENCE

Pestivirus persistence has already been introduced above, but understanding the unique strategy developed by these viruses during evolution depends on profound knowledge of their molecular biology provided in the previous sections. This chapter aims at summarizing our knowledge on the molecular biology of pestivirus persistence with a clear focus on BVDV, since most of the data were obtained for this system.

A variety of viruses are able to establish persistent infection in their hosts. In many cases, viral persistence is dependent on a delicate balance between the host's immune system and viral replication, often supported by immune evasion strategies. Other viruses rely on an almost complete shutdown of viral gene expression to avoid elimination of infected cells. The pestiviral strategy is very different since it aims at almost complete abrogation of the immune response enabling the viruses to replicate for long periods at high levels in their hosts without causing too much damage or increasing the susceptibility for other infectious diseases via general immunological repression. Virus/host interaction is manipulated at several levels in order to eliminate both the adaptive and innate immune response.

6.1 Elimination of the Adaptive Immune Response

Diaplacental infection of fetuses is an essential prerequisite for pestivirus persistence (Baker, 1987; Thiel et al., 1996). In BVDV, infection has to occur in the first trimester (between days 40 and 120 of pregnancy). The time point of infection is important since it should lead to establishment of an acquired immunotolerance for the infecting virus strain (Baker, 1987; Moennig & Plagemann, 1992; Peterhans & Schweizer, 2010; Thiel et al., 1996). It is generally accepted that self-reactive elements of the adaptive immune system are eliminated during early development of the immune system so that effector cells directed against the virus are removed from the fetus when BVDV infection occurs before this stage of establishment of self-tolerance. Accordingly, the animals cannot mount an adaptive immune response against the invading virus and this situation will not change during their lifetime. It is important to note that the immunotolerance established by this process is highly specific for the persisting virus (Bolin, 1988; Coria & McClurkin, 1978; Steck et al., 1980). Upon superinfection of PI calves with other BVDV strains, the animals are able to mount a normal adaptive response against the heterologous viruses. Sometimes, such superinfection scenarios resulted in induction of cross-reactive responses leading to breakage of tolerance (Brownlie & Clarke, 1993; Chase, 2013; Collen, Douglas, Paton, Zhang, & Morrison, 2000; Howard, 1990; Peterhans & Schweizer, 2010). Most importantly, these experiments proved that the PI animals are not generally immunocompromised but can eliminate invading pathogens with more or less normal efficiency which is a prerequisite for long-term survival despite persistent infection with BVDV. Moreover, PI calves often appear healthy, indicating that high-level virus replication does not necessarily cause significant damage to the host pointing at a role of the immune response for induction of (severe) symptoms in acutely BVDV-infected animals.

6.2 Inhibition of the Innate Immune Response

The elimination of the adaptive immune response via infection before self-/non-self-discrimination is a perfect strategy as long as the time point of infection is right. However, persistent infection as observed for BVDV would not be maintained for years without severe damage of the calves when the PI animal's innate immune system reacted against the virus with high efficiency. Accordingly, BVDV must also evade innate immune responses controlled by receptors recognizing PAMPs. Because innate immunity is crucially important to control viral infections, all viruses seem to express antagonists counteracting this system (Haller, Kochs, & Weber, 2006; Hengel et al., 2005; Pichlmair et al., 2012; Pichlmair & Reis e Sousa, 2007). For BVDV, three strategies or factors have been identified that work together.

6.2.1 Biotype: Control of Viral RNA Replication

As mentioned above, only infection of pregnant cattle with noncp BVDV may lead to persistent infection of the not yet immunocompetent fetus and the birth of PI offspring (Baker, 1987; Bolin et al., 1985; Brownlie & Clarke, 1993; Brownlie et al., 1984, 1989; Peterhans & Schweizer, 2010). In the light of our detailed knowledge of many aspects regarding the molecular basis of cytopathogenicity in pestiviruses (see above), it is obvious that cp viruses have significantly lower chances to escape the innate immune system due to their highly enhanced RNA replication rate (Lackner et al., 2004).

The increased amount of viral RNA and in consequence also the amount of double-stranded RNA present within the infected cell represents a much stronger trigger for the innate immune response compared to a noncp virus-infected cell. Moreover, the high number of dead and lysed cells will result in attraction of phagocytic cells and enhance the type 1 interferon response via cross-priming. Experimental evidence for a biotype-specific innate immune response in fetuses was obtained after direct infection of fetuses via laparotomy. The cp BVDV elicited a solid IFN-1 response in the fetuses, whereas IFN-1 could not be detected when the corresponding noncp BVDV from a virus pair was used (Charleston, Fray, Baigent, Carr, & Morrison, 2001). It therefore can be concluded that a noncp virus can also prevent induction of an IFN-1 response. Thus, the ability of noncp viruses to strictly control their RNA replication via the complex mechanism of regulating the NS2 autoprotease activity in dependence of the limited host factor Jiv is a prerequisite for long-term persistence.

6.2.2 Interference with the Innate Immune Response

Despite the control of RNA replication in noncp BVDV, a minimal amount of viral RNA must be present within the infected cell, inevitably leading to formation of dsRNA, a very potent trigger of the type 1 interferon response. Pestiviruses express two factors helping to control the interferon response, N^{pro} and E^{rns} RNase. The function of N^{pro} during acute infection of adult host animals is not clear. Its activity with regard to blocking the IFN-1 response is not a potent virulence factor and seems not to provide a significant advantage during acute infection (G. Meyers, unpublished; Ruggli et al., 2009; Tamura et al., 2014). Similarly, the detailed function of the E^{rns} RNase in the infected animal remains obscure. In this case, however, a clear disadvantage of RNase-negative viruses was detected in the natural host with decreased virus load in the infected animals and absence of virus transmission to contact animals (Von Freyburg, Ege, Saalmüller, & Meyers, 2004). However, elucidation of the mechanisms underlying this effect is still incomplete (Magkouras et al., 2008; Mätzener et al., 2009; Python et al., 2013; Zürcher et al., 2014). Interestingly, an adequate interferon response was observed in immunocompetent cattle after acute infection with noncp BVDV (Brackenbury et al., 2003; Charleston et al., 2002; Smirnova et al., 2008). In contrast, absence of detectable systemic levels of IFN-1 is typical in animals persistently infected with noncp BVDV and was proposed to play an important role in establishment and maintenance of persistent infection

(Charleston et al., 2001). Therefore, the most critical functions of N^{pro} and E^{rns} RNase may be to assist persistent infection of the fetus.

Evidence for a connection between N^{pro} and the E^{rns} RNase activity with the establishment of persistent BVDV infections was obtained in experiments with pregnant heifers. In fetuses directly infected with noncp BVDV mutants lacking either the E^{rns} RNase activity or N^{pro}, induction of a solid IFN-1 response was observed, with levels similar to those induced by the cp BVDV variant. Most importantly, the combination of N^{pro} deletion and RNase inactivation in one virus double mutant resulted in an extremely elevated IFN-1 level 7 days postinfection (Meyers et al., 2007). Although induction of an interferon response in the infected fetus is not directly connected to prevention of virus persistence, there is at least some correlation. The RNase or N^{pro} negative single mutants of BVDV could establish persistent infection, as documented by isolation of infectious virus from fetuses 2 months after infection of pregnant heifers (Meyers et al., 2007). In contrast, double mutants were never found in the fetus in these experiments. When such mutants were introduced directly into the fetus, abortion within the first 3–7 weeks after infection occurred in all animals. Therefore, it can be concluded that the double mutant cannot establish persistence, presumably because it induces a strong IFN-1 response, leading to severe innate immune reactions. The ongoing secretion of IFN-1 and the resulting downstream effects should lead to apoptosis of cells, ultimately causing massive damage to the fetus.

It is important to notice that even the combination of control of viral RNA replication with repression of innate immune reactions via N^{pro} and E^{rns} will not necessarily lead to a complete block of the innate response. In fact, published data indicate that low-level responses can be observed at least transiently (Hansen et al., 2010; Shoemaker et al., 2009; Smirnova et al., 2012, 2014). However, the innate immune reactions in noncp BVDV-infected fetuses are obviously reduced to a tolerable level in order to establish and maintain persistence.

According to the known features of N^{pro} and E^{rns}, it can be hypothesized that pestiviruses inhibit interferon induction on two levels. N^{pro} is used to suppress the interferon response through its effect on IRF3. In contrast, the secreted E^{rns} protein distributed by the blood stream could function on a more systemic level. It could be taken up by cells, specifically PDCs, through clathrin-dependent endocytosis (Zürcher et al., 2014) and degrade RNA that is directly transferred from infected cells and would be sensed in endosomal compartments by TLR7. Moreover, the finding of E^{rns} RNase blocking the IFN-1 response of PDCs in contact with infected cells (Python et al., 2013) should result in a decrease of the systemic interferon response mediated through PDCs. In this context, it is worth noting that E^{rns} has been shown to efficiently degrade pestiviral RNA and that its pH optimum is consistent with activity in an endosomal compartment (Schneider et al., 1993; Windisch et al., 1996).

One might expect that loss of either N^{pro} or E^{rns} RNase function would prevent establishment of persistence, because both are hypothesized to exert quite different activities on the innate immune response. However, because both seem to counteract the induction of IFN-1 expression, one could also imagine a certain degree of redundancy. Alternatively, it may be that the IFN-1 response must only be reduced below a threshold level, with just one of the two viral counteracting activities being sufficient to reach this level. In fact, experiments show that the presence of only one of the two functions is sufficient to allow establishment of persistence (Meyers et al., 2007; G. Meyers, unpublished results). However, it can be hypothesized that a reduction of the incidence and duration of persistence would be observed if a sufficiently high number of pregnant animals could be challenged with viruses lacking an active RNase or N^{pro}. It is also important to note that efficient control of viral replication is absolutely necessary, because cp viruses cannot prevent an IFN-1 response, even in the presence of normal amounts of functional N^{pro} and active E^{rns} RNase.

The fact that E^{rns} dimerization obviously also plays a role in its function cannot be explained in this model. E^{rns} dimerization has no effect on membrane association or secretion. Interestingly, E^{rns} structure–function analyses revealed differences in enzymatic activity between the monomeric and dimeric forms (Krey et al., 2012), notably an higher affinity for the substrate displayed by the dimer and a slower activity. CSFV E^{rns} has also been shown to bind glycosaminoglycans less in its monomeric form (van Gennip et al., 2005), but so far, the connection between these findings and the observed attenuation of dimerization-deficient viruses is not obvious. It is also still not clear why the presence of N^{pro} and E^{rns} RNase cannot prevent or at least reduce the systemic IFN-1 response in animals acutely infected with noncp BVDV (Brackenbury et al., 2003; Charleston et al., 2002; Smirnova et al., 2008), whereas the absence of increased levels of IFN-1 is seen not only in noncp BVDV-inoculated fetuses shortly after infection but also in persistently infected calves (Bryan Charleston, personal communication).

Persistence of other pestiviruses was not investigated in such detail as for BVDV, but differences are known (Moennig & Plagemann, 1992; Thiel et al., 1996). The similarity is in general higher within the group of viruses infecting ruminants compared to CSFV. However, it is important to notice that transplacental infection of fetuses is a general theme so that many aspects of the concept outlined for BVDV above will also be true for other pestiviruses, especially the need to control adaptive and innate immune responses.

Thus, E^{rns} RNase and N^{pro} work together in inhibition of the innate immune response. These viral factors are especially important for the establishment and maintenance of persistent infections. Since persistently infected animals play a crucial role in maintenance of pestiviruses within their host populations, these two factors are of major importance for these viruses and were therefore conserved during virus evolution.

7. PESTIVIRUS—HOST CELL INTERPLAY

The interference of pestiviruses with the innate immune response is only one aspect of the interplay between these viruses and their hosts. It is obvious that a virus expressing only 12 proteins requires cellular factors supporting its replication either as components of the replication machinery itself or as mediators inducing changes in cellular biology that indirectly promote virus propagation. One group of factors with a known direct function in virus replication are members of the NF90/NFAR protein group (Isken et al., 2003). These proteins interact with both the 5′- and 3′-NTR of the BVDV genome by binding to regulatory elements. Modification of the binding sites for these factors leads to RNAs that are deficient in replication. These mutations did not disrupt essential *cis*-acting structures of the RNA (Isken, Grassmann, Yu, & Behrens, 2004) but reduced RNA replication efficiency and NFAR protein association to a similar extent. This result together with the finding that downregulation of NFAR protein concentration via RNAi impaired replication supported the conclusion that NFAR proteins function as host factors in pestivirus replication. Detailed analysis of viral translation revealed that the interaction with these host factors is important for correct translation termination which again is coupled to the initiation of RNA replication (Isken et al., 2004). This concept fits nicely with the fact that most of the viral replication components have to be translated from the RNA molecule that serves as a template for negative-strand synthesis, and thus have to be provided *in cis*. The interaction of NFAR proteins with both the 5′- and 3′-NTR resulted in a model in which the viral RNA is pseudo-circularized via an NFAR protein complex. This arrangement can be hypothesized to be important for translation mimicking the structure of mRNAs resulting from interaction of poly(A)-binding protein with initiation factor eIF4F. Moreover, the pseudo-circular form could be crucial for coordination of translation and negative-strand RNA synthesis and thereby coordinate the switch from translation to RNA replication. The interaction with NFAR proteins was first analyzed for BVDV but can be supposed to occur with other pestiviral RNAs in an equivalent way. In fact, one member of this group of proteins, the cytoplasmic RNA helicase A (RHA), was found to associate with both NTRs of the CSFV genome (Sheng et al., 2013).

The NFAR proteins were identified via UV-cross-linking/label transfer experiments followed by mass spectrometry as factors that specifically interact with viral RNA. A variety of other approaches were used to identify host factors important for pestivirus replication. Among them, yeast two-hybrid (Y2H) searches with individual viral proteins are the most frequently used (Doceul et al., 2008; Gladue, Baker-Bransetter, et al., 2014; Gladue, O'Donnell, et al., 2014; Johns et al., 2010; Kang et al., 2012; Zahoor et al., 2010) followed by microarray techniques (Gladue et al., 2010; Yamane et al., 2009c). Other approaches relied on targeting possible interaction partners of viral proteins or RNA (Fu, Shi, Ren, et al., 2014; Fu, Shi, Shi, Meng, Bao, et al., 2014; Fu, Shi, Shi, Meng, Zhang, et al., 2014; Fu, Shi, Zhang, et al., 2014; Gladue et al., 2011, 2010; He et al., 2012; Jordan, Nikolaeva, et al., 2002; Lin et al., 2014; Mohamed et al., 2014; Sheng et al., 2015; Wang et al., 2011; Yamane et al., 2009a, 2009b). Some of this work needs further evaluation and elaboration to allow establishment of a hypothesis on the specific function of the host factors. Moreover, it is impossible to describe all these quite heterogeneous investigations in detail in this review. Therefore, only a few groups of factors or processes will be described below.

Several reports describe the induction of autophagy upon pestivirus infection of cells (Fu, Shi, Shi, Meng, Bao, et al., 2014; Fu, Shi, Zhang, et al., 2014; Pei et al., 2014). Autophagy represents an important cellular response to infection. It can be a mechanism of defense but can also support pathogen replication. For CSFV, enhancement of replication was described to follow induction of autophagic processes (Pei et al., 2014). Further cellular factors interacting with pestiviral proteins are Annexin A2 (interacts with NS5A and is involved in particle production) (Sheng et al., 2015), sphingosine kinase (interacts with NS3 and is thereby inhibited resulting in enhanced virus replication) (Yamane et al., 2009b), adenosine deaminase acting on RNA (ADAR) (interacts with NS4A, overexpression reduces BVDV replication) (Mohamed et al., 2014), and the porcine homologue of human antigen R (HuR) (binds to AU-rich sequences in the $3'$-NTR of CSFV RNA and might stabilize the RNA) (Nadar et al., 2011). Another group of factors interacts with the core protein, namely the ER-associated degradation pathway protein OS9 and the IQGAP1 protein. Inhibition of OS9 interaction by core mutagenesis led to decreased replication of CSFV in tissue culture but had no influence on virulence in pigs. In contrast, prevention of IQGAP1 protein/core interaction resulted in defective growth in swine macrophages and complete attenuation in pigs (Gladue, Holinka, et al., 2011; Gladue, O'Donnell, et al., 2014). One report shows that cp BVDV infection activates PERK and induces ER stress-mediated apoptosis but the underlying mechanisms are still not fully understood (Jordan, Nikolaeva, et al., 2002).

8. VACCINES

Vaccination represents a common and widely used strategy to prevent losses due to pestivirus infection in many countries worldwide (Blome, Meindl-Bohmer, Loeffen, Thuer, & Moennig, 2006; Dong & Chen, 2007; Greiser-Wilke & Moennig, 2004; Huang, Deng, Wang, Huang, & Chang, 2014; Kelling, 2004; van Oirschot, 1999, 2003; van Oirschot, Bruschke, & van Rijn, 1999). Killed and live attenuated vaccines against BVDV and CSFV are commercially available. The established live virus vaccines were attenuated via unspecific approaches like repeated cell culture passages of field isolates. For BVDV, clear proof of the attenuation is generally difficult to obtain since most field viruses do not induce significant symptoms. In general, the genetic basis of attenuation of these viruses is not known which implies an unpredictable risk of reversion to virulence. Attempts to identify the molecular markers for attenuation in the viral genomes did not provide final conclusions since genomic alterations often lead to undefined reduction of virus growth (Leifer, Ruggli, & Blome, 2013; Risatti, Borca, et al., 2005; Risatti, Holinka, Fernandez, Carrillo, Kutish, et al., 2007; Risatti, Holinka, Fernandez, Carrillo, Lu, et al., 2007; van Gennip et al., 2004). An even more important concern than the risk of reversion to virulence is the fact that all so far approved live pestivirus vaccines are problematic in pregnant animals. Dependent on the vaccine virus, vaccination of pregnant animals may lead to fetal damage, trigger abortion, or establish persistent infection (Becher et al., 2001). Accordingly, the use of live pestivirus vaccines is flawed by safety concerns despite their outstanding efficacy. To circumvent the safety issue, a variety of nonlive vaccines have been developed and approved, composed of killed viruses or heterologously expressed structural components of the viruses supplemented with adjuvants. These vaccine formulations are *per se* safe but considerably less efficient, especially concerning efficacy against diaplacental infection of fetuses in pregnant animals. Sometimes, a booster vaccination is necessary to achieve protective immunity. In addition to standard two-step immunization schemes using the same vaccine for both steps, prime boost regimes utilizing a killed vaccine as prime vaccination and a life virus vaccine for booster immunization have been developed to improve protection (Liang, van den Hurk, Babiuk, & van Drunen Littel-van den Hurk, 2006; Liang et al., 2008; Sun, Li, Li, Li, & Qiu, 2010; Voigt et al., 2007).

8.1 Life Attenuated Viruses

Keeping in mind the threat of pestiviruses for the fetus in pregnant animals, there is an obvious need for improved pestivirus vaccines. The use of such vaccines should be safe in pregnant animals and lead to a level of immunity that not only prevents development of clinical symptoms and virus shedding but also blocks transplacental transmission of field viruses to the fetus. The ideal vaccine would be a novel type of live attenuated virus that is not transmitted to the fetus but preserves the superior efficacy of a replicating virus.

The safety concern of novel vaccine approaches requires defined and effective knowledge about pestivirus attenuation. Mutations affecting genome replication, viral gene expression, the efficiency of target cell infection, or virus tropism in its natural host can reduce fitness and lead to attenuation. Different mutations reducing the efficacy of pestivirus gene expression have been identified. One of these changes is directly connected with deletion of N^{pro} that leads to reduced growth rates via impairment of viral protein translation (see Section 3.1). In addition, deletion of N^{pro} eliminates part of the ability of pestiviruses to interfere with the innate immunity of the host, which should also reduce its fitness for replication in the natural host even though loss of this function via mutation of the TRASH motif did not lead to significant attenuation (Ruggli et al., 2009).

IRES function can also be affected by mutations introduced into the NTRs of BVDV. The recovered virus mutants were viable and stable and exhibited reduced fitness in cell culture and attenuation in their animal host (Becher et al., 2000; Makoschey et al., 2004).

Similarly, alterations in the 3′-NTR were shown to attenuate pestiviruses. A 12-nucleotide insertion, first identified in a lapinized CSFV vaccine strain recovered after consecutive passage of pathogenic virus in rabbit cells, reduced the growth of the mutant virus in tissue culture by nearly 2 logs and conferred attenuation of the highly pathogenic strain Shimen (Wang et al., 2008). The stability of attenuation resulting from such inserted sequences is an important unanswered question. In theory, an insertion that hampers virus growth can easily be lost by recombination, restoring growth efficiency and probably also virulence.

Reduction of virus fitness at the level of target cell infection has been achieved by introducing mutations into structural protein-coding regions. A linear epitope in E2 used to differentiate CSFV from ruminant pestiviruses was changed in highly virulent CSFV Brescia to resemble the corresponding sequence of BVDV strain NADL. Two of the resulting viruses showed considerable growth retardation and proved to be attenuated in pigs (Risatti, Borca, et al., 2005; Risatti et al., 2006). In addition, a 19-amino-acid insertion introduced into the C-terminal region of E1 via transposon linker insertion mutagenesis resulted in an attenuated virus. In contrast to the mutants described above, the resulting virus grew about as well as wild-type virus in tissue culture (Risatti, Holinka, et al., 2005).

Other approaches to attenuation of CSFV by mutations in structural protein-coding regions rely on elimination of N-glycosylation sites as was done for E2, E^{rns}, and E1 (Risatti, Holinka, Fernandez, Carrillo, Lu, et al., 2007; Risatti, Holinka, et al., 2005; Sainz et al., 2008). In all cases, the mechanism underlying the observed attenuation is not fully understood and the long-term stability of the mutations has not yet been demonstrated.

In another approach, a cytopathic strain of CSFV was established by the insertion of Jiv-coding sequences into the viral genome (Gallei et al., 2008). This virus showed more efficient NS2–3 processing and upregulation of viral RNA synthesis, as observed previously for cp BVDV strains with Jiv insertions. In cell culture experiments, the cp CSFV mutant induced the expression of the interferon-regulated gene Mx, in contrast to the parental noncp CSFV strain. Accordingly, increased replication of the cp virus correlated with induction of an innate immune response. Importantly, the cp virus was found to be significantly attenuated in its natural host and induced high levels of neutralizing antibodies.

For several reasons, chimeric viruses that contain sequences from two different strains of one pestivirus species or even from members of two different species were established and tested as vaccine candidates. Several chimeras with sequences from the CSFV vaccine viruses "C-strain" or "CS-strain" introduced into the background of the highly pathogenic Brescia strain are attenuated in pigs (Gallei et al., 2008; Risatti, Borca, et al., 2005; Risatti, Holinka, Fernandez, Carrillo, Kutish, et al., 2007; van Gennip et al., 2004). In other approaches, specific fragments, such as the E^{rns}-coding region or part of the E2-coding sequence, in a CSFV genome were replaced by the corresponding BVDV sequences (de Smit et al., 2000; Gallei et al., 2008; Risatti, Borca, et al., 2005; Risatti, Holinka, Fernandez, Carrillo, Kutish, et al., 2007; van Gennip, van Rijn, Widjojoatmodjo, de Smit, & Moormann, 2000). These chimeras were shown to grow with acceptable or even high efficiency in porcine cells but proved to be non-pathogenic in pigs. Nevertheless, vaccination with these chimeras protected against a stringent CSFV challenge.

A different approach was chosen in constructing the chimeric virus, CP7_E2alf (Beer, Reimann, Hoffmann, & Depner, 2007; Koenig, Lange, Reimann, & Beer, 2007; Leifer et al., 2009). Here, in BVDV strain CP7, the E2-coding region was replaced by the corresponding sequence from CSFV strain Alfort 187. Efficient replication of this virus was basically restricted to porcine cells. Vaccination of pigs induced protective immunity against a stringent challenge. It is not clear whether the theoretically lower complexity of the CSFV-specific immune response due to the absence of CSFV E^{rns} as a second target for neutralizing antibodies (Weiland et al., 1992) and a complete set of T-cell epitopes in viral structural and non-structural proteins (Armengol et al., 2002) represent a relevant disadvantage of this vaccine with regard to cross-protection against different CSFV strains. In general, vaccination with interspecies chimeras bears an additional risk because new viruses are created which may spread in the field, but the risk connected with such an approach can be hypothesized to be low.

An interesting approach to attenuation via defined mutations relies on the elimination of viral factors blocking the innate immune response of the host. Recent studies demonstrated that a noncp BVDV double mutant with deletion of the N^{pro}-coding sequence and inactivation of the E^{rns} RNase fulfills all the requirements for a safe and effective pestivirus vaccine. When inoculated into a pregnant heifer, these mutants did not cross the placenta and were not detected in fetal tissue. Challenge infections of vaccinated pregnant animals did not lead to fetal infection either, demonstrating efficient protection by the vaccination. The observed high efficiency of these vaccine viruses is most likely due to the fact that they show almost no growth impairment compared to the wild-type viruses (Meyers et al., 2007).

Most of the recent efforts toward novel pestivirus vaccines have concentrated on BVDV which is probably based on the high commercial impact of BVDV, its nearly ubiquitous distribution, and the almost general acceptance of vaccination as one pillar for the control of this virus. As mentioned above, the two species BVDV-1 and BVDV-2 are found in the host population. Most of the currently available vaccines are derived from BVDV-1. Published data show that these vaccines provide at least partial protection also against a BVDV-2 challenge (Cortese, West, Hassard, Carman, & Ellis, 1998; Dean & Leyh, 1999; Fairbanks, Schnackel, & Chase, 2003; Makoschey, Janssen, Vrijenhoek, Korsten, & Marel, 2001). However, full protection against transmission of virus to the fetus in pregnant animals can obviously not be achieved by these vaccines (Paton, Sharp, & Ibata, 1999; Ridpath, 2005). Therefore, novel vaccines should include both BVDV-1 and BVDV-2 in order to provide full protection of cows against classical BVDV.

The development of novel vaccines against pestiviruses other than BVDV is pursued with lower intensity. The authors are not aware of ongoing efforts toward BDV vaccines, although border disease also has a significant economic impact. CSFV vaccines are available and are used in several countries. The most frequently administered CSFV vaccines consist of live viruses attenuated by serial passages in nonporcine cells. An example is the so-called C-strain developed in China by numerous passages in rabbits. These vaccines are safe and very efficient with rapid induction of a solid protective immunity. Nevertheless, vaccination is forbidden or at least restricted in many countries because of a general stamping-out policy for CSFV that cannot be combined with these vaccines. An important issue in this context is the absence of test systems able to discriminate vaccinated from field virus-infected animals. In the 1990s, CSFV vaccines based on baculovirus expressed E2 protein have been developed that fulfill the criteria of marker vaccines (Hulst, Westra, Wensvoort, & Moormann, 1993; Uttenthal, Le Potier, Romero, De Mia, & Floegel-Niesmann, 2001), but establishment of protective immunity after administration of these vaccines lasts rather long which lowers their suitability for emergency vaccination. Discussions about novel CSFV vaccines have been reactivated because CSF outbreaks in the last decades have required extensive culling campaigns, which are associated with serious ethical concerns and enormous costs. As for BVDV, safe and effective modified life vaccines with defined attenuating mutations would be reasonable for CSFV.

8.2 Replicons

A major disadvantage of killed vaccines is their defect with regard to induction of a cellular immune response due to the absence of *de novo* synthesis of viral antigens that could enter the MHC-associated presentation pathway. An interesting approach combining the advantages of killed and live vaccine is the use of autonomous replicons for immunization. Different replicon vaccine candidates have been established for BVDV and CSFV (Frey et al., 2006; Maurer, Stettler, Ruggli, Hofmann, & Tratschin, 2005; Reimann et al., 2007; van Gennip, Bouma, van Rijn, Widjojoatmodjo, & Moormann, 2002; Widjojoatmodjo et al., 2000). All of them contain deletions of essential sequences so that their propagation requires *in trans-* complementation. In theory, safety of such replicon-based vaccines is intrinsic since infection of a cell is a dead end, with no infectious virus being released. However, production of such replicons in complementing cell lines for vaccine purposes has to be controlled in a way that reintroduction of the deleted sequences into the genome via recombination between the replicon genome and RNA coding for the complementing proteins is prevented.

The efficacy of replicon-based vaccines is hampered by the fact that only a very limited number of cells will be infected and produce viral proteins, even when a high vaccine dose is applied. Thus, the trigger for the immune system is probably not very strong. Nevertheless, induction of protective immunity has been successfully demonstrated in stringent challenge models in several cases (Frey et al., 2006; Loy et al., 2013; Maurer et al., 2005; Sun et al., 2011; Yang et al., 2012; Zemke, Koenig, Mischkale, Reimann, & Beer, 2010). However, it is not yet clear whether a single vaccination with a replicon can consistently prevent fetal infection in a pregnant animal.

8.3 DIVA Vaccines

Control or even eradication of viruses can be facilitated by a combination of vaccination and culling of (persistently) infected animals. Marker vaccines allowing the *d*ifferentiation of field virus-*i*nfected from *v*accinated *a*nimals (DIVA) by serological techniques have been successfully used in other viruses. In pestiviruses, the benefit of a marker for control program has been questioned. The main reason for this debate is the specific role of PI animals. PI animals represent the main source of virus in the field, and their identification and elimination is the most important issue. As described above, PI animals do not produce antibodies so that infection cannot be identified by serological means in these animals requiring other solutions, such as a simple RT-PCR test, for their identification. A serological DIVA test could help in control programs for pestiviruses during later stages for monitoring reintroduction of field virus into regions from which the virus had been eradicated. Therefore, different groups pursue the aim of establishment of marker vaccines (recently reviewed for CSFV in Dong & Chen, 2007). High frequency of mutation and recombination as well as lack in diagnostic consistency more or less excludes the use of positive markers so that negative markers like deletion or mutation of conserved immunogenic regions is the preferred approach. Deletion of N^{pro}, the only nonessential protein of pestiviruses, would be a perfect marker, but N^{pro} is not sufficiently immunogenic. More promising approaches rely on chimeric pestiviruses since the absence of sequences replaced by sequences derived from different pestivirus species can be used for serological differentiation. Similarly, the absence of certain viral proteins from cells infected with replicons can serve as a basis for a marker system. In fact, vaccinated animals could be serologically differentiated from animals that had been infected with the corresponding field viruses when chimeras or deletion mutants were used as vaccines (Koenig et al., 2007; Leifer et al., 2009; Luo et al., 2012; Reimann et al., 2010). So far, it is not clear whether under field conditions the vaccination will not repress the replication of an infecting field virus so strongly that the amount of the DIVA target protein will be too low for induction of a solid immune response in a sufficiently high number of animals. Taken together, the search for a feasible marker for pestivirus vaccines remains an ongoing process still open for novel ideas.

9. CONCLUSION

Pestiviruses represent highly interesting and economically important pathogens. Molecular analyses during the last two decades elucidated many astonishing features of these viruses and added significantly to our view of virus/host interplay. Moreover, this work helped to establish promising approaches toward a new generation of safe and efficient vaccines that will help to improve control of pestiviruses in the future and reduce losses in food production. Many aspects of pestiviral biology are still obscure and await further work at the molecular level. It will be interesting to see where detailed analysis with a special emphasis on the structure of pestiviral proteins and the mechanisms underlying their interaction with the host organism will lead us.

REFERENCES

Abbas, Y. M., Pichlmair, A., Gorna, M. W., Superti-Furga, G., & Nagar, B. (2013). Structural basis for viral 5′-PPP-RNA recognition by human IFIT proteins. *Nature*, *494*(7435), 60–64. http://dx.doi.org/10.1038/nature11783.

Aberle, D., Muhle-Goll, C., Bürck, J., Wolf, M., Reisser, S., Luy, B., et al. (2014). Structure of the membrane anchor of pestivirus glycoprotein E(rns), a long tilted amphipathic helix. *PLoS Pathogens*, *10*(2), e1003973. http://dx.doi.org/10.1371/journal.ppat.1003973.

Agapov, E. V., Frolov, I., Lindenbach, B. D., Pragai, B. M., Schlesinger, S., & Rice, C. M. (1998). Noncytopathic sindbis virus RNA vectors for heterologous gene expression. *Proceedings of the National Academy of Sciences of the United States of America*, *95*, 12989–12994.

Agol, V. I. (1997). Recombination and other genomic rearrangements in picornaviruses. *Seminars in Virology*, *8*(2), 77–84. http://dx.doi.org/10.1006/smvy.1997.0112.

Ansari, I. H., Chen, L. M., Liang, D., Gil, L. H., Zhong, W., & Donis, R. O. (2004). Involvement of a bovine viral diarrhea virus NS5B locus in virion assembly. *Journal of Virology*, *78*(18), 9612–9623.

Aoki, H., Ishikawa, K., Sakoda, Y., Sekiguchi, H., Kodama, M., Suzuki, S., et al. (2001). Characterization of classical swine fever virus associated with defective interfering particles containing a cytopathogenic subgenomic RNA isolated from wild boar. *Journal of Veterinary Medical Science*, *63*(7), 751–758.

Aoki, H., Ishikawa, K., Sekiguchi, H., Suzuki, S., & Fukusho, A. (2003). Pathogenicity and kinetics of virus propagation in swine infected with the cytopathogenic classical swine fever virus containing defective interfering particles. *Archives of Virology*, *148*(2), 297–310. http://dx.doi.org/10.1007/s00705-002-0907-2.

Armengol, E., Wiesmüller, K. H., Wienhold, D., Büttner, M., Pfaff, E., Jung, G., et al. (2002). Identification of T-cell epitopes in the structural and non-structural proteins of classical swine fever virus. *Journal of General Virology*, *83*(Pt 3), 551–560.

Arnal, M., Fernandez-de-Luco, D., Riba, L., Maley, M., Gilray, J., Willoughby, K., et al. (2004). A novel pestivirus associated with deaths in Pyrenean chamois (Rupicapra pyrenaica pyrenaica). *Journal of General Virology*, *85*(Pt 12), 3653–3657. http://dx.doi.org/10.1099/vir.0.80235-0.

Austermann-Busch, S., & Becher, P. (2012). RNA structural elements determine frequency and sites of nonhomologous recombination in an animal plus-strand RNA virus. *Journal of Virology*, *86*(13), 7393–7402. http://dx.doi.org/10.1128/JVI.00864-12.

Avalos-Ramirez, R., Orlich, M., Thiel, H. J., & Becher, P. (2001). Evidence for the presence of two novel pestivirus species. *Virology*, *286*(2), 456–465.

Baginski, S. G., Pevear, D. C., Seipel, M., Sun, S. C., Benetatos, C. A., Chunduru, S. K., et al. (2000). Mechanism of action of a pestivirus antiviral compound. *Proceedings of the National academy of Sciences of the United States of America*, *97*(14), 7981–7986.

Baker, J. C. (1987). Bovine viral diarrhea virus: A review. *Journal of the American Veterinary Medical Association*, *190*, 1449–1458.

Balint, A., Baule, C., Kecskemeti, S., Kiss, I., & Belak, S. (2005). Cytopathogenicity markers in the genome of Hungarian cytopathic isolates of bovine viral diarrhoea virus. *Acta Veterinaria Hungarica*, *53*(1), 125–136. http://dx.doi.org/10.1556/AVet.53.2005.1.12.

Balint, A., Baule, C., Palfi, V., Dencso, L., Hornyak, A., & Belak, S. (2005). A 45-nucleotide insertion in the NS2 gene is responsible for the cytopathogenicity of a bovine viral diarrhoea virus strain. *Virus Genes, 31*(2), 135–144. http://dx.doi.org/10.1007/s11262-005-1785-y.

Balint, A., Palfi, V., Belak, S., & Baule, C. (2005). Viral sequence insertions and a novel cellular insertion in the NS2 gene of cytopathic isolates of bovine viral diarrhea virus as potential cytopathogenicity markers. *Virus Genes, 30*(1), 49–58. http://dx.doi.org/10.1007/s11262-004-4581-1.

Barlow, R. M. (1972). Experiments in border disease IV. Pathological changes in ewes. *Journal of Comparative Pathology, 82*, 151–157.

Baroth, M., Orlich, M., Thiel, H. J., & Becher, P. (2000). Insertion of cellular NEDD8 coding sequences in a pestivirus. *Virology, 278*(2), 456–466.

Barrett, A. D. T., & Dimmock, N. J. (1986). Defective interfering viruses and infections of animals. *Current Topics in Microbiology and Immunology, 128*, 55–84.

Bauermann, F. V., Ridpath, J. F., Weiblen, R., & Flores, E. F. (2013). HoBi-like viruses: An emerging group of pestiviruses. *Journal of Veterinary Diagnostic Investigation, 25*(1), 6–15. http://dx.doi.org/10.1177/1040638712473103.

Bauhofer, O., Summerfield, A., Sakoda, Y., Tratschin, J. D., Hofmann, M. A., & Ruggli, N. (2007). Classical swine fever virus Npro interacts with interferon regulatory factor 3 and induces its proteasomal degradation. *Journal of Virology, 81*(7), 3087–3096.

Bazan, J. F., & Fletterick, R. J. (1988). Viral cysteine proteases are homologous to the trypsin-like family of serine proteases: Structural and functional implications. *Proceedings of the National Academy of Sciences of the United States of America, 85*(21), 7872–7876.

Becher, P., Avalos, Ramirez R., Orlich, M., Cedillo, Rosales S., König, M., Schweizer, M., et al. (2003). Genetic and antigenic characterization of novel pestivirus genotypes: Implications for classification. *Virology, 311*(1), 96–104.

Becher, P., Fischer, N., Grundhoff, A., Stalder, H., Schweizer, M., & Postel, A. (2014). Complete genome sequence of bovine pestivirus strain PG-2, a second member of the tentative pestivirus species giraffe. *Genome Announcements, 2*(3), e00376. http://dx.doi.org/10.1128/genomeA.00376-14.

Becher, P., Meyers, G., Shannon, A. D., & Thiel, H. J. (1996). Cytopathogenicity of border disease virus is correlated with integration of cellular sequences into the viral genome. *Journal of Virology, 70*(5), 2992–2998.

Becher, P., Orlich, M., König, M., & Thiel, H. J. (1999). Nonhomologous RNA recombination in bovine viral diarrhea virus: Molecular characterization of a variety of subgenomic RNAs isolated during an outbreak of fatal mucosal disease. *Journal of Virology, 73*(7), 5646–5653.

Becher, P., Orlich, M., Kosmidou, A., König, M., Baroth, M., & Thiel, H. J. (1999). Genetic diversity of pestiviruses: Identification of novel groups and implications for classification. *Virology, 262*(1), 64–71.

Becher, P., Orlich, M., Shannon, A. D., Horner, G., König, M., & Thiel, H. J. (1997). Phylogenetic analysis of pestiviruses from domestic and wild ruminants. *Journal of General Virology, 78*(Pt 6), 1357–1366.

Becher, P., Orlich, M., & Thiel, H. J. (1998a). Complete genomic sequence of border disease virus, a pestivirus from sheep. *Journal of Virology, 72*(6), 5165–5173.

Becher, P., Orlich, M., & Thiel, H. J. (1998b). Ribosomal S27a coding sequences upstream of ubiquitin coding sequences in the genome of a pestivirus. *Journal of Virology, 72*(11), 8697–8704.

Becher, P., Orlich, M., & Thiel, H. J. (2000). Mutations in the 5' nontranslated region of bovine viral diarrhea virus result in altered growth characteristics. *Journal of Virology, 74*(17), 7884–7894.

Becher, P., Orlich, M., & Thiel, H. J. (2001). RNA recombination between persisting pestivirus and a vaccine strain: Generation of cytopathogenic virus and induction of lethal disease. *Journal of Virology, 75*(14), 6256–6264.

Becher, P., Schmeiser, S., Oguzoglu, T. C., & Postel, A. (2012). Complete genome sequence of a novel pestivirus from sheep. *Journal of Virology*, *86*(20), 11412. http://dx.doi.org/10.1128/JVI.01994-12.

Becher, P., & Tautz, N. (2011). RNA recombination in pestiviruses: Cellular RNA sequences in viral genomes highlight the role of host factors for viral persistence and lethal disease. *RNA Biology*, *8*(2), 216–224.

Becher, P., Thiel, H. J., Collins, M., Brownlie, J., & Orlich, M. (2002). Cellular sequences in pestivirus genomes encoding gamma-aminobutyric acid (A) receptor-associated protein and Golgi-associated ATPase enhancer of 16 kilodaltons. *Journal of Virology*, *76*(24), 13069–13076.

Beer, M., Reimann, I., Hoffmann, B., & Depner, K. (2007). Novel marker vaccines against classical swine fever. *Vaccine*, *25*(30), 5665–5670.

Behrens, S. E., Grassmann, C. W., Thiel, H. J., Meyers, G., & Tautz, N. (1998). Characterization of an autonomous subgenomic pestivirus RNA replicon. *Journal of Virology*, *72*(3), 2364–2372.

Bekal, S., Domier, L. L., Gonfa, B., McCoppin, N. K., Lambert, K. N., & Bhalerao, K. (2014). A novel flavivirus in the soybean cyst nematode. *Journal of General Virology*, *95*(Pt 6), 1272–1280. http://dx.doi.org/10.1099/vir.0.060889-0.

Bintintan, I., & Meyers, G. (2010). A new type of signal peptidase cleavage site identified in an RNA virus polyprotein. *Journal of Biological Chemistry*, *285*(12), 8572–8584.

Blome, S., Meindl-Bohmer, A., Loeffen, W., Thuer, B., & Moennig, V. (2006). Assessment of classical swine fever diagnostics and vaccine performance. *Revue Scientifique et Technique*, *25*(3), 1025–1038.

Bolin, S. R. (1988). Viral and viral protein specificity of antibodies induced in cows persistently infected with noncytopathic bovine viral diarrhea virus after vaccination with cytopathic bovine viral diarrhea virus. *American Journal of Veterinary Research*, *49*(7), 1040–1044.

Bolin, S. R., McClurkin, A. W., Cutlip, R. C., & Coria, M. F. (1985). Severe clinical disease induced in cattle persistently infected with noncytopathogenic bovine viral diarrhea virus by superinfection with cytopathogenic bovine viral diarrhea virus. *American Journal of Veterinary Research*, *46*, 573–576.

Brackenbury, L. S., Carr, B. V., & Charleston, B. (2003). Aspects of the innate and adaptive immune responses to acute infections with BVDV. *Veterinary Microbiology*, *96*(4), 337–344.

Branza-Nichita, N., Durantel, D., Carrouee-Durantel, S., Dwek, R. A., & Zitzmann, N. (2001). Antiviral effect of N-butyldeoxynojirimycin against bovine viral diarrhea virus correlates with misfolding of E2 envelope proteins and impairment of their association into E1-E2 heterodimers. *Journal of Virology*, *75*(8), 3527–3536. http://dx.doi.org/10.1128/JVI.75.8.3527-3536.2001.

Branza-Nichita, N., Lazar, C., Durantel, D., Dwek, R. A., & Zitzmann, N. (2002). Role of disulfide bond formation in the folding and assembly of the envelope glycoproteins of a pestivirus. *Biochemical and Biophysical Research Communications*, *296*(2), 470–476.

Branza-Nichita, N., Lazar, C., Dwek, R. A., & Zitzmann, N. (2004). Role of N-glycan trimming in the folding and secretion of the pestivirus protein E(rns). *Biochemical and Biophysical Research Communications*, *319*(2), 655–662.

Brass, V., Pal, Z., Sapay, N., Deleage, G., Blum, H. E., Penin, F., et al. (2007). Conserved determinants for membrane association of nonstructural protein 5A from hepatitis C virus and related viruses. *Journal of Virology*, *81*(6), 2745–2757. http://dx.doi.org/10.1128/JVI.01279-06.

Brock, K. V., Deng, R., & Riblet, S. M. (1992). Nucleotide sequencing of 5′ and 3′ termini of bovine viral diarrhea virus by RNA ligation and PCR. *Journal of Virological Methods*, *38*(1), 39–46.

Brownlie, J. (1991). The pathways for bovine virus diarrhoea virus biotypes in the pathogenesis of disease. *Archives of Virology. Supplementum, 3*, 79–96.

Brownlie, J., & Clarke, M. C. (1993). Experimental and spontaneous mucosal disease of cattle: A validation of Koch's postulates in the definition of pathogenesis. *Intervirology, 35*(1–4), 51–59.

Brownlie, J., Clarke, M. C., & Howard, C. J. (1984). Experimental production of fatal mucosal disease in cattle. *The Veterinary Record, 114*, 535–536.

Brownlie, J., Clarke, M. C., & Howard, C. J. (1989). Experimental infection of cattle in early pregnancy with a cytopathic strain of bovine virus diarrhoea virus. *Research in Veterinary Science, 46*(3), 307–311.

Bruschke, C. J., Moormann, R. J., van Oirschot, J. T., & van Rijn, P. A. (1997). A subunit vaccine based on glycoprotein E2 of bovine virus diarrhea virus induces fetal protection in sheep against homologous challenge. *Vaccine, 15*(17–18), 1940–1945.

Brzozka, K., Finke, S., & Conzelmann, K. K. (2006). Inhibition of interferon signaling by rabies virus phosphoprotein P: Activation-dependent binding of STAT1 and STAT2. *Journal of Virology, 80*(6), 2675–2683. http://dx.doi.org/10.1128/JVI.80.6.2675-2683.2006.

Burrack, S., Aberle, D., Burck, J., Ulrich, A. S., & Meyers, G. (2012). A new type of intracellular retention signal identified in a pestivirus structural glycoprotein. *The FASEB Journal, 26*(8), 3292–3305. http://dx.doi.org/10.1096/fj.12-207191.

Charini, W. A., Todd, S., Gutman, G. A., & Semler, B. L. (1994). Transduction of a human RNA sequence by poliovirus. *Journal of Virology, 68*(10), 6547–6552.

Charleston, B., Brackenbury, L. S., Carr, B. V., Fray, M. D., Hope, J. C., Howard, C. J., et al. (2002). Alpha/beta and gamma interferons are induced by infection with noncytopathic bovine viral diarrhea virus in vivo. *Journal of Virology, 76*(2), 923–927.

Charleston, B., Fray, M. D., Baigent, S., Carr, B. V., & Morrison, W. I. (2001). Establishment of persistent infection with non-cytopathic bovine viral diarrhoea virus in cattle is associated with a failure to induce type I interferon. *Journal of General Virology, 82*(Pt 8), 1893–1897.

Chase, C. C. (2013). The impact of BVDV infection on adaptive immunity. *Biologicals, 41*(1), 52–60. http://dx.doi.org/10.1016/j.biologicals.2012.09.009.

Chen, Z., Rijnbrand, R., Jangra, R. K., Devaraj, S. G., Qu, L., Ma, Y., et al. (2007). Ubiquitination and proteasomal degradation of interferon regulatory factor-3 induced by Npro from a cytopathic bovine viral diarrhea virus. *Virology, 366*(2), 277–292.

Chen, Y., Xiao, J., Xiao, J., Sheng, C., Wang, J., Jia, L., et al. (2012). Classical swine fever virus NS5A regulates viral RNA replication through binding to NS5B and 3'UTR. *Virology, 432*(2), 376–388. http://dx.doi.org/10.1016/j.virol.2012.04.014.

Choi, C., & Chae, C. (2002). Localization of classical swine fever virus in male gonads during subclinical infection. *Journal of General Virology, 83*(Pt 11), 2717–2721.

Choi, K. H., Gallei, A., Becher, P., & Rossmann, M. G. (2006). The structure of bovine viral diarrhea virus RNA-dependent RNA polymerase and its amino-terminal domain. *Structure, 14*(7), 1107–1113. http://dx.doi.org/10.1016/j.str.2006.05.020.

Choi, K. H., Groarke, J. M., Young, D. C., Kuhn, R. J., Smith, J. L., Pevear, D. C., et al. (2004). The structure of the RNA-dependent RNA polymerase from bovine viral diarrhea virus establishes the role of GTP in de novo initiation. *Proceedings of the National Academy of Sciences of the United States of America, 101*(13), 4425–4430. http://dx.doi.org/10.1073/pnas.0400660101.

Choi, H. K., Tong, L., Minor, W., Dumas, P., Boege, U., Rossmann, M. G., et al. (1991). Structure of Sindbis virus core protein reveals a chymotrypsin-like serine proteinase and the organization of the virion. *Nature, 354*(6348), 37–43.

Chon, S. K., Perez, D. R., & Donis, R. O. (1998). Genetic analysis of the internal ribosome entry segment of bovine viral diarrhea virus. *Virology, 251*(2), 370–382.

Cocquerel, L., De, Op, Beeck, A., Lambot, M., Roussel, J., Delgrange, D., et al. (2002). Topological changes in the transmembrane domains of hepatitis C virus envelope glycoproteins. *The EMBO Journal, 21*(12), 2893–2902.

Collen, T., Douglas, A. J., Paton, D. J., Zhang, G., & Morrison, W. I. (2000). Single amino acid differences are sufficient for CD4(+) T-cell recognition of a heterologous virus by cattle persistently infected with bovine viral diarrhea virus. *Virology, 276*(1), 70–82.

Collett, M. S., Anderson, D. K., & Retzel, E. (1988). Comparisons of the Pestivirus bovine viral diarrhoea virus with members of the Flaviviridae. *Journal of General Virology, 69*, 2637–2643.

Collett, M. S., Larson, R., Belzer, S., & Retzel, E. (1988). Proteins encoded by bovine viral diarrhea virus: The genome organization of a pestivirus. *Virology, 165*, 200–208.

Collett, M. S., Larson, R., Gold, C., Strick, D., Anderson, D. K., & Purchio, A. F. (1988). Molecular cloning and nucleotide sequence of the pestivirus bovine viral diarrhea virus. *Virology, 165*, 191–199.

Collett, M. S., Moennig, V., & Horzinek, M. C. (1989). Recent advances in pestivirus research. *Journal of General Virology, 70*, 253–266.

Collett, M. S., Wiskerchen, M. A., Welniak, E., & Belzer, S. K. (1991). Bovine viral diarrhea virus genomic organization. *Archives of Virology. Supplementum, 3*, 19–27.

Corapi, W. V., Donis, R. O., & Dubovi, E. J. (1988). Monoclonal antibody analyses of cytopathic and noncytopathic viruses from fatal bovine viral diarrhea infections. *Journal of Virology, 62*, 2823–2827.

Corapi, W. V., Elliott, R. D., French, T. W., & Dubovi, E. J. (1990). Thrombocytopenia and hemorrhages in veal calves infected with bovine viral diarrhea virus. *Journal of the American Veterinary Medical Association, 196*, 590–596.

Corapi, W. V., French, T. W., & Dubovi, E. J. (1989). Severe thrombocytopenia in young calves experimentally infected with noncytopathic bovine viral diarrhea virus. *Journal of Virology, 63*(9), 3934–3943.

Coria, M. F., & McClurkin, A. W. (1978). Specific immunotolerance in an apparently healthy bull persistently infected with BVD virus. *Journal of the American Veterinary Medical Association, 172*, 449–451.

Cortese, V. S., West, K. H., Hassard, L. E., Carman, S., & Ellis, J. A. (1998). Clinical and immunologic responses of vaccinated and unvaccinated calves to infection with a virulent type-II isolate of bovine viral diarrhea virus. *Journal of the American Veterinary Medical Association, 213*(9), 1312–1319.

Darbyshire, J. H. (1969). A serological relationship between swine fever and mucosal disease of cattle. *The Veterinary Record, 72*, 331–333.

Darweesh, M. F., Rajput, M. K., Braun, L. J., Ridpath, J. F., Neill, J. D., & Chase, C. C. (2015). Characterization of the cytopathic BVDV strains isolated from 13 mucosal disease cases arising in a cattle herd. *Virus Research, 195*, 141–147. http://dx.doi.org/10.1016/j.virusres.2014.09.015.

de Breyne, S., Yu, Y., Pestova, T. V., & Hellen, C. U. (2008). Factor requirements for translation initiation on the Simian picornavirus internal ribosomal entry site. *RNA, 14*(2), 367–380. http://dx.doi.org/10.1261/rna.696508.

de Moerlooze, L., Desport, M., Renard, A., Lecomte, C., Brownlie, J., & Martial, J. A. (1990). The coding region for the 54-kDa protein of several pestiviruses lacks host insertions but reveals a "zinc finger-like" domain. *Virology, 177*, 812–815.

de Smit, A. J., van Gennip, H. G., Miedema, G. K., van Rijn, P. A., Terpstra, C., & Moormann, R. J. (2000). Recombinant classical swine fever (CSF) viruses derived from the Chinese vaccine strain (C-strain) of CSF virus retain their avirulent and immunogenic characteristics. *Vaccine, 18*(22), 2351–2358.

Dean, H. J., & Leyh, R. (1999). Cross-protective efficacy of a bovine viral diarrhea virus (BVDV) type 1 vaccine against BVDV type 2 challenge. *Vaccine, 17*(9–10), 1117–1124.

Decaro, N., Mari, V., Pinto, P., Lucente, M. S., Sciarretta, R., Cirone, F., et al. (2012). Hobi-like pestivirus: Both biotypes isolated from a diseased animal. *Journal of General Virology, 93*(Pt 9), 1976–1983. http://dx.doi.org/10.1099/vir.0.044552-0.

Deng, R., & Brock, K. V. (1993). 5′ and 3′ untranslated regions of pestivirus genome: Primary and secondary structure anayses. *Nucleic Acids Research, 21*, 1949–1957.

Depner, K., Bauer, T., & Liess, B. (1992). Thermal and pH stability of pestiviruses. *Revue Scientifique et Technique, 11*(3), 885–893.

Deregt, D., & Loewen, K. G. (1995). Bovine viral diarrhea virus: Biotypes and disease. *The Canadian Veterinary Journal, 36*(6), 371–378.

Deshpande, R. A., & Shankar, V. (2002). Ribonucleases from T2 family. *Critical Reviews in Microbiology, 28*(2), 79–122. http://dx.doi.org/10.1080/1040-840291046704.

Desport, M., Collins, M. E., & Brownlie, J. (1998). Genome instability in BVDV: An examination of the sequence and structural influences on RNA recombination. *Virology, 246*, 352–361.

Diderholm, H., & Dinter, Z. (1966). Interference between strains of bovine virus diarrhea virus and their capacity to suppress interferon of a heterologous virus. *Proceedings of the Society for Experimental Biology and Medicine, 121*(3), 976–980.

Ding, Q., von Schaewen, M., & Ploss, A. (2014). The impact of hepatitis C virus entry on viral tropism. *Cell Host & Microbe, 16*(5), 562–568. http://dx.doi.org/10.1016/j.chom.2014.10.009.

Dixit, E., & Kagan, J. C. (2013). Intracellular pathogen detection by RIG-I-like receptors. In Alt, F. (Ed.), *Advances in immunology: Vol. 117* (pp. 99–125): Academic Press, Elsevier B.V.

Doceul, V., Charleston, B., Crooke, H., Reid, E., Powell, P. P., & Seago, J. (2008). The Npro product of classical swine fever virus interacts with I{kappa} B{alpha}, the NF-{kappa}B inhibitor. *Journal of General Virology, 89*(Pt 8), 1881–1889.

Dong, X. N., & Chen, Y. H. (2007). Marker vaccine strategies and candidate CSFV marker vaccines. *Vaccine, 25*(2), 205–230. http://dx.doi.org/10.1016/j.vaccine.2006.07.033.

Donis, R. O., & Dubovi, E. J. (1987). Characterization of bovine diarrhoea-mucosal disease virus-specific proteins in bovine cells. *Journal of General Virology, 68*, 1597–1605.

Dougherty, W. G., & Semler, B. L. (1993). Expression of virus-encoded proteinases: Functional and structural similarities with cellular enzymes. *Microbiological Reviews, 57*(4), 781–822.

Dräger, C., Beer, M., & Blome, S. (2015). Porcine complement regulatory protein CD46 and heparan sulfates are the major factors for classical swine fever virus attachment in vitro. *Archives of Virology, 160*, 739–746. http://dx.doi.org/10.1007/s00705-014-2313-y.

Dunne, H. W. (1973). Hog cholera (European swine fever). *Advances in Veterinary Science and Comparative Medicine, 17*, 315–359.

Einav, S., Elazar, M., Danieli, T., & Glenn, J. S. (2004). A nucleotide binding motif in hepatitis C virus (HCV) NS4B mediates HCV RNA replication. *Journal of Virology, 78*(20), 11288–11295. http://dx.doi.org/10.1128/JVI.78.20.11288-11295.2004.

El Omari, K., Iourin, O., Harlos, K., Grimes, J. M., & Stuart, D. I. (2013). Structure of a pestivirus envelope glycoprotein E2 clarifies its role in cell entry. *Cell Reports, 3*(1), 30–35. http://dx.doi.org/10.1016/j.celrep.2012.12.001.

Elbers, K., Tautz, N., Becher, P., Rümenapf, T., & Thiel, H. J. (1996). Processing in the Pestivirus E2-NS2 region: Identification of the nonstructural proteins p7 and E2p7. *Journal of Virology, 70*(6), 4131–4135.

Ettema, T. J., Huynen, M. A., de Vos, W. M., & van der Oost, J. (2003). TRASH: A novel metal-binding domain predicted to be involved in heavy-metal sensing, trafficking and resistance. *Trends in Biochemical Sciences, 28*(4), 170–173. http://dx.doi.org/10.1016/S0968-0004(03)00037-9.

Eymann-Hani, R., Leifer, I., McCullough, K. C., Summerfield, A., & Ruggli, N. (2011). Propagation of classical swine fever virus in vitro circumventing heparan sulfate-adaptation. *Journal of Virological Methods, 176*(1–2), 85–95.

Fairbanks, K., Schnackel, J., & Chase, C. C. (2003). Evaluation of a modified live virus type-1a bovine viral diarrhea virus vaccine (Singer strain) against a type-2 (strain 890) challenge. *Veterinary Therapeutics, 4*(1), 24–34.

Fan, Z. C., & Bird, R. C. (2010). The extra 16-amino-acid peptide at C-terminal NS2 of the hypervirulent type-2 bovine viral diarrhea viruses has no effect on viral replication and NS2-3 processing of type-1 virus. *Virus Genes, 41*(2), 218–223. http://dx.doi.org/10.1007/s11262-010-0503-6.

Fernandez-Sainz, I. J., Largo, E., Gladue, D. P., Fletcher, P., O'Donnell, V., Holinka, L. G., et al. (2014). Effect of specific amino acid substitutions in the putative fusion peptide of structural glycoprotein E2 on Classical Swine Fever Virus replication. *Virology, 456–457,* 121–130. http://dx.doi.org/10.1016/j.virol.2014.03.005.

Fetzer, C., Tews, B. A., & Meyers, G. (2005). The carboxy-terminal sequence of the pestivirus glycoprotein E(rns) represents an unusual type of membrane anchor. *Journal of Virology, 79*(18), 11901–11913.

Fiebach, A. R., Guzylack-Piriou, L., Python, S., Summerfield, A., & Ruggli, N. (2011). Classical Swine Fever virus npro limits type I interferon induction in plasmacytoid dendritic cells by interacting with interferon regulatory factor 7. *Journal of Virology, 85*(16), 8002–8011.

Firth, C., Bhat, M., Firth, M. A., Williams, S. H., Frye, M. J., Simmonds, P., et al. (2014). Detection of zoonotic pathogens and characterization of novel viruses carried by commensal Rattus norvegicus in New York City. *MBio. 5*(5). http://dx.doi.org/10.1128/mBio.01933-14, e01933–01914.

Firth, A. E., & Brierley, I. (2012). Non-canonical translation in RNA viruses. *Journal of General Virology, 93*(Pt 7), 1385–1409. http://dx.doi.org/10.1099/vir.0.042499-0.

Fletcher, S. R., Ali, I. K., Kaminski, A., Digard, P., & Jackson, R. J. (2002). The influence of viral coding sequences on pestivirus IRES activity reveals further parallels with translation initiation in prokaryotes. *RNA, 8*(12), 1558–1571.

Fletcher, S. P., & Jackson, R. J. (2002). Pestivirus internal ribosome entry site (IRES) structure and function: Elements in the 5′ untranslated region important for IRES function. *Journal of Virology, 76*(10), 5024–5033.

Foster, T. L., Verow, M., Wozniak, A. L., Bentham, M. J., Thompson, J., Atkins, E., et al. (2011). Resistance mutations define specific antiviral effects for inhibitors of the hepatitis C virus p7 ion channel. *Hepatology, 54*(1), 79–90. http://dx.doi.org/10.1002/hep.24371.

Fray, M. D., Clarke, M. C., Thomas, L. H., McCauley, J. W., & Charleston, B. (1998). Prolonged nasal shedding and viraemia of cytopathogenic bovine virus diarrhoea virus in experimental late-onset mucosal disease. *The Veterinary Record, 143*(22), 608–611.

Frey, C. F., Bauhofer, O., Ruggli, N., Summerfield, A., Hofmann, M. A., & Tratschin, J. D. (2006). Classical swine fever virus replicon particles lacking the Erns gene: A potential marker vaccine for intradermal application. *Veterinary Research, 37*(5), 655–670. http://dx.doi.org/10.1051/vetres:2006028.

Fricke, J., Gunn, M., & Meyers, G. (2001). A family of closely related bovine viral diarrhea virus recombinants identified in an animal suffering from mucosal disease: New insights into the development of a lethal disease in cattle. *Virology, 291*(1), 77–90.

Fricke, J., Voss, C., Thumm, M., & Meyers, G. (2004). Processing of a pestivirus protein by a cellular protease specific for light chain 3 of microtubule-associated proteins. *Journal of Virology, 78*(11), 5900–5912.

Fritzemeier, J., Greiser-Wilke, I., Haas, L., Pituco, E., Moennig, V., & Liess, B. (1995). Experimentally induced "late-onset" mucosal disease—Characterization of the cytopathogenic viruses isolated. *Veterinary Microbiology, 46*, 285–294.

Fritzemeier, J., Haas, L., Liebler, E., Moennig, V., & Greiser-Wilke, I. (1997). The development of early vs. late onset mucosal disease is a consequence of two different pathogenic mechanisms. *Archives of Virology, 142*(7), 1335–1350.

Frolov, I., McBride, M. S., & Rice, C. M. (1998). cis-acting RNA elements required for replication of bovine viral diarrhea virus-hepatitis C virus 5′ nontranslated region chimeras. *RNA, 4*(11), 1418–1435.

Fu, Q., Shi, H., Ren, Y., Guo, F., Ni, W., Qiao, J., et al. (2014). Bovine viral diarrhea virus infection induces autophagy in MDBK cells. *Journal of Microbiology, 52*(7), 619–625. http://dx.doi.org/10.1007/s12275-014-3479-4.

Fu, Q., Shi, H., Shi, M., Meng, L., Bao, H., Zhang, G., et al. (2014). Roles of bovine viral diarrhea virus envelope glycoproteins in inducing autophagy in MDBK cells. *Microbial Pathogenesis, 76*, 61–66. http://dx.doi.org/10.1016/j.micpath.2014.09.011.

Fu, Q., Shi, H., Shi, M., Meng, L., Zhang, H., Ren, Y., et al. (2014). bta-miR-29b attenuates apoptosis by directly targeting caspase-7 and NAIF1 and suppresses bovine viral diarrhea virus replication in MDBK cells. *Canadian Journal of Microbiology, 60*(7), 455–460. http://dx.doi.org/10.1139/cjm-2014-0277.

Fu, Q., Shi, H., Zhang, H., Ren, Y., Guo, F., Qiao, J., et al. (2014). Autophagy during early stages contributes to bovine viral diarrhea virus replication in MDBK cells. *Journal of Basic Microbiology, 54*(10), 1044–1052. http://dx.doi.org/10.1002/jobm.201300750.

Gallei, A., Blome, S., Gilgenbach, S., Tautz, N., Moennig, V., & Becher, P. (2008). Cytopathogenicity of classical Swine Fever virus correlates with attenuation in the natural host. *Journal of Virology, 82*(19), 9717–9729.

Gallei, A., Orlich, M., Thiel, H. J., & Becher, P. (2005). Noncytopathogenic pestivirus strains generated by nonhomologous RNA recombination: Alterations in the NS4A/NS4B coding region. *Journal of Virology, 79*(22), 14261–14270.

Gallei, A., Pankraz, A., Thiel, H. J., & Becher, P. (2004). RNA recombination in vivo in the absence of viral replication. *Journal of Virology, 78*(12), 6271–6281.

Gamlen, T., Richards, K. H., Mankouri, J., Hudson, L., McCauley, J., Harris, M., et al. (2010). Expression of the NS3 protease of cytopathogenic bovine viral diarrhea virus results in the induction of apoptosis but does not block activation of the beta interferon promoter. *Journal of General Virology, 91*(Pt 1), 133–144.

Gil, L. H., Ansari, I. H., Vassilev, V., Liang, D., Lai, V. C., Zhong, W., et al. (2006). The amino-terminal domain of bovine viral diarrhea virus Npro protein is necessary for alpha/beta interferon antagonism. *Journal of Virology, 80*(2), 900–911.

Gladue, D. P., Baker-Bransetter, R., Holinka, L. G., Fernandez-Sainz, I. J., O'Donnell, V., Fletcher, P., et al. (2014). Interaction of CSFV E2 protein with swine host factors as detected by yeast two-hybrid system. *PLoS One, 9*(1), e85324. http://dx.doi.org/10.1371/journal.pone.0085324.

Gladue, D. P., Gavrilov, B. K., Holinka, L. G., Fernandez-Sainz, I. J., Vepkhvadze, N. G., Rogers, K., et al. (2011). Identification of an NTPase motif in classical swine fever virus NS4B protein. *Virology, 411*(1), 41–49.

Gladue, D. P., Holinka, L. G., Fernandez-Sainz, I. J., Prarat, M. V., O'Donell, V., Vepkhvadze, N., et al. (2010). Effects of the interactions of classical swine fever virus Core protein with proteins of the SUMOylation pathway on virulence in swine. *Virology, 407*(1), 129–136. http://dx.doi.org/10.1016/j.virol.2010.07.040.

Gladue, D. P., Holinka, L. G., Fernandez-Sainz, I. J., Prarat, M. V., O'Donnell, V., Vepkhvadze, N. G., et al. (2011). Interaction between core protein of classical swine fever virus with cellular IQGAP1 protein appears essential for virulence in swine. *Virology, 412*(1), 68–74. http://dx.doi.org/10.1016/j.virol.2010.12.060.

Gladue, D. P., Holinka, L. G., Largo, E., Fernandez Sainz, I., Carrillo, C., O'Donnell, V., et al. (2012). Classical swine fever virus p7 protein is a viroporin involved in virulence in swine. *Journal of Virology*, *86*(12), 6778–6791. http://dx.doi.org/10.1128/JVI.00560-12.

Gladue, D. P., O'Donnell, V., Fernandez-Sainz, I. J., Fletcher, P., Baker-Branstetter, R., Holinka, L. G., et al. (2014). Interaction of structural core protein of classical swine fever virus with endoplasmic reticulum-associated degradation pathway protein OS9. *Virology*, *460–461*, 173–179. http://dx.doi.org/10.1016/j.virol.2014.05.008.

Gladue, D. P., Zhu, J., Holinka, L. G., Fernandez-Sainz, I., Carrillo, C., Prarat, M. V., et al. (2010). Patterns of gene expression in swine macrophages infected with classical swine fever virus detected by microarray. *Virus Research*, *151*(1), 10–18. http://dx.doi.org/10.1016/j.virusres.2010.03.007.

Gmyl, A. P., Belousov, E. V., Maslova, S. V., Khitrina, E. V., Chetverin, A. B., & Agol, V. I. (1999). Nonreplicative RNA recombination in poliovirus. *Journal of Virology*, *73*(11), 8958–8965.

Gonzalez, M. E., & Carrasco, L. (2003). Viroporins. *FEBS Letters*, *552*(1), 28–34.

Gorbalenya, A. E., Donchenko, A. P., Koonin, E. V., & Blinov, V. M. (1989). N-terminal domains of putative helicases of flavi- and pestiviruses may be serine proteases. *Nucleic Acids Research*, *17*, 3889–3897.

Gorbalenya, A. E., Koonin, E. V., Donchenko, A. P., & Blinov, V. M. (1989). Two related superfamilies of putative helicases involved in replication, recombination, repair and expression of DNA and RNA genomes. *Nucleic Acids Research*, *17*, 4713–4729.

Gottipati, K., Acholi, S., Ruggli, N., & Choi, K. H. (2014). Autocatalytic activity and substrate specificity of the pestivirus N-terminal protease Npro. *Virology*, *452–453*, 303–309. http://dx.doi.org/10.1016/j.virol.2014.01.026.

Gottipati, K., Ruggli, N., Gerber, M., Tratschin, J. D., Benning, M., Bellamy, H., et al. (2013). The structure of classical swine fever virus N(pro): A novel cysteine Autoprotease and zinc-binding protein involved in subversion of type I interferon induction. *PLoS Pathogens*, *9*(10), e1003704. http://dx.doi.org/10.1371/journal.ppat.1003704.

Grakoui, A., McCourt, D. W., Wychowski, C., Feinstone, S. M., & Rice, C. (1993). A second hepatitis C virus-encoded proteinase. *Proceedings of the National Academy of Sciences of the United States of America*, *90*, 10583–10587.

Grassmann, C. W., Isken, O., & Behrens, S. E. (1999). Assignment of the multifunctional NS3 protein of bovine viral diarrhea virus during RNA replication: An in vivo and in vitro study. *Journal of Virology*, *73*(11), 9196–9205.

Grassmann, C. W., Isken, O., Tautz, N., & Behrens, S. E. (2001). Genetic analysis of the pestivirus nonstructural coding region: Defects in the NS5A unit can be complemented in trans. *Journal of Virology*, *75*(17), 7791–7802.

Grassmann, C. W., Yu, H., Isken, O., & Behrens, S. E. (2005). Hepatitis C virus and the related bovine viral diarrhea virus considerably differ in the functional organization of the 5′ non-translated region: Implications for the viral life cycle. *Virology*, *333*(2), 349–366. http://dx.doi.org/10.1016/j.virol.2005.01.007.

Greiser-Wilke, I., Liebler, E., Haas, L., Liess, B., Pohlenz, J., & Moennig, V. (1993). Distribution of cytopathogenic and noncytopathogenic bovine virus diarrhea virus in tissues from a calf with experimentally induced mucosal disease using antigenic and genetic markers. *Archives of Virology. Supplementum*, *7*, 295–302.

Greiser-Wilke, I., & Moennig, V. (2004). Vaccination against classical swine fever virus: limitations and new strategies. *Animal Health Research Reviews*, *5*(2), 223–226.

Griffin, S. D., Beales, L. P., Clarke, D. S., Worsfold, O., Evans, S. D., Jaeger, J., et al. (2003). The p7 protein of hepatitis C virus forms an ion channel that is blocked by the antiviral drug, Amantadine. *FEBS Letters*, *535*(1–3), 34–38.

The Molecular Biology of Pestiviruses   139

Griffin, S. D., Harvey, R., Clarke, D. S., Barclay, W. S., Harris, M., & Rowlands, D. J. (2004). A conserved basic loop in hepatitis C virus p7 protein is required for amantadine-sensitive ion channel activity in mammalian cells but is dispensable for localization to mitochondria. *Journal of General Virology, 85*(Pt 2), 451–461.
Grooms, D. L. (2004). Reproductive consequences of infection with bovine viral diarrhea virus. *The Veterinary Clinics of North America. Food Animal Practice, 20*(1), 5–19. http://dx.doi.org/10.1016/j.cvfa.2003.11.006.
Grummer, B., Beer, M., Liebler-Tenorio, E., & Greiser-Wilke, I. (2001). Localization of viral proteins in cells infected with bovine viral diarrhoea virus. *Journal of General Virology, 82*(Pt 11), 2597–2605.
Grummer, B., Grotha, S., & Greiser-Wilke, I. (2004). Bovine viral diarrhoea virus is internalized by clathrin-dependent receptor-mediated endocytosis. *Journal of Veterinary Medicine. B, Infectious Disease and Veterinary Public Health, 51*(10), 427–432.
Haller, O., Kochs, G., & Weber, F. (2006). The interferon response circuit: Induction and suppression by pathogenic viruses. *Virology, 344*(1), 119–130. http://dx.doi.org/10.1016/j.virol.2005.09.024.
Hamers, C., Dehan, P., Couvreur, B., Letellier, C., Kerkhofs, P., & Pastoret, P. P. (2001). Diversity among bovine pestiviruses. *The Veterinary Journal, 161*(2), 112–122. http://dx.doi.org/10.1053/tvjl.2000.0504.
Hansen, T. R., Smirnova, N. P., Van Campen, H., Shoemaker, M. L., Ptitsyn, A. A., & Bielefeldt-Ohmann, H. (2010). Maternal and fetal response to fetal persistent infection with bovine viral diarrhea virus. *American Journal of Reproductive Immunology, 64*(4), 295–306. http://dx.doi.org/10.1111/j.1600-0897.2010.00904.x.
Harada, T., Tautz, N., & Thiel, H. J. (2000). E2-p7 region of the bovine viral diarrhea virus polyprotein: Processing and functional studies. *Journal of Virology, 74*(20), 9498–9506.
Harasawa, R., Giangaspero, M., Ibata, G., & Paton, D. J. (2000). Giraffe strain of pestivirus: Its taxonomic status based on the 5′-untranslated region. *Microbiology and Immunology, 44*(11), 915–921.
Hashem, Y., des Georges, A., Dhote, V., Langlois, R., Liao, H. Y., Grassucci, R. A., et al. (2013). Hepatitis-C-virus-like internal ribosome entry sites displace eIF3 to gain access to the 40S subunit. *Nature, 503*(7477), 539–543. http://dx.doi.org/10.1038/nature12658.
Hausmann, Y., Roman-Sosa, G., Thiel, H. J., & Rümenapf, T. (2004). Classical swine fever virus glycoprotein E rns is an endoribonuclease with an unusual base specificity. *Journal of Virology, 78*(10), 5507–5512.
He, L., Zhang, Y. M., Lin, Z., Li, W. W., Wang, J., & Li, H. L. (2012). Classical swine fever virus NS5A protein localizes to endoplasmic reticulum and induces oxidative stress in vascular endothelial cells. *Virus Genes, 45*(2), 274–282. http://dx.doi.org/10.1007/s11262-012-0773-2.
Heimann, M., Roman-Sosa, G., Martoglio, B., Thiel, H. J., & Rümenapf, T. (2006). Core protein of pestiviruses is processed at the C terminus by signal peptide peptidase. *Journal of Virology, 80*(4), 1915–1921.
Hellen, C. U., & de Breyne, S. (2007). A distinct group of hepacivirus/pestivirus-like internal ribosomal entry sites in members of diverse picornavirus genera: Evidence for modular exchange of functional noncoding RNA elements by recombination. *Journal of Virology, 81*(11), 5850–5863. http://dx.doi.org/10.1128/JVI. 02403-06.
Hellen, C. U., & Pestova, T. V. (1999). Translation of hepatitis C virus RNA. *Journal of Viral Hepatitis, 6*(2), 79–87.
Hemelaar, J., Lelyveld, V. S., Kessler, B. M., & Ploegh, H. L. (2003). A single protease, Apg4B, is specific for the autophagy-related ubiquitin-like proteins GATE-16, MAP1-LC3, GABARAP, and Apg8L. *Journal of Biological Chemistry, 278*(51), 51841–51850.

Hengel, H., Koszinowski, U. H., & Conzelmann, K. K. (2005). Viruses know it all: New insights into IFN networks. *Trends in Immunology, 26*(7), 396–401. http://dx.doi.org/10.1016/j.it.2005.05.004.

Hijikata, M., Mizushima, H., Akagi, T., Mori, S., Kakiuchi, N., Kato, N., et al. (1993). Two distinct proteinase activities required for the processing of a putative nonstructural precursor protein of hepatitis C virus. *Journal of Virology, 67*, 4665–4675.

Hilton, L., Moganeradj, K., Zhang, G., Chen, Y. H., Randall, R. E., McCauley, J. W., et al. (2006). The NPro product of bovine viral diarrhea virus inhibits DNA binding by interferon regulatory factor 3 and targets it for proteasomal degradation. *Journal of Virology, 80*(23), 11723–11732.

Hinnebusch, A. G. (2014). The scanning mechanism of eukaryotic translation initiation. *Annual Review of Biochemistry, 83*, 779–812. http://dx.doi.org/10.1146/annurev-biochem-060713-035802.

Hinnebusch, A. G., & Lorsch, J. R. (2012). The mechanism of eukaryotic translation initiation: New insights and challenges. *Cold Spring Harbor Perspectives in Biology, 4*(10). http://dx.doi.org/10.1101/cshperspect.a011544.

Hoff, H. S., & Donis, R. O. (1997). Induction of apoptosis and cleavage of poly(ADP-ribose) polymerase by cytopathic bovine viral diarrhea virus infection. *Virus Research, 49*(1), 101–113.

Howard, C. J. (1990). Immunological responses to bovine virus diarrhoea virus infections. *Revue Scientifique et Technique, 9*(1), 95–103.

Howard, C. J., Brownlie, J., & Clarke, M. C. (1987). Comparison by the neutralisation assay of pairs of non-cytopathogenic and cytopathogenic strains of bovine virus diarrhoea virus isolated from cases of mucosal disease. *Veterinary Microbiology, 19*, 13–21.

Hsu, W. L., Chen, C. L., Huang, S. W., Wu, C. C., Chen, I. H., Nadar, M., et al. (2014). The untranslated regions of classic swine fever virus RNA trigger apoptosis. *PLoS One, 9*(2), e88863. http://dx.doi.org/10.1371/journal.pone.0088863.

Huang, Y. L., Deng, M. C., Wang, F. I., Huang, C. C., & Chang, C. Y. (2014). The challenges of classical swine fever control: Modified live and E2 subunit vaccines. *Virus Research, 179*, 1–11. http://dx.doi.org/10.1016/j.virusres.2013.10.025.

Hulst, M. M., Himes, G., Newbigin, E., & Moormann, R. J. M. (1994). Glycoprotein E2 of classical swine fever virus: Expression in insect cells and identification as a ribonuclease. *Virology, 200*, 558–565.

Hulst, M. M., & Moormann, R. J. (1997). Inhibition of pestivirus infection in cell culture by envelope proteins E(rns) and E2 of classical swine fever virus: E(rns) and E2 interact with different receptors. *Journal of General Virology, 78*(Pt 11), 2779–2787.

Hulst, M. M., & Moormann, R. J. (2001). Erns protein of pestiviruses. *Methods in Enzymology, 342*, 431–440.

Hulst, M. M., Panoto, F. E., Hoekman, A., van Gennip, H. G., & Moormann, R. J. (1998). Inactivation of the RNase activity of glycoprotein E rns of classical swine fever virus results in a cytopathogenic virus. *Journal of Virology, 72*, 151–157.

Hulst, M. M., van Gennip, H. G., & Moormann, R. J. (2000). Passage of classical swine fever virus in cultured swine kidney cells selects virus variants that bind to heparan sulfate due to a single amino acid change in envelope protein E(rns). *Journal of Virology, 74*(20), 9553–9561.

Hulst, M. M., van Gennip, H. G., Vlot, A. C., Schooten, E., de Smit, A. J., & Moormann, R. J. (2001). Interaction of classical swine fever virus with membrane-associated heparan sulfate: Role for virus replication in vivo and virulence. *Journal of Virology, 75*(20), 9585–9595.

Hulst, M. M., Westra, D. F., Wensvoort, G., & Moormann, R. J. (1993). Glycoprotein E1 of hog cholera virus expressed in insect cells protects swine from hog cholera. *Journal of Virology, 67*(9), 5435–5442.

Inaba, Y., Omori, T., & Kumagai, T. (1963). Detection and measurement of non-cytopathogenic strains of virus diarrhea of cattle by the end method (Brief Report). *Archiv für die Gesamte Virusforschung*, *13*, 425–429.

Iqbal, M., Flick-Smith, H., & McCauley, J. W. (2000). Interactions of bovine viral diarrhoea virus glycoprotein E(rns) with cell surface glycosaminoglycans. *Journal of General Virology*, *81*(Pt 2), 451–459.

Iqbal, M., & McCauley, J. W. (2002). Identification of the glycosaminoglycan-binding site on the glycoprotein E(rns) of bovine viral diarrhoea virus by site-directed mutagenesis. *Journal of General Virology*, *83*(Pt 9), 2153–2159.

Iqbal, M., Poole, E., Goodbourn, S., & McCauley, J. W. (2004). Role for bovine viral diarrhea virus Erns glycoprotein in the control of activation of beta interferon by double-stranded RNA. *Journal of Virology*, *78*(1), 136–145.

Irie, M., & Ohgi, K. (2001). Ribonuclease T2. *Methods in Enzymology*, *341*, 42–55.

Isken, O., Grassmann, C. W., Sarisky, R. T., Kann, M., Zhang, S., Grosse, F., et al. (2003). Members of the NF90/NFAR protein group are involved in the life cycle of a positive-strand RNA virus. *EMBO Journal*, *22*(21), 5655–5665. http://dx.doi.org/10.1093/emboj/cdg562.

Isken, O., Grassmann, C. W., Yu, H., & Behrens, S. E. (2004). Complex signals in the genomic 3′ nontranslated region of bovine viral diarrhea virus coordinate translation and replication of the viral RNA. *RNA*, *10*(10), 1637–1652. http://dx.doi.org/10.1261/rna.7290904.

Isken, O., Langerwisch, U., Schönherr, R., Lamp, B., Schröder, K., Duden, R., et al. (2014). Functional characterization of bovine viral diarrhea virus nonstructural protein 5A by reverse genetic analysis and live cell imaging. *Journal of Virology*, *88*(1), 82–98. http://dx.doi.org/10.1128/JVI.01957-13.

Ivanyi-Nagy, R., Lavergne, J. P., Gabus, C., Ficheux, D., & Darlix, J. L. (2008). RNA chaperoning and intrinsic disorder in the core proteins of Flaviviridae. *Nucleic Acids Research*, *36*(3), 712–725. http://dx.doi.org/10.1093/nar/gkm1051.

Jackson, R. J. (2005). Alternative mechanisms of initiating translation of mammalian mRNAs. *Biochemical Society Transactions*, *33*(Pt 6), 1231–1241.

Jenckel, M., Höper, D., Schirrmeier, H., Reimann, I., Goller, K. V., Hoffmann, B., et al. (2014). Mixed triple: Allied viruses in unique recent isolates of highly virulent type 2 bovine viral diarrhea virus detected by deep sequencing. *Journal of Virology*, *88*(12), 6983–6992. http://dx.doi.org/10.1128/JVI.00620-14.

Ji, H., Fraser, C. S., Yu, Y., Leary, J., & Doudna, J. A. (2004). Coordinated assembly of human translation initiation complexes by the hepatitis C virus internal ribosome entry site RNA. *Proceedings of the National Academy of Sciences of the United States of America*, *101*(49), 16990–16995. http://dx.doi.org/10.1073/pnas.0407402101.

Jirasko, V., Montserret, R., Appel, N., Janvier, A., Eustachi, L., Brohm, C., et al. (2008). Structural and functional characterization of nonstructural protein 2 for its role in hepatitis C virus assembly. *Journal of Biological Chemistry*, *283*(42), 28546–28562. http://dx.doi.org/10.1074/jbc.M803981200.

Johns, H. L., Doceul, V., Everett, H., Crooke, H., Charleston, B., & Seago, J. (2010). The classical swine fever virus N-terminal protease N(pro) binds to cellular HAX-1. *Journal of General Virology*, *91*(Pt 11), 2677–2686. http://dx.doi.org/10.1099/vir.0.022897-0.

Johnson, C. M., Perez, D. R., French, R., Merrick, W. C., & Donis, R. O. (2001). The NS5A protein of bovine viral diarrhoea virus interacts with the alpha subunit of translation elongation factor-1. *Journal of General Virology*, *82*(Pt 12), 2935–2943.

Jones, C. T., Murray, C. L., Eastman, D. K., Tassello, J., & Rice, C. M. (2007). Hepatitis C virus p7 and NS2 proteins are essential for production of infectious virus. *Journal of Virology*, *81*(16), 8374–8383. http://dx.doi.org/10.1128/JVI.00690-07.

Jordan, R., Nikolaeva, O. V., Wang, L., Conyers, B., Mehta, A., Dwek, R. A., et al. (2002). Inhibition of host ER glucosidase activity prevents Golgi processing of virion-associated bovine viral diarrhea virus E2 glycoproteins and reduces infectivity of secreted virions. *Virology, 295*(1), 10–19. http://dx.doi.org/10.1006/viro. 2002.1370.

Jordan, R., Wang, L., Graczyk, T. M., Block, T. M., & Romano, P. R. (2002). Replication of a cytopathic strain of bovine viral diarrhea virus activates perk and induces endoplasmic reticulum stress-mediated apoptosis of MDBK cells. *Journal of Virology, 76*(19), 9588–9599.

Jubin, R., Vantuno, N. E., Kieft, J. S., Murray, M. G., Doudna, J. A., Lau, J. Y., et al. (2000). Hepatitis C virus internal ribosome entry site (IRES) stem loop IIId contains a phylogenetically conserved GGG triplet essential for translation and IRES folding. *Journal of Virology, 74*(22), 10430–10437.

Kameyama, K., Sakoda, Y., Matsuno, K., Ito, A., Tajima, M., Nakamura, S., et al. (2008). Cleavage of the NS2-3 protein in the cells of cattle persistently infected with noncytopathogenic bovine viral diarrhea virus. *Microbiology and Immunology, 52*(5), 277–282. http://dx.doi.org/10.1111/j.1348-0421.2008.00013.x.

Kang, K., Guo, K., Tang, Q., Zhang, Y., Wu, J., Li, W., et al. (2012). Interactive cellular proteins related to classical swine fever virus non-structure protein 2 by yeast two-hybrid analysis. *Molecular Biology Reports, 39*(12), 10515–10524. http://dx.doi.org/10.1007/s11033-012-1936-x.

Kao, C. C., Del Vecchio, A. M., & Zong, W. (1999). De novo initiation of RNA synthesis by a recombinant Flaviviridae RNA-dependent RNA polymerase. *Virology, 253*, 1–7.

Kelling, C. L. (2004). Evolution of bovine viral diarrhea virus vaccines. *The Veterinary Clinics of North America. Food Animal Practice, 20*(1), 115–129. http://dx.doi.org/10.1016/j.cvfa.2003.11.001.

Khan, D., Bhat, P., & Das, S. (2014). HCV-like IRESs sequester eIF3: Advantage virus. *Trends in Microbiology, 22*(2), 57–58. http://dx.doi.org/10.1016/j.tim.2013.12.009.

Khan, A. G., Whidby, J., Miller, M. T., Scarborough, H., Zatorski, A. V., Cygan, A., et al. (2014). Structure of the core ectodomain of the hepatitis C virus envelope glycoprotein 2. *Nature, 509*(7500), 381–384. http://dx.doi.org/10.1038/nature13117.

Khatchikian, D., Orlich, M., & Rott, R. (1989). Increased viral pathogenicity after insertion of a 28S ribosomal RNA sequence into the haemagglutinin gene of an influenza virus. *Nature, 340*, 156–157.

King, A. M., McCahon, D., Slade, W. R., & Newman, J. W. (1982). Recombination in RNA. *Cell, 29*(3), 921–928.

Kirkegaard, K., & Baltimore, D. (1986). The mechanism of RNA recombination in poliovirus. *Cell, 47*(3), 433–443.

Kirkland, P. D., Frost, M. J., Finlaison, D. S., King, K. R., Ridpath, J. F., & Gu, X. (2007). Identification of a novel virus in pigs—Bungowannah virus: A possible new species of pestivirus. *Virus Research, 129*(1), 26–34. http://dx.doi.org/10.1016/j.virusres. 2007.05.002.

Koenig, P., Lange, E., Reimann, I., & Beer, M. (2007). CP7_E2alf: A safe and efficient marker vaccine strain for oral immunisation of wild boar against Classical swine fever virus (CSFV). *Vaccine, 25*(17), 3391–3399.

Köhl, W., Zimmer, G., Greiser-Wilke, I., Haas, L., Moennig, V., & Herrler, G. (2004). The surface glycoprotein E2 of bovine viral diarrhoea virus contains an intracellular localization signal. *Journal of General Virology, 85*(Pt 5), 1101–1111.

Kolupaeva, V. G., Pestova, T. V., & Hellen, C. U. (2000a). An enzymatic footprinting analysis of the interaction of 40S ribosomal subunits with the internal ribosomal entry site of hepatitis C virus. *Journal of Virology, 74*(14), 6242–6250.

Kolupaeva, V. G., Pestova, T. V., & Hellen, C. U. (2000b). Ribosomal binding to the internal ribosomal entry site of classical swine fever virus. *RNA*, *6*(12), 1791–1807.

Kolykhalov, A. A., Mihalik, K., Feinstone, S. M., & Rice, C. M. (2000). Hepatitis C virus-encoded enzymatic activities and conserved RNA elements in the 3′ nontranslated region are essential for virus replication in vivo. *Journal of Virology*, *74*(4), 2046–2051.

Kong, L., Giang, E., Nieusma, T., Kadam, R. U., Cogburn, K. E., Hua, Y., et al. (2013). Hepatitis C virus E2 envelope glycoprotein core structure. *Science*, *342*(6162), 1090–1094. http://dx.doi.org/10.1126/science.1243876.

Kosmidou, A., Ahl, R., Thiel, H. J., & Weiland, E. (1995). Differentiation of classical swine fever virus (CSFV) strains using monoclonal antibodies against structural glycoproteins. *Veterinary Microbiology*, *47*, 111–118.

Kosmidou, A., Büttner, M., & Meyers, G. (1998). Isolation and characterization of cytopathogenic classical swine fever virus (CSFV). *Archives of Virology*, *143*(7), 1295–1309.

Krey, T., Bontems, F., Vonrhein, C., Vaney, M. C., Bricogne, G., Rümenapf, T., et al. (2012). Crystal structure of the pestivirus envelope glycoprotein E(rns) and mechanistic analysis of its ribonuclease activity. *Structure*, *20*(5), 862–873. http://dx.doi.org/10.1016/j.str.2012.03.018.

Krey, T., Himmelreich, A., Heimann, M., Menge, C., Thiel, H. J., Maurer, K., et al. (2006). Function of bovine CD46 as a cellular receptor for bovine viral diarrhea virus is determined by complement control protein 1. *Journal of Virology*, *80*(8), 3912–3922.

Krey, T., Thiel, H. J., & Rümenapf, T. (2005). Acid-resistant bovine pestivirus requires activation for pH-triggered fusion during entry. *Journal of Virology*, *79*(7), 4191–4200.

Kumagai, T., Shimizu, T., & Matumoto, M. (1958). Detection of hog cholera virus by its effect on Newcastle disease virus in swine tissue culture. *Science*, *128*(3320), 366.

Kümmerer, B. M., & Meyers, G. (2000). Correlation between point mutations in NS2 and the viability and cytopathogenicity of bovine viral diarrhea virus strain Oregon analyzed with an infectious cDNA clone. *Journal of Virology*, *74*(1), 390–400.

Kümmerer, B. M., Stoll, D., & Meyers, G. (1998). Bovine viral diarrhea virus strain Oregon: A novel mechanism for processing of NS2-3 based on point mutations. *Journal of Virology*, *72*, 4127–4138.

Kupfermann, H., Thiel, H. J., Dubovi, E. J., & Meyers, G. (1996). Bovine viral diarrhea virus: Characterization of a cytopathogenic defective interfering particle with two internal deletions. *Journal of Virology*, *70*(11), 8175–8181.

La Rocca, S. A., Herbert, R. J., Crooke, H., Drew, T. W., Wileman, T. E., & Powell, P. P. (2005). Loss of interferon regulatory factor 3 in cells infected with classical swine fever virus involves the N-terminal protease Npro. *Journal of Virology*, *79*(11), 7239–7247.

Lackner, T., Müller, A., König, M., Thiel, H. J., & Tautz, N. (2005). Persistence of bovine viral diarrhea virus is determined by a cellular cofactor of a viral autoprotease. *Journal of Virology*, *79*(15), 9746–9755.

Lackner, T., Müller, A., Pankraz, A., Becher, P., Thiel, H. J., Gorbalenya, A. E., et al. (2004). Temporal modulation of an autoprotease is crucial for replication and pathogenicity of an RNA virus. *Journal of Virology*, *78*(19), 10765–10775.

Lackner, T., Thiel, H. J., & Tautz, N. (2006). Dissection of a viral autoprotease elucidates a function of a cellular chaperone in proteolysis. *Proceedings of the National academy of Sciences of the United States of America*, *103*(5), 1510–1515.

Lai, V. C., Kao, C. C., Ferrari, E., Park, J., Uss, A. S., Wright-Minogue, J., et al. (1999). Mutational analysis of bovine viral diarrhea virus RNA-dependent RNA polymerase. *Journal of Virology*, *73*(12), 10129–10136.

Lambot, M., Hanon, E., Lecomte, C., Hamers, C., Letesson, J. J., & Pastoret, P. P. (1998). Bovine viral diarrhoea virus induces apoptosis in blood mononuclear cells by a mechanism largely dependent on monocytes. *Journal of General Virology*, *79*(Pt 7), 1745–1749.

Lamp, B., Riedel, C., Roman-Sosa, G., Heimann, M., Jacobi, S., Becher, P., et al. (2011). Biosynthesis of classical swine fever virus nonstructural proteins. *Journal of Virology*, *85*(7), 3607–3620.

Lamp, B., Riedel, C., Wentz, E., Tortorici, M. A., & Rümenapf, T. (2013). Autocatalytic cleavage within classical swine fever virus NS3 leads to a functional separation of protease and helicase. *Journal of Virology*, *87*(21), 11872–11883. http://dx.doi.org/10.1128/JVI.00754-13.

Lancaster, A. M., Jan, E., & Sarnow, P. (2006). Initiation factor-independent translation mediated by the hepatitis C virus internal ribosome entry site. *RNA*, *12*(5), 894–902.

Langedijk, J. P., van Veelen, P. A., Schaaper, W. M., de Ru, A. H., Meloen, R. H., & Hulst, M. M. (2002). A structural model of pestivirus E(rns) based on disulfide bond connectivity and homology modeling reveals an extremely rare vicinal disulfide. *Journal of Virology*, *76*(20), 10383–10392.

Largo, E., Gladue, D. P., Huarte, N., Borca, M. V., & Nieva, J. L. (2014). Pore-forming activity of pestivirus p7 in a minimal model system supports genus-specific viroporin function. *Antiviral Research*, *101*, 30–36. http://dx.doi.org/10.1016/j.antiviral.2013.10.015.

Lattwein, E., Klemens, O., Schwindt, S., Becher, P., & Tautz, N. (2012). Pestivirus virion morphogenesis in the absence of uncleaved nonstructural protein 2-3. *Journal of Virology*, *86*(1), 427–437. http://dx.doi.org/10.1128/JVI.06133-11.

Laude, H., & Gelfi, J. (1979). Properties of border disease virus as studied in a sheep cell line. *Archives of Virology*, *62*, 342–346.

Leifer, I., Lange, E., Reimann, I., Blome, S., Juanola, S., Duran, J. P., et al. (2009). Modified live marker vaccine candidate CP7_E2alf provides early onset of protection against lethal challenge infection with classical swine fever virus after both intramuscular and oral immunization. *Vaccine*, *27*, 6522–6529.

Leifer, I., Ruggli, N., & Blome, S. (2013). Approaches to define the viral genetic basis of classical swine fever virus virulence. *Virology*, *438*(2), 51–55. http://dx.doi.org/10.1016/j.virol.2013.01.013.

Li, Y., Wang, J., Kanai, R., & Modis, Y. (2013). Crystal structure of glycoprotein E2 from bovine viral diarrhea virus. *Proceedings of the National Academy of Sciences of the United States of America*, *110*(17), 6805–6810. http://dx.doi.org/10.1073/pnas.1300524110.

Li, X., Wang, L., Zhao, D., Zhang, G., Luo, J., Deng, R., et al. (2011). Identification of host cell binding peptide from an overlapping peptide library for inhibition of classical swine fever virus infection. *Virus Genes*, *43*(1), 33–40. http://dx.doi.org/10.1007/s11262-011-0595-7.

Liang, D., Chen, L., Ansari, I. H., Gil, L. H., Topliff, C. L., Kelling, C. L., et al. (2009). A replicon trans-packaging system reveals the requirement of nonstructural proteins for the assembly of bovine viral diarrhea virus (BVDV) virion. *Virology*, *387*(2), 331–340. http://dx.doi.org/10.1016/j.virol.2009.02.019.

Liang, D., Sainz, I. F., Ansari, I. H., Gil, L. H., Vassilev, V., & Donis, R. O. (2003). The envelope glycoprotein E2 is a determinant of cell culture tropism in ruminant pestiviruses. *Journal of General Virology*, *84*(Pt 5), 1269–1274.

Liang, R., van den Hurk, J. V., Babiuk, L. A., & van Drunen Littel-van den Hurk, S. (2006). Priming with DNA encoding E2 and boosting with E2 protein formulated with CpG oligodeoxynucleotides induces strong immune responses and protection from Bovine viral diarrhea virus in cattle. *Journal of General Virology*, *87*(Pt 10), 2971–2982. http://dx.doi.org/10.1099/vir.0.81737-0.

Liang, R., van den Hurk, J. V., Landi, A., Lawman, Z., Deregt, D., Townsend, H., et al. (2008). DNA prime protein boost strategies protect cattle from bovine viral diarrhea virus type 2 challenge. *Journal of General Virology*, *89*(Pt 2), 453–466. http://dx.doi.org/10.1099/vir.0.83251-0.

Liebler, E. M., Waschbusch, J., Pohlenz, J. F., Moennig, V., & Liess, B. (1991). Distribution of antigen of noncytopathogenic and cytopathogenic bovine virus diarrhea virus biotypes in the intestinal tract of calves following experimental production of mucosal disease. *Archives of Virology. Supplementum*, *3*, 109–124.

Liebler-Tenorio, E. M., Greiser-Wilke, I., & Pohlenz, J. F. (1997). Organ and tissue distribution of the antigen of the cytopathogenic bovine viral diarrhea virus in the early and advanced phase of experimental mucosal disease. *Archives of Virology*, *142*, 1613–1634.

Liebler-Tenorio, E. M., Lanwehr, A., Greiser-Wilke, I., Loehr, B. I., & Pohlenz, J. F. (2000). Comparative investigation of tissue alterations and distribution of BVD-viral antigen in cattle with early onset versus late onset mucosal disease. *Veterinary Microbiology*, *77*, 163–174.

Liebler-Tenorio, E. M., Ridpath, J. E., & Neill, J. D. (2002). Distribution of viral antigen and development of lesions after experimental infection with highly virulent bovine viral diarrhea virus type 2 in calves. *American Journal of Veterinary Research*, *63*(11), 1575–1584.

Liebler-Tenorio, E. M., Ridpath, J. F., & Neill, J. D. (2003a). Distribution of viral antigen and development of lesions after experimental infection of calves with a BVDV 2 strain of low virulence. *Journal of Veterinary Diagnostic Investigation*, *15*(3), 221–232.

Liebler-Tenorio, E. M., Ridpath, J. F., & Neill, J. D. (2003b). Lesions and tissue distribution of viral antigen in severe acute versus subclinical acute infection with BVDV2. *Biologicals*, *31*(2), 119–122.

Liebler-Tenorio, E. M., Ridpath, J. E., & Neill, J. D. (2004). Distribution of viral antigen and tissue lesions in persistent and acute infection with the homologous strain of noncytopathic bovine viral diarrhea virus. *Journal of Veterinary Diagnostic Investigation*, *16*(5), 388–396.

Liess, B., Frey, H. R., Kittsteiner, H., Baumann, F., & Neumann, W. (1974). Bovine mucosal disease, an immunobiological explainable late stage of BVD-MD virus infection with criteria of a "slow virus infection" *Deutsche Tierärztliche Wochenschrift*, *81*(20), 481–487.

Lin, Z., Liang, W., Kang, K., Li, H., Cao, Z., & Zhang, Y. (2014). Classical swine fever virus and p7 protein induce secretion of IL-1beta in macrophages. *Journal of General Virology*, *95*(Pt 12), 2693–2699. http://dx.doi.org/10.1099/vir.0.068502-0.

Lin, C., Lindenbach, B. D., Pragal, B. M., McCourt, D. W., & Rice, C. M. (1994). Processing in the Hepatitis C Virus E2-NS2 region: Identification of p7 and two distinct E2- specific products with different C termini. *Journal of Virology*, *68*, 5063–5073.

Lindberg, A. L. (2003). Bovine viral diarrhoea virus infections and its control. A review. *The Veterinary Quarterly*, *25*(1), 1–16. http://dx.doi.org/10.1080/01652176.2003.9695140.

Lindenbach, B. D., Evans, M. J., Syder, A. J., Wolk, B., Tellinghuisen, T. L., Liu, C. C., et al. (2005). Complete replication of hepatitis C virus in cell culture. *Science*, *309*(5734), 623–626. http://dx.doi.org/10.1126/science.1114016.

Lindenbach, B. D., Murray, C. L., Thiel, H. J., & Rice, C. M. (2013). Flaviviridae. In D. M. Knipe & P. M. Howley (Eds.), *Vol. 6*. *Fields virology* (pp. 712–746). Philadelphia: Lippincott Williams & Wilkins.

Lindenbach, B. D., & Rice, C. M. (2013). The ins and outs of hepatitis C virus entry and assembly. *Nature Reviews. Microbiology*, *11*(10), 688–700. http://dx.doi.org/10.1038/nrmicro3098.

Liu, L., Xia, H., Wahlberg, N., Belak, S., & Baule, C. (2009). Phylogeny, classification and evolutionary insights into pestiviruses. *Virology*, *385*(2), 351–357. http://dx.doi.org/10.1016/j.virol.2008.12.004.

Locker, N., Easton, L. E., & Lukavsky, P. J. (2007). HCV and CSFV IRES domain II mediate eIF2 release during 80S ribosome assembly. *The EMBO Journal*, *26*(3), 795–805. http://dx.doi.org/10.1038/sj.emboj.7601549.

Lohmann, V., Overton, H., & Bartenschlager, R. (1999). Selective stimulation of hepatitis C virus and pestivirus NS5B RNA polymerase activity by GTP. *Journal of Biological Chemistry*, *274*(16), 10807–10815.

Lopez-Lastra, M., Ramdohr, P., Letelier, A., Vallejos, M., Vera-Otarola, J., & Valiente-Echeverria, F. (2010). Translation initiation of viral mRNAs. *Reviews in Medical Virology*, *20*(3), 177–195. http://dx.doi.org/10.1002/rmv.649.

Lorenz, I. C., Marcotrigiano, J., Dentzer, T. G., & Rice, C. M. (2006). Structure of the catalytic domain of the hepatitis C virus NS2-3 protease. *Nature*, *442*(7104), 831–835. http://dx.doi.org/10.1038/nature04975.

Loy, J. D., Gander, J., Mogler, M., Vander Veen, R., Ridpath, J., Harris, D. H., et al. (2013). Development and evaluation of a replicon particle vaccine expressing the E2 glycoprotein of bovine viral diarrhea virus (BVDV) in cattle. *Virology Journal*, *10*, 35. http://dx.doi.org/10.1186/1743-422x-10-35.

Lukavsky, P. J., Otto, G. A., Lancaster, A. M., Sarnow, P., & Puglisi, J. D. (2000). Structures of two RNA domains essential for hepatitis C virus internal ribosome entry site function. *Natural Structural Biology*, *7*(12), 1105–1110.

Lunardi, M., Headley, S. A., Lisboa, J. A., Amude, A. M., & Alfieri, A. A. (2008). Outbreak of acute bovine viral diarrhea in Brazilian beef cattle: Clinicopathological findings and molecular characterization of a wild-type BVDV strain subtype 1b. *Research in Veterinary Science*, *85*(3), 599–604. http://dx.doi.org/10.1016/j.rvsc.2008.01.002.

Luo, Y., Li, S., Sun, Y., & Qiu, H. J. (2014). Classical swine fever in China: A minireview. *Veterinary Microbiology*, *172*(1–2), 1–6. http://dx.doi.org/10.1016/j.vetmic.2014.04.004.

Luo, Y., Yuan, Y., Ankenbauer, R. G., Nelson, L. D., Witte, S. B., Jackson, J. A., et al. (2012). Construction of chimeric bovine viral diarrhea viruses containing glycoprotein E rns of heterologous pestiviruses and evaluation of the chimeras as potential marker vaccines against BVDV. *Vaccine*, *30*(26), 3843–3848. http://dx.doi.org/10.1016/j.vaccine.2012.04.016.

Luscombe, C. A., Huang, Z., Murray, M. G., Miller, M., Wilkinson, J., & Ewart, G. D. (2010). A novel Hepatitis C virus p7 ion channel inhibitor, BIT225, inhibits bovine viral diarrhea virus in vitro and shows synergism with recombinant interferon-alpha-2b and nucleoside analogues. *Antiviral Research*, *86*(2), 144–153. http://dx.doi.org/10.1016/j.antiviral.2010.02.312.

Luzzago, C., Lauzi, S., Ebranati, E., Giammarioli, M., Moreno, A., Cannella, V., et al. (2014). Extended genetic diversity of bovine viral diarrhea virus and frequency of genotypes and subtypes in cattle in Italy between 1995 and 2013. *BioMed Research International*, *2014*, 147145. http://dx.doi.org/10.1155/2014/147145.

Ma, Y., Yates, J., Liang, Y., Lemon, S. M., & Yi, M. (2008). NS3 helicase domains involved in infectious intracellular hepatitis C virus particle assembly. *Journal of Virology*, *82*(15), 7624–7639. http://dx.doi.org/10.1128/JVI.00724-08.

Macovei, A., Zitzmann, N., Lazar, C., Dwek, R. A., & Branza-Nichita, N. (2006). Brefeldin A inhibits pestivirus release from infected cells, without affecting its assembly and infectivity. *Biochemical and Biophysical Research Communications*, *346*(3), 1083–1090. http://dx.doi.org/10.1016/j.bbrc.2006.06.023.

Magkouras, I., Mätzener, P., Rümenapf, T., Peterhans, E., & Schweizer, M. (2008). RNase-dependent inhibition of extracellular, but not intracellular, dsRNA-induced interferon synthesis by Erns of pestiviruses. *Journal of General Virology*, *89*(Pt 10), 2501–2506.

Makoschey, B., Becher, P., Janssen, M. G., Orlich, M., Thiel, H. J., & Lutticken, D. (2004). Bovine viral diarrhea virus with deletions in the 5′-nontranslated region: Reduction of replication in calves and induction of protective immunity. *Vaccine*, *22*(25–26), 3285–3294.

Makoschey, B., Janssen, M. G., Vrijenhoek, M. P., Korsten, J. H., & Marel, P. (2001). An inactivated bovine virus diarrhoea virus (BVDV) type 1 vaccine affords clinical protection against BVDV type 2. *Vaccine*, *19*(23–24), 3261–3268.

Martin, G. S. (1970). Rous sarcoma virus: A function required for the maintenance of the transformed state. *Nature*, *227*(5262), 1021–1023.

Mätzener, P., Magkouras, I., Rümenapf, T., Peterhans, E., & Schweizer, M. (2009). The viral RNase E(rns) prevents IFN type-I triggering by pestiviral single- and double-stranded RNAs. *Virus Research*, *140*(1–2), 15–23.

Maurer, K., Krey, T., Moennig, V., Thiel, H. J., & Rümenapf, T. (2004). CD46 is a cellular receptor for bovine viral diarrhea virus. *Journal of Virology*, *78*(4), 1792–1799.

Maurer, R., Stettler, P., Ruggli, N., Hofmann, M. A., & Tratschin, J. D. (2005). Oronasal vaccination with classical swine fever virus (CSFV) replicon particles with either partial or complete deletion of the E2 gene induces partial protection against lethal challenge with highly virulent CSFV. *Vaccine*, *23*(25), 3318–3328.

Mayer, D., Hofmann, M. A., & Tratschin, J. D. (2004). Attenuation of classical swine fever virus by deletion of the viral N(pro) gene. *Vaccine*, *22*(3–4), 317–328.

Mayer, D., Thayer, T. M., Hofmann, M. A., & Tratschin, J. D. (2003). Establishment and characterisation of two cDNA-derived strains of classical swine fever virus, one highly virulent and one avirulent. *Virus Research*, *98*(2), 105–116.

McClure, M. A., & Perrault, J. (1985). Poliovirus genome RNA hybridizes specifically to higher eukaryotic rRNAs. *Nucleic Acids Research*, *13*(19), 6797–6816.

McClurkin, A. W., Coria, M. F., & Bolin, S. R. (1985). Isolation of cytopathic and non-cytopathic bovine viral diarrhea virus from the spleen of cattle acutely and chronically affected with bovine viral diarrhea. *Journal of the American Veterinary Medical Association*, *186*, 568–569.

Mendez, E., Ruggli, N., Collett, M. S., & Rice, C. M. (1998). Infectious bovine viral diarrhea virus (strain NADL) RNA from stable cDNA clones: A cellular insert determines NS3 production and viral cytopathogenicity. *Journal of Virology*, *72*, 4737–4745.

Meyer, C., Von Freyburg, M., Elbers, K., & Meyers, G. (2002). Recovery of virulent and RNase-negative attenuated type 2 bovine viral diarrhea viruses from infectious cDNA clones. *Journal of Virology*, *76*(16), 8494–8503.

Meyers, G., Ege, A., Fetzer, C., von Freyburg, M., Elbers, K., Carr, V., et al. (2007). Bovine viral diarrhoea virus: Prevention of persistent foetal infection by a combination of two mutations affecting the Erns RNase and the Npro protease. *Journal of Virology*, *81*(7), 3327–3338.

Meyers, G., Rümenapf, T., & Thiel, H. J. (1989a). Molecular cloning and nucleotide sequence of the genome of hog cholera virus. *Virology*, *171*, 555–567.

Meyers, G., Rümenapf, T., & Thiel, H. J. (1989b). Ubiquitin in a togavirus. *Nature*, *341*, 491.

Meyers, G., Rümenapf, T., & Thiel, H. J. (1990). Insertion of ubiquitin-coding sequence identified in the RNA genome of a Togavirus. In M. A. Brinton & F. X. Heinz (Eds.), *New aspects of positive strand RNA viruses* (pp. 25–29).Washington, DC: American Society for Microbiology, Reprinted from: Not in file.

Meyers, G., Saalmüller, A., & Büttner, M. (1999). Mutations abrogating the RNase activity in glycoprotein e(rns) of the pestivirus classical swine fever virus lead to virus attenuation. *Journal of Virology*, *73*(12), 10224–10235.

Meyers, G., Tautz, N., Becher, P., Thiel, H. J., & Kümmerer, B. (1996). Recovery of cytopathogenic and noncytopathogenic bovine viral diarrhea viruses from cDNA constructs. *Journal of Virology*, *70*(12), 8606–8613.

Meyers, G., Tautz, N., Dubovi, E. J., & Thiel, H. J. (1991). Viral cytopathogenicity correlated with integration of ubiquitin-coding sequences. *Virology*, *180*, 602–616.

Meyers, G., Tautz, N., Stark, R., Brownlie, J., Dubovi, E. J., Collett, M. S., et al. (1992). Rearrangement of viral sequences in cytopathogenic pestiviruses. *Virology, 191,* 368–386.

Meyers, G., & Thiel, H. J. (1995). Cytopathogenicity of classical swine fever virus caused by defective interfering particles. *Journal of Virology, 69*(6), 3683–3689.

Meyers, G., & Thiel, H. J. (1996). Molecular characterization of pestiviruses. *Advances in Virus Research, 47,* 53–117.

Meyling, A., Houe, H., & Jensen, A. M. (1990). Epidemiology of bovine virus diarrhoea virus. *Revue Scientifique et Technique, 9*(1), 75–93.

Mittelholzer, C., Moser, C., Tratschin, J. D., & Hofmann, M. A. (1997). Generation of cytopathogenic subgenomic RNA of classical swine fever virus in persistently infected porcine cell lines. *Virus Research, 51*(2), 125–137.

Moennig, V. (2000). Introduction to classical swine fever: Virus, disease and control policy. *Veterinary Microbiology, 73*(2–3), 93–102.

Moennig, V., Floegel-Niesmann, G., & Greiser-Wilke, I. (2003). Clinical signs and epidemiology of classical swine fever: A review of new knowledge. *The Veterinary Journal, 165*(1), 11–20.

Moennig, V., Frey, H. R., Liebler, E., Polenz, P., & Liess, B. (1990). Reproduction of mucosal disease with cytopathogenic bovine viral diarrhoea virus selected in vitro. *The Veterinary Record, 127,* 200–203.

Moennig, V., Greiser-Wilke, I., Frey, H. R., Haas, L., Liebler, E., Pohlenz, J., et al. (1993). Prolonged persistence of cytopathogenic bovine viral diarrhea virus (BVDV) in a persistently viremic cattle. *Zentralblatt für Veterinärmedizin Reihe B, 40*(5), 371–377.

Moennig, V., Houe, H., & Lindberg, A. (2005). BVD control in Europe: Current status and perspectives. *Animal Health Research Reviews, 6*(1), 63–74.

Moennig, V., & Plagemann, P. G. W. (1992). The pestiviruses. *Advances in Virus Research, 41,* 53–98.

Moes, L., & Wirth, M. (2007). The internal initiation of translation in bovine viral diarrhea virus RNA depends on the presence of an RNA pseudoknot upstream of the initiation codon. *Virology Journal, 4,* 124. http://dx.doi.org/10.1186/1743-422X-4-124.

Mohamed, Y. M., Bangphoomi, N., Yamane, D., Suda, Y., Kato, K., Horimoto, T., et al. (2014). Physical interaction between bovine viral diarrhea virus nonstructural protein 4A and adenosine deaminase acting on RNA (ADAR). *Archives of Virology, 159*(7), 1735–1741. http://dx.doi.org/10.1007/s00705-014-1997-3.

Monroe, S. S., & Schlesinger, S. (1983). RNAs from two independently isolated defective interfering particles of Sindbis virus contain a cellular tRNA sequence at their 5′ ends. *Proceedings of the National Academy of Sciences of the United States of America, 80*(11), 3279–3283.

Moormann, R. J. M., Warmerdam, P. A. M., Van der Meer, B., Schaaper, W. M. M., Wensvoort, G., & Hulst, M. M. (1990). Molecular cloning and nucleotide sequence of hog cholera virus strain brescia and mapping of the genomic region encoding envelope glycoprotein E1. *Virology, 177,* 184–198.

Moulin, H. R., Seuberlich, T., Bauhofer, O., Bennett, L. C., Tratschin, J. D., Hofmann, M. A., et al. (2007). Nonstructural proteins NS2-3 and NS4A of classical swine fever virus: Essential features for infectious particle formation. *Virology, 365*(2), 376–389.

Müller, A., Rinck, G., Thiel, H. J., & Tautz, N. (2003). Cell-derived sequences in the N-terminal region of the polyprotein of a cytopathogenic pestivirus. *Journal of Virology, 77*(19), 10663–10669.

Murray, C. L., Jones, C. T., & Rice, C. M. (2008). Architects of assembly: Roles of Flaviviridae non-structural proteins in virion morphogenesis. *Nature Reviews. Microbiology, 6*(9), 699–708. http://dx.doi.org/10.1038/nrmicro1928.

Murray, C. L., Marcotrigiano, J., & Rice, C. M. (2008). Bovine viral diarrhea virus core is an intrinsically disordered protein that binds RNA. *Journal of Virology, 82*(3), 1294–1304.

Myers, T. M., Kolupaeva, V. G., Mendez, E., Baginski, S. G., Frolov, I., Hellen, C. U., et al. (2001). Efficient translation initiation is required for replication of bovine viral diarrhea virus subgenomic replicons. *Journal of Virology, 75*(9), 4226–4238.

Nadar, M., Chan, M. Y., Huang, S. W., Huang, C. C., Tseng, J. T., & Tsai, C. H. (2011). HuR binding to AU-rich elements present in the 3' untranslated region of Classical swine fever virus. *Virology Journal, 8,* 340. http://dx.doi.org/10.1186/1743-422X-8-340.

Nagai, M., Sakoda, Y., Mori, M., Hayashi, M., Kida, H., & Akashi, H. (2003). Insertion of cellular sequence and RNA recombination in the structural protein coding region of cytopathogenic bovine viral diarrhoea virus. *Journal of General Virology, 84*(Pt 2), 447–452.

Nagy, P. D., & Bujarski, J. J. (1997). Engineering of homologous recombination hotspots with AU-rich sequences in Brome Mosaic virus. *Journal of Virology, 71*(5), 3799–3810.

Nettleton, P. F., Gilmour, J. S., Herring, J. A., & Sinclair, J. A. (1992). The production and survival of lambs persistently infected with border disease virus. *Comparative Immunology, Microbiology and Infectious Diseases, 15,* 179–188.

Nieva, J. L., Madan, V., & Carrasco, L. (2012). Viroporins: Structure and biological functions. *Nature Reviews. Microbiology, 10*(8), 563–574. http://dx.doi.org/10.1038/nrmicro2820.

Nishida, T., Kaneko, F., Kitagawa, M., & Yasuda, H. (2001). Characterization of a novel mammalian SUMO-1/Smt3-specific isopeptidase, a homologue of rat axam, which is an axin-binding protein promoting beta-catenin degradation. *Journal of Biological Chemistry, 276*(42), 39060–39066. http://dx.doi.org/10.1074/jbc.M103955200.

Pan, C. H., Jong, M. H., Huang, T. S., Liu, H. F., Lin, S. Y., & Lai, S. S. (2005). Phylogenetic analysis of classical swine fever virus in Taiwan. *Archives of Virology, 150*(6), 1101–1119. http://dx.doi.org/10.1007/s00705-004-0485-6.

Pankraz, A., Preis, S., Thiel, H. J., Gallei, A., & Becher, P. (2009). A single point mutation in nonstructural protein NS2 of bovine viral diarrhea virus results in temperature-sensitive attenuation of viral cytopathogenicity. *Journal of Virology, 83*(23), 12415–12423.

Pankraz, A., Thiel, H. J., & Becher, P. (2005). Essential and nonessential elements in the 3' nontranslated region of Bovine viral diarrhea virus. *Journal of Virology, 79*(14), 9119–9127.

Passler, T., & Walz, P. H. (2010). Bovine viral diarrhea virus infections in heterologous species. *Animal Health Research Reviews, 11*(2), 191–205. http://dx.doi.org/10.1017/s1466252309990065.

Passler, T., Walz, H. L., Ditchkoff, S. S., van Santen, E., Brock, K. V., & Walz, P. H. (2012). Distribution of bovine viral diarrhoea virus antigen in persistently infected white-tailed deer (Odocoileus virginianus). *Journal of Comparative Pathology, 147*(4), 533–541. http://dx.doi.org/10.1016/j.jcpa.2012.02.008.

Patkar, C. G., & Kuhn, R. J. (2008). Yellow Fever virus NS3 plays an essential role in virus assembly independent of its known enzymatic functions. *Journal of Virology, 82*(7), 3342–3352. http://dx.doi.org/10.1128/JVI.02447-07.

Paton, D. J., Lowings, J. P., & Barrett, A. D. T. (1990). Epitope mapping of the gp53 envelope protein of bovine viral diarrhea virus. *Virology, 190,* 763–772.

Paton, D. J., Lowings, J. P., & Barrett, A. D. (1992). Epitope mapping of the gp53 envelope protein of bovine viral diarrhea virus. *Virology, 190*(2), 763–772.

Paton, D. J., McGoldrick, A., Greiser-Wilke, I., Parchariyanon, S., Song, J. Y., Liou, P. P., et al. (2000). Genetic typing of classical swine fever virus. *Veterinary Microbiology, 73*(2–3), 137–157.

Paton, D. J., Sharp, G., & Ibata, G. (1999). Foetal cross-protection experiments between type 1 and type 2 bovine viral diarrhoea virus in pregnant ewes. *Veterinary Microbiology, 64*(2–3), 185–196.

Pei, J., Zhao, M., Ye, Z., Gou, H., Wang, J., Yi, L., et al. (2014). Autophagy enhances the replication of classical swine fever virus in vitro. *Autophagy, 10*(1), 93–110. http://dx.doi.org/10.4161/auto.26843.

Perler, L., Schweizer, M., Jungi, T. W., & Peterhans, E. (2000). Bovine viral diarrhoea virus and bovine herpesvirus-1 prime uninfected macrophages for lipopolysaccharide-triggered apoptosis by interferon-dependent and -independent pathways. *Journal of General Virology, 81*(Pt 4), 881–887.

Pestova, T. V., de Breyne, S., Pisarev, A. V., Abaeva, I. S., & Hellen, C. U. (2008). eIF2-dependent and eIF2-independent modes of initiation on the CSFV IRES: A common role of domain II. *The EMBO Journal, 27*(7), 1060–1072. http://dx.doi.org/10.1038/emboj.2008.49.

Pestova, T. V., Shatsky, I. N., Fletcher, S. P., Jackson, R. J., & Hellen, C. U. (1998). A prokaryotic-like mode of cytoplasmic eukaryotic ribosome binding to the initiation codon during internal translation initiation of hepatitis C and classical swine fever virus RNAs. *Genes and Development, 12*(1), 67–83.

Peterhans, E., & Schweizer, M. (2010). Pestiviruses: How to outmaneuver your hosts. *Veterinary Microbiology, 142*(1–2), 18–25. http://dx.doi.org/10.1016/j.vetmic.2009.09.038.

Pichlmair, A., Kandasamy, K., Alvisi, G., Mulhern, O., Sacco, R., Habjan, M., et al. (2012). Viral immune modulators perturb the human molecular network by common and unique strategies. *Nature, 487*(7408), 486–490. http://dx.doi.org/10.1038/nature11289.

Pichlmair, A., & Reis e Sousa, C. (2007). Innate recognition of viruses. *Immunity, 27*(3), 370–383. http://dx.doi.org/10.1016/j.immuni.2007.08.012.

Pisarev, A. V., Shirokikh, N. E., & Hellen, C. U. (2005). Translation initiation by factor-independent binding of eukaryotic ribosomes to internal ribosomal entry sites. *Comptes Rendus Biologies, 328*(7), 589–605.

Pocock, D. H., Howard, C. J., Clarke, M. C., & Brownlie, J. (1987). Variation in the intra-cellular polypeptide profiles from different isolates of bovine viral diarrhea virus. *Archives of Virology, 94*, 43–53.

Poole, T. L., Wang, C., Popp, R. A., Potgieter, L. N. D., Siddiqui, A., & Collett, M. S. (1995). Pestivirus translation occurs by internal ribosome entry. *Virology, 206*, 750–754.

Postel, A., Schmeiser, S., Bernau, J., Meindl-Boehmer, A., Pridotkas, G., Dirbakova, Z., et al. (2012). Improved strategy for phylogenetic analysis of classical swine fever virus based on full-length E2 encoding sequences. *Veterinary Research, 43*, 50. http://dx.doi.org/10.1186/1297-9716-43-50.

Potgieter, L. N. (1995). Immunology of bovine viral diarrhea virus. *The Veterinary Clinics of North America. Food Animal Practice, 11*(3), 501–520.

Python, S., Gerber, M., Suter, R., Ruggli, N., & Summerfield, A. (2013). Efficient sensing of infected cells in absence of virus particles by plasmacytoid dendritic cells is blocked by the viral ribonuclease E(rns). *PLoS Pathogens, 9*(6), e1003412. http://dx.doi.org/10.1371/journal.ppat.1003412.

Qi, F., Ridpath, J. F., & Berry, E. S. (1998). Insertion of a bovine SMT3B gene in NS4B and duplication of NS3 in a bovine viral diarrhea virus genome correlated with the cytopath-ogenicity of the virus. *Virus Research, 57*, 1–9.

Radostits, O. M., & Littlejohns, I. R. (1988). New concepts in the pathogenesis, diagnosis and control of diseases caused by the bovine viral diarrhea virus. *The Canadian Veterinary Journal, 29*(6), 513–528.

Rawlings, N. D., Barrett, A. J., & Bateman, A. (2012). MEROPS: The database of proteo-lytic enzymes, their substrates and inhibitors. *Nucleic Acids Research, 40*(Database issue), D343–D350. http://dx.doi.org/10.1093/nar/gkr987.

Rebhuhn, W. C., French, T. W., Perdrizet, J. A., Dubovi, E. J., Dill, S. G., & Karcher, L. F. (1989). Thrombocytopenia associated with acute bovine virus diarrhea infection in cattle. *Journal of Veterinary Internal Medicine, 3*, 42–46.

Reed, K. E., Gorbalenya, A. E., & Rice, C. M. (1998). The NS5A/NS5 proteins of viruses from three genera of the family flaviviridae are phosphorylated by associated serine/threonine kinases. *Journal of Virology*, *72*(7), 6199–6206.

Reimann, I., Depner, K., Trapp, S., & Beer, M. (2004). An avirulent chimeric Pestivirus with altered cell tropism protects pigs against lethal infection with classical swine fever virus. *Virology*, *322*(1), 143–157.

Reimann, I., Depner, K., Utke, K., Leifer, I., Lange, E., & Beer, M. (2010). Characterization of a new chimeric marker vaccine candidate with a mutated antigenic E2-epitope. *Veterinary Microbiology*, *142*(1–2), 45–50. http://dx.doi.org/10.1016/j.vetmic.2009.09.042.

Reimann, I., Semmler, I., & Beer, M. (2007). Packaged replicons of bovine viral diarrhea virus are capable of inducing a protective immune response. *Virology*, *366*(2), 377–386.

Reizis, B., Bunin, A., Ghosh, H. S., Lewis, K. L., & Sisirak, V. (2011). Plasmacytoid dendritic cells: Recent progress and open questions. *Annual Review of Immunology*, *29*, 163–183. http://dx.doi.org/10.1146/annurev-immunol-031210-101345.

Renard, A., Dino, D., & Martial, J. (1987). Vaccines and diagnostics derived from bovine diarrhea virus. *European patent Application number 86870095.6*, publication number 02.08672.

Reusken, C. B., Dalebout, T. J., Eerligh, P., Bredenbeek, P. J., & Spaan, W. J. (2003). Analysis of hepatitis C virus/classical swine fever virus chimeric 5′NTRs: Sequences within the hepatitis C virus IRES are required for viral RNA replication. *Journal of General Virology*, *84*(Pt 7), 1761–1769.

Ridpath, J. F. (2003). BVDV genotypes and biotypes: Practical implications for diagnosis and control. *Biologicals*, *31*(2), 127–131.

Ridpath, J. F. (2005). Practical significance of heterogeneity among BVDV strains: Impact of biotype and genotype on U.S. control programs. *Preventive Veterinary Medicine*, *72*(1–2), 17–30. http://dx.doi.org/10.1016/j.prevetmed.2005.08.003, discussion 215–219.

Ridpath, J. (2010). The contribution of infections with bovine viral diarrhea viruses to bovine respiratory disease. *The Veterinary Clinics of North America. Food Animal Practice*, *26*(2), 335–348. http://dx.doi.org/10.1016/j.cvfa.2010.04.003.

Ridpath, J. F., & Bolin, S. R. (1995). The genomic sequence of a virulent bovine viral diarrhea virus (BVDV) from the type 2 genotype: Detection of a large genomic insertion in a noncytopathic BVDV. *Virology*, *212*, 39–46.

Ridpath, J. F., & Neill, J. D. (2000). Detection and characterization of genetic recombination in cytopathic type 2 bovine viral diarrhea viruses. *Journal of Virology*, *74*(18), 8771–8774.

Riedel, C., Lamp, B., Heimann, M., König, M., Blome, S., Moennig, V., et al. (2012). The core protein of classical Swine Fever virus is dispensable for virus propagation in vitro. *PLoS Pathogens*, *8*(3), e1002598. http://dx.doi.org/10.1371/journal.ppat.1002598.

Riedel, C., Lamp, B., Heimann, M., & Rümenapf, T. (2010). Characterization of essential domains and plasticity of the classical Swine Fever virus Core protein. *Journal of Virology*, *84*(21), 11523–11531.

Rieder, M., Finke, S., & Conzelmann, K. K. (2012). Interferon in lyssavirus infection. *Berliner und Münchener Tierärztliche Wochenschrift*, *125*(5–6), 209–218.

Rijnbrand, R., Bredenbeek, P. J., Haasnoot, P. C., Kieft, J. S., Spaan, W. J., & Lemon, S. M. (2001). The influence of downstream protein-coding sequence on internal ribosome entry on hepatitis C virus and other flavivirus RNAs. *RNA*, *7*(4), 585–597.

Rijnbrand, R., van der Straaten, T., van Rijn, P. A., Spaan, W. J., & Bredenbeek, P. (1997). Internal entry of ribosomes is directed by the 5′ noncoding region of classical swine fever virus and is dependent on the presence of an RNA pseudoknot upstream of the initiation codon. *Journal of Virology*, *71*(1), 451–457.

Rinck, G., Birghan, C., Harada, T., Meyers, G., Thiel, H. J., & Tautz, N. (2001). A cellular J-domain protein modulates polyprotein processing and cytopathogenicity of a pestivirus. *Journal of Virology, 75*(19), 9470–9482.

Risager, P. C., Fahnoe, U., Gullberg, M., Rasmussen, T. B., & Belsham, G. J. (2013). Analysis of classical swine fever virus RNA replication determinants using replicons. *Journal of General Virology, 94*(Pt 8), 1739–1748. http://dx.doi.org/10.1099/vir.0.052688-0.

Risatti, G. R., Borca, M. V., Kutish, G. F., Lu, Z., Holinka, L. G., French, R. A., et al. (2005). The E2 glycoprotein of classical swine fever virus is a virulence determinant in swine. *Journal of Virology, 79*(6), 3787–3796.

Risatti, G. R., Holinka, L. G., Carrillo, C., Kutish, G. F., Lu, Z., Tulman, E. R., et al. (2006). Identification of a novel virulence determinant within the E2 structural glyco-protein of classical swine fever virus. *Virology, 355*(1), 94–101.

Risatti, G. R., Holinka, L. G., Fernandez, Sainz I., Carrillo, C., Kutish, G. F., Lu, Z., et al. (2007). Mutations in the carboxyl terminal region of E2 glycoprotein of classical swine fever virus are responsible for viral attenuation in swine. *Virology, 364*(2), 371–382.

Risatti, G. R., Holinka, L. G., Fernandez, Sainz I., Carrillo, C., Lu, Z., & Borca, M. V. (2007). N-linked glycosylation status of classical swine fever virus strain Brescia E2 glycoprotein influences virulence in swine. *Journal of Virology, 81*(2), 924–933.

Risatti, G. R., Holinka, L. G., Lu, Z., Kutish, G. F., Tulman, E. R., French, R. A., et al. (2005). Mutation of E1 glycoprotein of classical swine fever virus affects viral virulence in swine. *Virology, 343*(1), 116–127.

Ronecker, S., Zimmer, G., Herrler, G., Greiser-Wilke, I., & Grummer, B. (2008). Forma-tion of bovine viral diarrhea virus E1-E2 heterodimers is essential for virus entry and depends on charged residues in the transmembrane domains. *Journal of General Virology, 89*(Pt 9), 2114–2121. http://dx.doi.org/10.1099/vir.0.2008/001792-0.

Rosamilia, A., Grattarola, C., Caruso, C., Peletto, S., Gobbi, E., Tarello, V., et al. (2014). Detection of border disease virus (BDV) genotype 3 in Italian goat herds. *The Veterinary Journal, 199*(3), 446–450. http://dx.doi.org/10.1016/j.tvjl.2013.12.006.

Rose, I. A., & Warms, J. V. (1983). An enzyme with ubiquitin carboxy-terminal esterase activity from reticulocytes. *Biochemistry, 22*(18), 4234–4237.

Ruggli, N., Bird, B. H., Liu, L., Bauhofer, O., Tratschin, J. D., & Hofmann, M. A. (2005). N(pro) of classical swine fever virus is an antagonist of double-stranded RNA-mediated apoptosis and IFN-alpha/beta induction. *Virology, 340*(2), 265–276.

Ruggli, N., Summerfield, A., Fiebach, A. R., Guzylack-Piriou, L., Bauhofer, O., Lamm, C. G., et al. (2009). Classical swine fever virus can remain virulent after specific elimination of the interferon regulatory factor 3-degrading function of Npro. *Journal of Virology, 83*(2), 817–829.

Ruggli, N., Tratschin, J. D., Schweizer, M., McCullough, K. C., Hofmann, M. A., & Summerfield, A. (2003). Classical swine fever virus interferes with cellular antiviral defense: Evidence for a novel function of N(pro). *Journal of Virology, 77*(13), 7645–7654.

Rümenapf, T., Stark, R., Heimann, M., & Thiel, H. J. (1998). N-terminal protease of pestiviruses: Identification of putative catalytic residues by site-directed mutagenesis. *Journal of Virology, 72*(3), 2544–2547.

Rümenapf, T., & Thiel, H.-J. (2014). Molecular biology of pestiviruses. In T. C. Mettenleiter & F. Sobrino (Eds.), *Animal viruses: Molecular biology* (pp. 39–96). Norfolk, UK: Caister Academic Press.

Rümenapf, T., Unger, G., Strauss, J. H., & Thiel, H. J. (1993). Processing of the envelope glycoproteins of pestiviruses. *Journal of Virology, 67*, 3288–3295.

Sainz, I. F., Holinka, L. G., Lu, Z., Risatti, G. R., & Borca, M. V. (2008). Removal of a N-linked glycosylation site of classical swine fever virus strain Brescia Erns glycoprotein affects virulence in swine. *Virology, 370*(1), 122–129.

Schechter, I., & Berger, A. (1967). On the size of the active site in proteases I. Papain. *Biochemical and Biophysical Research Communications, 27*(2), 157–162.

Schelp, C., Greiser-Wilke, I., Wolf, G., Beer, M., Moennig, V., & Liess, B. (1995). Identification of cell membrane proteins linked to susceptibility to bovine viral diarrhoea virus infection. *Archives of Virology, 140*(11), 1997–2009.

Scherer, C. F., Flores, E. F., Weiblen, R., Caron, L., Irigoyen, L. F., Neves, J. P., et al. (2001). Experimental infection of pregnant ewes with bovine viral diarrhea virus type-2 (BVDV-2): Effects on the pregnancy and fetus. *Veterinary Microbiology, 79*(4), 285–299.

Schirrmeier, H., Strebelow, G., Depner, K., Hoffmann, B., & Beer, M. (2004). Genetic and antigenic characterization of an atypical pestivirus isolate, a putative member of a novel pestivirus species. *Journal of General Virology, 85*(Pt 12), 3647–3652. http://dx.doi.org/10.1099/vir.0.80238-0.

Schmeiser, S., Mast, J., Thiel, H. J., & König, M. (2014). Morphogenesis of pestiviruses: New insights from ultrastructural studies of strain Giraffe-1. *Journal of Virology, 88*(5), 2717–2724. http://dx.doi.org/10.1128/JVI.03237-13.

Schneider, R., Unger, G., Stark, R., Schneider-Scherzer, E., & Thiel, H. J. (1993). Identification of a structural glycoprotein of an RNA virus as a ribonuclease. *Science, 261*, 1169–1171.

Schweizer, M., & Peterhans, E. (1999). Oxidative stress in cells infected with bovine viral diarrhoea virus: A crucial step in the induction of apoptosis. *Journal of General Virology, 80*(Pt 5), 1147–1155.

Seago, J., Hilton, L., Reid, E., Doceul, V., Jeyatheesan, J., Moganeradj, K., et al. (2007). The Npro product of classical swine fever virus and bovine viral diarrhea virus uses a conserved mechanism to target interferon regulatory factor-3. *Journal of General Virology, 88*(Pt 11), 3002–3006.

Shaw, I. G., Winkler, C. E., Gibbons, D. F., Terlecki, S., Hebert, C. N., Patterson, D. S., et al. (1969). Border disease of sheep: Vaccination of ewes before mating. *The Veterinary Record, 84*(6), 147–148.

Sheng, C., Chen, Y., Xiao, J., Xiao, J., Wang, J., Li, G., et al. (2012). Classical swine fever virus NS5A protein interacts with 3'-untranslated region and regulates viral RNA synthesis. *Virus Research, 163*(2), 636–643. http://dx.doi.org/10.1016/j.virusres.2012.01.004.

Sheng, C., Liu, X., Jiang, Q., Xu, B., Zhou, C., Wang, Y., et al. (2015). Annexin A2 is involved in the production of classical swine fever virus infectious particles. *Journal of General Virology*, in press. http://dx.doi.org/10.1099/vir.0.000048.

Sheng, C., Wang, J., Xiao, J., Xiao, J., Chen, Y., Jia, L., et al. (2012). Classical swine fever virus NS5B protein suppresses the inhibitory effect of NS5A on viral translation by binding to NS5A. *Journal of General Virology, 93*(Pt 5), 939–950. http://dx.doi.org/10.1099/vir.0.039495-0.

Sheng, C., Yao, Y., Chen, B., Wang, Y., Chen, J., & Xiao, M. (2013). RNA helicase is involved in the expression and replication of classical swine fever virus and interacts with untranslated region. *Virus Research, 171*(1), 257–261. http://dx.doi.org/10.1016/j.virusres.2012.11.014.

Shimizu, Y., Furuuchi, S., Kumagai, T., & Sasahara, J. (1970). A mutant of hog cholera virus inducing interference in swine testicle cell cultures. *American Journal of Veterinary Research, 31*(10), 1787–1794.

Shimizu, M., Satou, K., Nishioka, N., Yoshino, T., Momotani, E., & Ishikawa, Y. (1989). Serological characterization of viruses isolated from experimental mucosal disease. *Veterinary Microbiology, 19*(1), 13–21.

Shine, J., & Dalgarno, L. (1975). Determinant of cistron specificity in bacterial ribosomes. *Nature, 254*(5495), 34–38.

Shoemaker, M. L., Smirnova, N. P., Bielefeldt-Ohmann, H., Austin, K. J., van Olphen, A., Clapper, J. A., et al. (2009). Differential expression of the type I interferon pathway during persistent and transient bovine viral diarrhea virus infection. *Journal of Interferon and Cytokine Research, 29*(1), 23–35. http://dx.doi.org/10.1089/jir.2008.0033.

Simmonds, P., Becher, P., Collett, M. S., Gould, E. A., Heinz, F. X., Meyers, G., et al. (2012). Flaviviridae. In A. M. Q. King, E. Lefkowitz, M. J. Adams, E. B. Carstens, & C. M. Fauquet (Eds.), *Virus taxonomy. Ninth report of the international committee on taxonomy of viruses* (pp. 1003–1020).San Diego, USA: Academic Press, Reprinted from: Not in file.

Sizova, D. V., Kolupaeva, V. G., Pestova, T. V., Shatsky, I. N., & Hellen, C. U. (1998). Specific interaction of eukaryotic translation initiation factor 3 with the 5′ nontranslated regions of hepatitis C virus and classical swine fever virus RNAs. *Journal of Virology, 72*(6), 4775–4782.

Smirnova, N. P., Bielefeldt-Ohmann, H., Van Campen, H., Austin, K. J., Han, H., Montgomery, D. L., et al. (2008). Acute non-cytopathic bovine viral diarrhea virus infection induces pronounced type I interferon response in pregnant cows and fetuses. *Virus Research, 132*(1–2), 49–58. http://dx.doi.org/10.1016/j.virusres.2007.10.011.

Smirnova, N. P., Webb, B. T., Bielefeldt-Ohmann, H., Van Campen, H., Antoniazzi, A. Q., Morarie, S. E., et al. (2012). Development of fetal and placental innate immune responses during establishment of persistent infection with bovine viral diarrhea virus. *Virus Research, 167*(2), 329–336. http://dx.doi.org/10.1016/j.virusres.2012.05.018.

Smirnova, N. P., Webb, B. T., McGill, J. L., Schaut, R. G., Bielefeldt-Ohmann, H., Van Campen, H., et al. (2014). Induction of interferon-gamma and downstream pathways during establishment of fetal persistent infection with bovine viral diarrhea virus. *Virus Research, 183*, 95–106. http://dx.doi.org/10.1016/j.virusres.2014.02.002.

Sonenberg, N., & Hinnebusch, A. G. (2009). Regulation of translation initiation in eukaryotes: Mechanisms and biological targets. *Cell, 136*(4), 731–745. http://dx.doi.org/10.1016/j.cell.2009.01.042.

Stahl, K., Kampa, J., Alenius, S., Persson Wadman, A., Baule, C., Aiumlamai, S., et al. (2007). Natural infection of cattle with an atypical 'HoBi'-like pestivirus—Implications for BVD control and for the safety of biological products. *Veterinary Research, 38*(3), 517–523. http://dx.doi.org/10.1051/vetres:2007012.

Stalder, H. P., Meier, P., Pfaffen, G., Wageck-Canal, C., Rufenacht, J., Schaller, P., et al. (2005). Genetic heterogeneity of pestiviruses of ruminants in Switzerland. *Preventive Veterinary Medicine, 72*(1–2), 37–41. http://dx.doi.org/10.1016/j.prevetmed.2005.01.020, discussion 215–219.

Stapleton, J. T., Foung, S., Muerhoff, A. S., Bukh, J., & Simmonds, P. (2011). The GB viruses: A review and proposed classification of GBV-A, GBV-C (HGV), and GBV-D in genus Pegivirus within the family Flaviviridae. *Journal of General Virology, 92*(Pt 2), 233–246. http://dx.doi.org/10.1099/vir.0.027490-0.

Stark, R., Meyers, G., Rümenapf, T., & Thiel, H. J. (1993). Processing of pestivirus polyprotein: Cleavage site between autoprotease and nucleocapsid protein of classical swine fever virus. *Journal of Virology, 67*, 7088–7095.

Stark, R., Rümenapf, T., Meyers, G., & Thiel, J. H. (1990). Genomic localization of hog cholera virus glycoproteins. *Virology, 174*, 286–289.

Steck, F., Lazary, S., Fey, H., Wandeler, A., Huggler, C., Oppliger, G., et al. (1980). Immune responsiveness in cattle fatally affected by bovine virus diarrhea–mucosal disease. *Zentralblatt für Veterinärmedizin Reihe B, 27*(6), 429–445.

Steffen, D. J., Topliff, C. L., Schmitz, J. A., Kammerman, J. R., Henningson, J. N., Eskridge, K. M., et al. (2014). Distribution of lymphoid depletion and viral antigen in alpacas experimentally infected with Bovine viral diarrhea virus 1. *Journal of Veterinary Diagnostic Investigation, 26*(1), 35–41. http://dx.doi.org/10.1177/1040638713509626.

Steffens, S., Thiel, H. J., & Behrens, S. E. (1999). The RNA-dependent RNA polymerases of different members of the family Flaviviridae exhibit similar properties in vitro. *Journal of General Virology, 80*, 2583–2590.

Stehelin, D., Varmus, H. E., Bishop, J. M., & Vogt, P. K. (1976). DNA related to the transforming gene(s) of avian sarcoma viruses is present in normal avian DNA. *Nature, 260*(5547), 170–173.

Steinmann, E., Penin, F., Kallis, S., Patel, A. H., Bartenschlager, R., & Pietschmann, T. (2007). Hepatitis C virus p7 protein is crucial for assembly and release of infectious virions. *PLoS Pathogens, 3*(7), e103. http://dx.doi.org/10.1371/journal.ppat.0030103.

Stokstad, M., Collins, M., Sorby, R., Barboni, P., Meyers, G., Loken, T., et al. (2004). The role of the defective interfering particle DI9c in mucosal disease in cattle. *Archives of Virology, 149*(3), 571–582.

Sun, Y., Li, N., Li, H. Y., Li, M., & Qiu, H. J. (2010). Enhanced immunity against classical swine fever in pigs induced by prime-boost immunization using an alphavirus replicon-vectored DNA vaccine and a recombinant adenovirus. *Veterinary Immunology and Immunopathology, 137*(1–2), 20–27. http://dx.doi.org/10.1016/j.vetimm.2010.04.005.

Sun, Y., Li, H. Y., Tian, D. Y., Han, Q. Y., Zhang, X., Li, N., et al. (2011). A novel alphavirus replicon-vectored vaccine delivered by adenovirus induces sterile immunity against classical swine fever. *Vaccine, 29*(46), 8364–8372. http://dx.doi.org/10.1016/j.vaccine.2011.08.085.

Szymanski, M. R., Fiebach, A. R., Tratschin, J. D., Gut, M., Ramanujam, V. M., Gottipati, K., et al. (2009). Zinc binding in pestivirus N(pro) is required for interferon regulatory factor 3 interaction and degradation. *Journal of Molecular Biology, 391*(2), 438–449. http://dx.doi.org/10.1016/j.jmb.2009.06.040.

Tamura, T., Nagashima, N., Ruggli, N., Summerfield, A., Kida, H., & Sakoda, Y. (2014). Npro of classical swine fever virus contributes to pathogenicity in pigs by preventing type I interferon induction at local replication sites. *Veterinary Research, 45*, 47. http://dx.doi.org/10.1186/1297-9716-45-47.

Tamura, T., Sakoda, Y., Yoshino, F., Nomura, T., Yamamoto, N., Sato, Y., et al. (2012). Selection of classical swine fever virus with enhanced pathogenicity reveals synergistic virulence determinants in E2 and NS4B. *Journal of Virology, 86*(16), 8602–8613. http://dx.doi.org/10.1128/JVI.00551-12.

Tamura, J. K., Warrener, P., & Collett, M. S. (1993). RNA-stimulated NTPase activity associated with the p80 protein of the pestivirus bovine viral diarrhea virus. *Virology, 193*, 1–10.

Tao, J., Liao, J., Wang, Y., Zhang, X., Wang, J., & Zhu, G. (2013). Bovine viral diarrhea virus (BVDV) infections in pigs. *Veterinary Microbiology, 165*(3–4), 185–189. http://dx.doi.org/10.1016/j.vetmic.2013.03.010.

Tautz, N., Elbers, K., Stoll, D., Meyers, G., & Thiel, H. J. (1997). Serine protease of pestiviruses: Determination of cleavage sites. *Journal of Virology, 71*(7), 5415–5422.

Tautz, N., Harada, T., Kaiser, A., Rinck, G., Behrens, S. E., & Thiel, H. J. (1999). Establishment and characterization of cytopathogenic and noncytopathogenic pestivirus replicons. *Journal of Virology, 73*(11), 9422–9432.

Tautz, N., Kaiser, A., & Thiel, H. J. (2000). NS3 serine protease of bovine viral diarrhea virus: Characterization of active site residues, NS4A cofactor domain, and protease-cofactor interactions. *Virology, 273*(2), 351–363.

Tautz, N., Meyers, G., Stark, R., Dubovi, E. J., & Thiel, H. J. (1996). Cytopathogenicity of a pestivirus correlated with a 27 nucleotide insertion. *Journal of Virology, 70*(11), 7851–7858.

Tautz, N., Meyers, G., & Thiel, H. J. (1993). Processing of poly-ubiquitin in the polyprotein of an RNA virus. *Virology, 197*, 74–85.

Tautz, N., & Thiel, H. J. (2003). Cytopathogenicity of pestiviruses: Cleavage of bovine viral diarrhea virus NS2-3 has to occur at a defined position to allow viral replication. *Archives of Virology, 148*(7), 1405–1412.

Tautz, N., Thiel, H. J., Dubovi, E. J., & Meyers, G. (1994). Pathogenesis of mucosal disease: A cytopathogenic pestivirus generated by internal deletion. *Journal of Virology, 68*, 3289–3297.

Tellinghuisen, T. L., Paulson, M. S., & Rice, C. M. (2006). The NS5A protein of bovine viral diarrhea virus contains an essential zinc-binding site similar to that of the hepatitis C virus NS5A protein. *Journal of Virology, 80*(15), 7450–7458.

Terpstra, C. (1981). Border disease: Virus persistence, antibody response and transmission studies. *Research in Veterinary Science, 30*, 185–191.

Tews, B. A., & Meyers, G. (2007). The pestivirus glycoprotein Erns is anchored in plane in the membrane via an amphipathic helix. *Journal of Biological Chemistry, 282*(45), 32730–32741.

Tews, B. A., Schürmann, E. M., & Meyers, G. (2009). Mutation of cysteine 171 of pestivirus E rns RNase prevents homodimer formation and leads to attenuation of classical swine fever virus. *Journal of Virology, 83*(10), 4823–4834.

Thabti, F., Letellier, C., Hammami, S., Pepin, M., Ribiere, M., Mesplede, A., et al. (2005). Detection of a novel border disease virus subgroup in Tunisian sheep. *Archives of Virology, 150*(2), 215–229. http://dx.doi.org/10.1007/s00705-004-0427-3.

Thiel, H. J., Plagemann, P. G. W., & Moennig, V. (1996). Pestiviruses. In B. N. Fields, D. M. Knipe, & P. M. Howley (Eds.), *Fields virology: Vol. 3* (pp. 1059–1073). Philadelphia, New York: Lippincott-Raven Publishers, Reprinted from: Not in file.

Thiel, H. J., Stark, R., Weiland, E., Rümenapf, T., & Meyers, G. (1991). Hog cholera virus: Molecular composition of virions from a pestivirus. *Journal of Virology, 65*, 4705–4712.

Thurner, C., Witwer, C., Hofacker, I. L., & Stadler, P. F. (2004). Conserved RNA secondary structures in Flaviviridae genomes. *Journal of General Virology, 85*(Pt 5), 1113–1124.

Toba, M., & Matumoto, M. (1969). Role of interferon in enhanced replication of Newcastle disease virus in swine cells infected with hog cholera virus. *Japanese Journal of Microbiology, 13*(3), 303–305.

Tratschin, J. D., Moser, C., Ruggli, N., & Hofmann, M. A. (1998). Classical swine fever virus leader proteinase Npro is not required for viral replication in cell culture. *Journal of Virology, 72*(9), 7681–7684.

Tscherne, D. M., Evans, M. J., Macdonald, M. R., & Rice, C. M. (2008). Transdominant inhibition of bovine viral diarrhea virus entry. *Journal of Virology, 82*(5), 2427–2436. http://dx.doi.org/10.1128/JVI.02158-07.

Tsiang, M., Monroe, S. S., & Schlesinger, S. (1985). Studies of defective interfering RNAs of Sindbis virus with and without tRNAAsp sequences at their 5′ termini. *Journal of Virology, 54*(1), 38–44.

Uttenthal, A., Le Potier, M. F., Romero, L., De Mia, G. M., & Floegel-Niesmann, G. (2001). Classical swine fever (CSF) marker vaccine Trial I. Challenge studies in weaner pigs. *Veterinary Microbiology, 83*(2), 85–106.

van Gennip, H. G., Bouma, A., van Rijn, P. A., Widjojoatmodjo, M. N., & Moormann, R. J. (2002). Experimental non-transmissible marker vaccines for classical swine fever (CSF) by trans-complementation of E(rns) or E2 of CSFV. *Vaccine, 20*(11–12), 1544–1556.

van Gennip, H. G., Hesselink, A. T., Moormann, R. J., & Hulst, M. M. (2005). Dimerization of glycoprotein E(rns) of classical swine fever virus is not essential for viral replication and infection. *Archives of Virology, 150*(11), 2271–2286.

van Gennip, H. G., van Rijn, P. A., Widjojoatmodjo, M. N., de Smit, A. J., & Moormann, R. J. (2000). Chimeric classical swine fever viruses containing envelope protein E(RNS) or E2 of bovine viral diarrhoea virus protect pigs against challenge with CSFV and induce a distinguishable antibody response. *Vaccine, 19*(4–5), 447–459.

van Gennip, H. G., Vlot, A. C., Hulst, M. M., de Smit, A. J., & Moormann, R. J. (2004). Determinants of virulence of classical swine fever virus strain Brescia. *Journal of Virology, 78*(16), 8812–8823.

van Oirschot, J. T. (1999). Diva vaccines that reduce virus transmission. *Journal of Biotechnology, 73*(2–3), 195–205.

van Oirschot, J. T. (2003). Vaccinology of classical swine fever: From lab to field. *Veterinary Microbiology, 96*(4), 367–384.

van Oirschot, J. T., Bruschke, C. J., & van Rijn, P. A. (1999). Vaccination of cattle against bovine viral diarrhoea. *Veterinary Microbiology, 64*(2–3), 169–183.

van Rijn, P. A., Miedema, G. K. W., Wensvoort, G., van Gennip, H. G. P., & Moormann, R. J. M. (1994). Antigenic structure of envelope glycoprotein E1 of hog cholera virus. *Journal of Virology, 68*, 3934–3942.

van Rijn, P. A., van Gennip, H. G., de Meijer, E. J., & Moormann, R. J. (1993). Epitope mapping of envelope glycoprotein E1 of hog cholera virus strain Brescia. *Journal of General Virology, 74*(Pt 10), 2053–2060.

Vaney, M. C., & Rey, F. A. (2011). Class II enveloped viruses. *Cellular Microbiology, 13*(10), 1451–1459. http://dx.doi.org/10.1111/j.1462-5822.2011.01653.x.

Vantsis, J. T., Barlow, R. M., Fraser, J., Rennie, J. C., & Mould, D. L. (1976). Experiments in border disease. VIII. Propagation and properties of a cytopathic virus. *Journal of Comparative Pathology, 86*, 111–120.

Vantsis, J. T., Linklater, K. A., Rennie, J. C., & Barlow, R. M. (1979). Experimental challenge infection of ewes following a field outbreak of Border disease. *Journal of Comparative Pathology, 89*(3), 331–339.

Vassilev, V. B., & Donis, R. O. (2000). Bovine viral diarrhea virus induced apoptosis correlates with increased intracellular viral RNA accumulation. *Virus Research, 69*(2), 95–107.

Vilcek, S., Greiser-Wilke, I., Nettleton, P., & Paton, D. J. (2000). Cellular insertions in the NS2-3 genome region of cytopathic bovine viral diarrhoea virus (BVDV) isolates. *Veterinary Microbiology, 77*(1–2), 129–136.

Vilcek, S., & Nettleton, P. F. (2006). Pestiviruses in wild animals. *Veterinary Microbiology, 116*(1–3), 1–12. http://dx.doi.org/10.1016/j.vetmic.2006.06.003.

Vilcek, S., Ridpath, J. F., Van Campen, H., Cavender, J. L., & Warg, J. (2005). Characterization of a novel pestivirus originating from a pronghorn antelope. *Virus Research, 108*(1–2), 187–193. http://dx.doi.org/10.1016/j.virusres.2004.09.010.

Voigt, H., Merant, C., Wienhold, D., Braun, A., Hutet, E., Le Potier, M. F., et al. (2007). Efficient priming against classical swine fever with a safe glycoprotein E2 expressing Orf virus recombinant (ORFV VrV-E2). *Vaccine, 25*(31), 5915–5926. http://dx.doi.org/10.1016/j.vaccine.2007.05.035.

Von Freyburg, M., Ege, A., Saalmüller, A., & Meyers, G. (2004). Comparison of the effects of RNase-negative and wild-type classical swine fever virus on peripheral blood cells of infected pigs. *Journal of General Virology, 85*(Pt 7), 1899–1908.

von Heijne, G. (1990). The signal peptide. *Journal of Membrane Biology, 115*(3), 195–201.

Wakita, T., Pietschmann, T., Kato, T., Date, T., Miyamoto, M., Zhao, Z., et al. (2005). Production of infectious hepatitis C virus in tissue culture from a cloned viral genome. *Nature Medicine, 11*(7), 791–796. http://dx.doi.org/10.1038/nm1268.

Walker, J. E., Saraste, M., Runswick, M. J., & Gay, N. J. (1982). Distantly related sequences in the alpha- and beta-subunits of ATP synthase, myosin, kinases and other ATP-requiring enzymes and a common nucleotide binding fold. *The EMBO Journal, 1*(8), 945–951.

Wang, J., Liu, B., Wang, N., Lee, Y. M., Liu, C., & Li, K. (2011). TRIM56 is a virus- and interferon-inducible E3 ubiquitin ligase that restricts pestivirus infection. *Journal of Virology, 85*(8), 3733–3745.

Wang, Z., Nie, Y., Wang, P., Ding, M., & Deng, H. (2004). Characterization of classical swine fever virus entry by using pseudotyped viruses: E1 and E2 are sufficient to mediate viral entry. *Virology, 330*(1), 332–341.

Wang, Y., Wang, Q., Lu, X., Zhang, C., Fan, X., Pan, Z., et al. (2008). 12-nt insertion in 3′ untranslated region leads to attenuation of classic swine fever virus and protects host against lethal challenge. *Virology, 374*(2), 390–398. http://dx.doi.org/10.1016/j.virol.2008.01.008.

Warrener, P., & Collett, M. S. (1995). Pestivirus NS3 (p80) protein possesses RNA helicase activity. *Journal of Virology, 69*(3), 1720–1726.

Wegelt, A., Reimann, I., Zemke, J., & Beer, M. (2009). New insights into processing of bovine viral diarrhea virus glycoproteins E(rns) and E1. *Journal of General Virology, 90*(Pt 10), 2462–2467.

Weiland, E., Ahl, R., Stark, R., Weiland, F., & Thiel, H. J. (1992). A second envelope glycoprotein mediates neutralization of a pestivirus, hog cholera virus. *Journal of Virology, 66*(6), 3677–3682.

Weiland, E., Stark, R., Haas, B., Rümenapf, T., Meyers, G., & Thiel, H. J. (1990). Pestivirus glycoprotein which induces neutralizing antibodies forms part of a disulfide linked heterodimer. *Journal of Virology, 64*(8), 3563–3569.

Weiland, F., Weiland, E., Unger, G., Saalmüller, A., & Thiel, H. J. (1999). Localization of pestiviral envelope proteins E(rns) and E2 at the cell surface and on isolated particles. *Journal of General Virology, 80*(Pt 5), 1157–1165.

Weiskircher, E., Aligo, J., Ning, G., & Konan, K. V. (2009). Bovine viral diarrhea virus NS4B protein is an integral membrane protein associated with Golgi markers and rearranged host membranes. *Virology Journal, 6*, 185. http://dx.doi.org/10.1186/1743-422X-6-185.

Welbourn, S., Green, R., Gamache, I., Dandache, S., Lohmann, V., Bartenschlager, R., et al. (2005). Hepatitis C virus NS2/3 processing is required for NS3 stability and viral RNA replication. *Journal of Biological Chemistry, 280*(33), 29604–29611. http://dx.doi.org/10.1074/jbc.M505019200.

Wensvoort, G., & Terpstra, C. (1985). Swine fever: A changing clinical picture. *Tijdschrift voor Diergeneeskunde, 110*(7), 263–269.

Widjojoatmodjo, M. N., van Gennip, H. G., Bouma, A., van Rijn, P. A., & Moormann, R. J. (2000). Classical swine fever virus E(rns) deletion mutants: Trans-complementation and potential use as nontransmissible, modified, live-attenuated marker vaccines. *Journal of Virology, 74*(7), 2973–2980.

Windisch, J. M., Schneider, R., Stark, R., Weiland, E., Meyers, G., & Thiel, H. J. (1996). RNase of classical swine fever virus: Biochemical characterization and inhibition by virus-neutralizing monoclonal antibodies. *Journal of Virology, 70*, 352–358.

Wiskerchen, M., Belzer, S. K., & Collett, M. S. (1991). Pestivirus gene expression: The first protein product of the bovine viral diarrhea virus large open reading frame, p20, possesses proteolytic activity. *Journal of Virology, 65*, 4508–4514.

Wiskerchen, M., & Collett, M. S. (1991). Pestivirus gene expression: Protein p80 of bovine viral diarrhea virus is a proteinase involved in polyprotein processing. *Virology, 184*(1), 341–350.

Wozniak, A. L., Griffin, S., Rowlands, D., Harris, M., Yi, M., Lemon, S. M., et al. (2010). Intracellular proton conductance of the hepatitis C virus p7 protein and its contribution to infectious virus production. *PLoS Pathogens, 6*(9), e1001087. http://dx.doi.org/10.1371/journal.ppat.1001087.

Wu, Z., Ren, X., Yang, L., Hu, Y., Yang, J., He, G., et al. (2012). Virome analysis for identification of novel mammalian viruses in bat species from Chinese provinces. *Journal of Virology, 86*(20), 10999–11012. http://dx.doi.org/10.1128/JVI.01394-12.

Wu, K., Yamoah, K., Dolios, G., Gan-Erdene, T., Tan, P., Chen, A., et al. (2003). DEN1 is a dual function protease capable of processing the C terminus of Nedd8 and deconjugating hyper-neddylated CUL1. *Journal of Biological Chemistry*, *278*(31), 28882–28891. http://dx.doi.org/10.1074/jbc.M302888200.

Xiao, M., Li, H., Wang, Y., Wang, X., Wang, W., Peng, J., et al. (2006). Characterization of the N-terminal domain of classical swine fever virus RNA-dependent RNA polymerase. *Journal of General Virology*, *87*(Pt 2), 347–356. http://dx.doi.org/10.1099/vir.0.81385-0.

Xiao, M., Wang, Y., Zhu, Z., Ding, C., Yu, J., Wan, L., et al. (2011). Influence of the 5′-proximal elements of the 5′-untranslated region of classical swine fever virus on translation and replication. *Journal of General Virology*, *92*(Pt 5), 1087–1096. http://dx.doi.org/10.1099/vir.0.027870-0.

Xiao, M., Wang, Y., Zhu, Z., Yu, J., Wan, L., & Chen, J. (2009). Influence of NS5A protein of classical swine fever virus (CSFV) on CSFV internal ribosome entry site-dependent translation. *Journal of General Virology*, *90*(Pt 12), 2923–2928. http://dx.doi.org/10.1099/vir.0.014472-0.

Xu, J., Mendez, E., Caron, P. R., Lin, C., Murcko, M. A., Collett, M. S., et al. (1997). Bovine viral diarrhea virus NS3 serine proteinase: Polyprotein cleavage sites, cofactor requirements, and molecular model of an enzyme essential for pestivirus replication. *Journal of Virology*, *71*(7), 5312–5322.

Yamane, D., Kato, K., Tohya, Y., & Akashi, H. (2006). The double-stranded RNA-induced apoptosis pathway is involved in the cytopathogenicity of cytopathogenic Bovine viral diarrhea virus. *Journal of General Virology*, *87*(Pt 10), 2961–2970. http://dx.doi.org/10.1099/vir.0.81820-0.

Yamane, D., Zahoor, M. A., Mohamed, Y. M., Azab, W., Kato, K., Tohya, Y., et al. (2009a). Activation of extracellular signal-regulated kinase in MDBK cells infected with bovine viral diarrhea virus. *Archives of Virology*, *154*(9), 1499–1503. http://dx.doi.org/10.1007/s00705-009-0453-2.

Yamane, D., Zahoor, M. A., Mohamed, Y. M., Azab, W., Kato, K., Tohya, Y., et al. (2009b). Inhibition of sphingosine kinase by bovine viral diarrhea virus NS3 is crucial for efficient viral replication and cytopathogenesis. *Journal of Biological Chemistry*, *284*(20), 13648–13659. http://dx.doi.org/10.1074/jbc.M807498200.

Yamane, D., Zahoor, M. A., Mohamed, Y. M., Azab, W., Kato, K., Tohya, Y., et al. (2009c). Microarray analysis reveals distinct signaling pathways transcriptionally activated by infection with bovine viral diarrhea virus in different cell types. *Virus Research*, *142*(1–2), 188–199. http://dx.doi.org/10.1016/j.virusres.2009.02.015.

Yang, Z., Wu, R., Li, R. W., Li, L., Xiong, Z., Zhao, H., et al. (2012). Chimeric classical swine fever (CSF)-Japanese encephalitis (JE) viral replicon as a non-transmissible vaccine candidate against CSF and JE infections. *Virus Research*, *165*(1), 61–70. http://dx.doi.org/10.1016/j.virusres.2012.01.007.

Yesilbag, K., Forster, C., Ozyigit, M. O., Alpay, G., Tuncer, P., Thiel, H. J., et al. (2014). Characterisation of bovine viral diarrhoea virus (BVDV) isolates from an outbreak with haemorrhagic enteritis and severe pneumonia. *Veterinary Microbiology*, *169*(1–2), 42–49. http://dx.doi.org/10.1016/j.vetmic.2013.12.005.

Yu, H., Grassmann, C. W., & Behrens, S. E. (1999). Sequence and structural elements at the 3′ terminus of bovine viral diarrhea virus genomic RNA: Functional role during RNA replication. *Journal of Virology*, *73*(5), 3638–3648.

Yu, H., Isken, O., Grassmann, C. W., & Behrens, S. E. (2000). A stem-loop motif formed by the immediate 5′ terminus of the bovine viral diarrhea virus genome modulates translation as well as replication of the viral RNA. *Journal of Virology*, *74*(13), 5825–5835.

Zahoor, M. A., Yamane, D., Mohamed, Y. M., Suda, Y., Kobayashi, K., Kato, K., et al. (2010). Bovine viral diarrhea virus non-structural protein 5A interacts with NIK- and IKKbeta-binding protein. *Journal of General Virology*, *91*(Pt 8), 1939–1948. http://dx.doi.org/10.1099/vir.0.020990-0.

Zemke, J., Koenig, P., Mischkale, K., Reimann, I., & Beer, M. (2010). Novel BVDV-2 mutants as new candidates for modified-live vaccines. *Veterinary Microbiology*, *142*(1–2), 69–80. http://dx.doi.org/10.1016/j.vetmic.2009.09.045.

Zhong, J., Gastaminza, P., Cheng, G., Kapadia, S., Kato, T., Burton, D. R., et al. (2005). Robust hepatitis C virus infection in vitro. *Proceedings of the National Academy of Sciences of the United States of America*, *102*(26), 9294–9299. http://dx.doi.org/10.1073/pnas.0503596102.

Zhong, W., Gutshall, L. L., & Del Vecchio, A. M. (1998). Identification and characterization of an RNA-dependent RNA polymerase activity within the nonstructural protein 5B region of bovine viral diarrhea virus. *Journal of Virology*, *72*(11), 9365–9369.

Zögg, T., Sponring, M., Schindler, S., Koll, M., Schneider, R., Brandstetter, H., et al. (2013). Crystal structures of the viral protease Npro imply distinct roles for the catalytic water in catalysis. *Structure*, *21*(6), 929–938. http://dx.doi.org/10.1016/j.str.2013.04.003.

Zürcher, C., Sauter, K. S., Mathys, V., Wyss, F., & Schweizer, M. (2014). Prolonged activity of the pestiviral RNase Erns as an interferon antagonist after uptake by clathrin-mediated endocytosis. *Journal of Virology*, *88*(13), 7235–7243. http://dx.doi.org/10.1128/JVI.00672-14.

CHAPTER THREE

Cyprinid Herpesvirus 3: An Archetype of Fish Alloherpesviruses

Maxime Boutier*, Maygane Ronsmans*, Krzysztof Rakus*, Joanna Jazowiecka-Rakus*, Catherine Vancsok*, Léa Morvan*, Ma. Michelle D. Peñaranda*, David M. Stone[†], Keith Way[†], Steven J. van Beurden[‡], Andrew J. Davison[§], Alain Vanderplasschen*,[1]

*Immunology-Vaccinology (B43b), Department of Infectious and Parasitic Diseases, Fundamental and Applied Research for Animals & Health (FARAH), Faculty of Veterinary Medicine, University of Liège, Liège, Belgium
[†]The Centre for Environment, Fisheries and Aquaculture Science, Weymouth Laboratory, Weymouth, Dorset, United Kingdom
[‡]Department of Pathobiology, Faculty of Veterinary Medicine, Utrecht University, Utrecht, The Netherlands
[§]MRC-University of Glasgow Centre for Virus Research, Glasgow, United Kingdom
[1]Corresponding author: e-mail address: a.vdplasschen@ulg.ac.be

Contents

Abstract

The order *Herpesvirales* encompasses viruses that share structural, genetic, and biological properties. However, members of this order infect hosts ranging from molluscs to humans. It is currently divided into three phylogenetically related families. The *Alloherpesviridae* family contains viruses infecting fish and amphibians. There are 12 alloherpesviruses described to date, 10 of which infect fish. Over the last decade, cyprinid herpesvirus 3 (CyHV-3) infecting common and koi carp has emerged as the archetype of fish alloherpesviruses. Since its first description in the late 1990s, this virus has induced important economic losses in common and koi carp worldwide. It has also

Advances in Virus Research, Volume 93
ISSN 0065-3527
http://dx.doi.org/10.1016/bs.aivir.2015.03.001

161

had negative environmental implications by affecting wild carp populations. These negative impacts and the importance of the host species have stimulated studies aimed at developing diagnostic and prophylactic tools. Unexpectedly, the data generated by these applied studies have stimulated interest in CyHV-3 as a model for fundamental research. This review intends to provide a complete overview of the knowledge currently available on CyHV-3.

1. INTRODUCTION

The order *Herpesvirales* contains a large number of viruses that share structural, genetic, and biological properties. It is divided into three phylogenetically related families infecting a wide range of hosts (Pellett et al., 2011b). The *Herpesviridae* family encompasses viruses infecting mammals, birds, or reptiles. It is by far the most important, both in terms of the number of its members and the volume of studies that have been devoted to them. The *Malacoherpesviridae* family comprises viruses infecting molluscs. Finally, the *Alloherpesviridae* family encompasses viruses infecting fish and amphibians. Twelve alloherpesviruses have been described to date, ten of them infecting fish (Hanson, Dishon, & Kotler, 2011; Waltzek et al., 2009).

Over the last decade, an increasing number of studies have been devoted to alloherpesviruses that infect fish. Scientific interest in a specific virus tends to originate from its impact on wildlife, the economic losses it causes to the aquaculture industry, or its importance as a fundamental research object. On rare occasions, all three of these reasons apply. This is the case for cyprinid herpesvirus 3 (CyHV-3), also known as koi herpesvirus (KHV), which has emerged as the archetype of fish alloherpesviruses (Adamek, Steinhagen, et al., 2014; Rakus et al., 2013).

Since its emergence in the late 1990s, CyHV-3 has had an ecological impact and induced severe economic losses in the common and koi carp industries (Bondad-Reantaso et al., 2005; Perelberg et al., 2003; Rakus et al., 2013). The common carp (*Cyprinus carpio*) is one of the oldest cultivated freshwater fish species (Balon, 1995) and is now one of the most economically valuable species in aquaculture. It is widely cultivated for human consumption, with a worldwide production of 3.8 million tons in 2012 representing US$5.2 billion (FAO, 2012). Furthermore, its colorful ornamental varieties (koi carp), grown for personal pleasure and competitive exhibitions, represent one of the most expensive markets for individual freshwater fish. The economic importance of CyHV-3 has rapidly stimulated research efforts aimed at building essential knowledge for the development of diagnostic and prophylactic tools (Ilouze, Dishon, & Kotler, 2006; Rakus et al., 2013). In addition, these studies have stimulated interest in CyHV-3 as an object of fundamental research. As a result, CyHV-3 can be considered today as the archetype of fish alloherpesviruses and is the subject of an increasing number of studies. Most of the present review is devoted to this virus.

This review consists of two sections. In the first part, we describe an up-to-date phylogenetic analysis of the family *Alloherpesviridae* as a component of the order *Herpesvirales*. We also summarize the main properties of herpesviruses and the specific properties of fish alloherpesviruses. In the second and main part, we provide a full overview of the knowledge currently available on CyHV-3.

2. THE ORDER *HERPESVIRALES*

2.1 Phylogeny

2.1.1 Phylogeny of the Order Herpesvirales

In historical terms, recognition of an agent as a herpesvirus has rested on morphology: a linear, double-stranded DNA genome packed into a $T=16$ icosahedral capsid, embedded in a complex protein layer known as the tegument, wrapped in a glycoprotein-containing lipid membrane, yielding a spherical virion. However, extensive understanding of the genetic structure of herpesviruses, especially in relation to conserved genes, now allows these features to be inferred rather than demonstrated directly. As a result, classification of an entity as a herpesvirus and determination of its detailed taxonomy depend principally on the interpretation of primary sequence data.

For many years, the International Committee on Taxonomy of Viruses (ICTV) counted several fish pathogens as being likely members of the family *Herpesviridae* based on morphology. In 1998, the first species of fish herpesvirus was founded in the family, namely *Ictalurid herpesvirus 1* (ictalurid herpesvirus 1 [IcHV-1], also known as channel catfish virus). The genus in which this species was placed adopted the name *Ictalurivirus*. However, it had been clear for some years that this virus was only very distantly related to mammalian herpesviruses (Davison, 1992). In 2008, this, as well as other considerations, led to the adoption of the order *Herpesvirales* (Davison et al., 2009; Pellett et al., 2011b). This order was established to contain three families: the already existing family *Herpesviridae*, which now contains herpesviruses of mammals, birds, and reptiles (Pellett et al., 2011c), the new families *Alloherpesviridae*, encompassing herpesviruses of amphibians and fish (Pellett et al., 2011a), and *Malacoherpesviridae*, containing herpesviruses of invertebrates (Pellett et al., 2011d). The assignment of herpesviruses of certain hosts to these families is descriptive rather than prescriptive.

The ICTV (http://www.ictvonline.org) currently lists 87 species in the family *Herpesviridae* distributed among the three subfamilies, *Alphaherpesvirinae*, *Betaherpesvirinae*, and *Gammaherpesvirinae*, plus one unassigned species. The subfamilies contain five, four, and four genera, respectively. Establishment of this taxonomical structure has been fostered by an extensively researched phylogeny (McGeoch, Dolan, & Ralph, 2000; McGeoch & Gatherer, 2005; McGeoch, Rixon, & Davison, 2006). A phylogenetic description of 65 viruses classified in this family, based on the complete sequence of the highly conserved viral gene encoding DNA polymerase, is shown in Fig. 1A. The overall genetic coherence of the family is apparent from the fact that 43 genes are conserved among members of the family. These genes are presumed to have been present in the last common ancestor, which has been inferred to have existed 400 million years ago (McGeoch et al., 2006).

A description of the phylogeny of the family *Alloherpesviridae*, to which CyHV-3 belongs, is given in Section 2.1.2. The third family, *Malacoherpesviridae*, consists of two genera, *Aurivirus*, which contains the species *Haliotid herpesvirus 1* (haliotid herpesvirus 1 or abalone herpesvirus), and *Ostreavirus*, which includes the species *Ostreid herpesvirus 1* (ostreid herpesvirus 1 or oyster herpesvirus).

Since herpesviruses continue to be identified, it seems likely that more members of the order *Herpesvirales* remain to be discovered. Although the coherence of the order is apparent from structural conservation of the virion, particularly the capsid, among the three families (Booy, Trus, Davison, & Steven, 1996; Davison et al., 2005), detectable genetic similarities are very few. The most convincingly conserved gene is that encoding DNA packaging terminase subunit 1, a subunit of an enzyme complex responsible for incorporating genomes into preformed capsids. Conservation of the predicted amino-acid sequence of this protein in herpesviruses and tailed bacteriophages (Davison, 1992), as well as the existence of conserved structural elements in other proteins (Rixon & Schmid, 2014), points to an origin of all herpesviruses from ancient precursors having existed in bacteria.

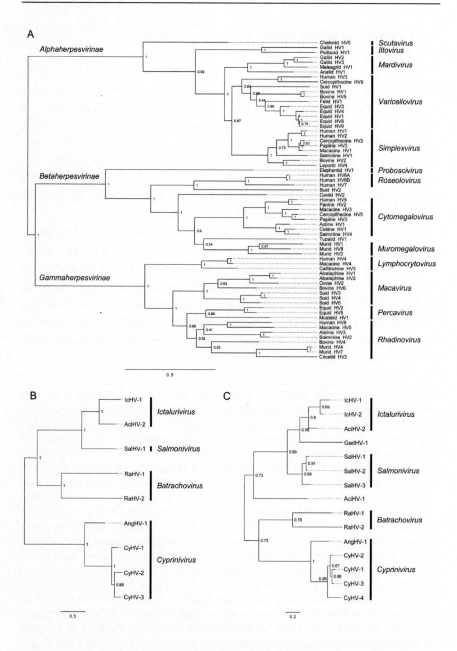

Figure 1 Phylogenetic analysis of the *Herpesviridae* and *Alloherpesviridae* families. Unrooted phylogenetic tree based on (A) the full-length DNA polymerases of members of the family *Herpesviridae*, (B) the full-length DNA polymerases of members of the family *Alloherpesviridae*, and (C) partial DNA polymerases of members or potential members
(Continued)

2.1.2 Phylogeny of the Family Alloherpesviridae

Shortly after the first formal reports of its discovery (Ariav, Tinman, Paperna, & Bejerano, 1999; Bretzinger, Fischer-Scherl, Oumouma, Hoffmann, & Truyen, 1999), CyHV-3 was characterized as a herpesvirus based on virion morphology (Hedrick et al., 2000). Although there was some suggestion, based on early DNA sequence data, that this assignment might not be correct (Hutoran et al., 2005; Ronen et al., 2003), the initial characterization was soon shown to be sound (Waltzek et al., 2005). The subsequent accumulation of extensive sequence data for a range of fish and amphibian herpesviruses provided a solid understanding of the phylogeny and evolution of the family *Alloherpesviridae*.

The ICTV currently lists twelve species in the family *Alloherpesviridae*, distributed among four genera, of which three contain fish viruses (*Cyprinivirus, Ictalurivirus*, and *Salmonivirus*, with CyHV-3 in the genus *Cyprinivirus*) and one contains amphibian viruses (*Batrachovirus*) (Table 1). Full genome sequences are available for seven of these viruses, representing three genera (Table 2). Partial sequence data are available for the other five classified, and also several unclassified, fish herpesviruses. A phylogenetic

---

Figure 1—Cont'd of the family *Alloherpesviridae*. For (A), the sequences (996–1357 amino-acid residues in length) were derived from relevant GenBank accessions. Virus names are aligned at the branch tips in the style that mirrors the species names (e.g., chelonid herpesvirus 5 (Chelonid HV5) is in the species *Chelonid herpesvirus 5*). The names of subfamilies and genera are marked on the left and right, respectively. The branching order in the genus *Rhadinovirus* is typically difficult to determine (McGeoch et al., 2006). For (B), the sequences (1507–1720 residues in length) were derived from the GenBank accessions listed in Table 1 and also from FJ815289.2 (Doszpoly, Somogyi, LaPatra, & Benko, 2011) for AciHV-2 and AAC59316.1 (Davison, 1998) and unpublished data (A.J.D.) for SalHV-1. Abbreviated virus names are shown at the branch tips (see Table 1), and the names of genera are marked on the right. For (C), partial sequences (134–158 residues in length; some truncated from longer sequences) located between the highly conserved DF(A/T/S)(S/A)(L/M)YP and GDTDS(V/T/I)M motifs were derived from EF685904.1 (Kelley et al., 2005) for AciHV-1, HQ857783.1 (Marcos-Lopez et al., 2012) for GadHV-1, KM357278.1 (Doszpoly et al., 2015) for CyHV-4, FJ641907.1 (Doszpoly et al., 2008; Waltzek et al., 2009) for IcHV-2, FJ641908.1 (Waltzek et al., 2009) for SalHV-2, and EU349277.1 (Waltzek et al., 2009) for SalHV-3. Abbreviated virus names are shown at the branch tips (see Table 1), and the names of genera are marked on the right. For (A), (B), and (C), the sequences were aligned by using Clustal Omega (Sievers & Higgins, 2014), and the tree was calculated by using MEGA6 (Tamura, Stecher, Peterson, Filipski, & Kumar, 2013) under an LG+G+I model with 100 bootstraps (values shown at the branch nodes). The scale in each panel shows the number of changes per site.

Table 1 Classification of the Family *Alloherpesviridae*

Genus Name	Species Name	Virus Name and Abbreviation	Alternative Virus Name[a]
Batrachovirus	*Ranid herpesvirus 1*	ranid herpesvirus 1 (RaHV-1)	Lucké tumor herpesvirus
	Ranid herpesvirus 2	ranid herpesvirus 2 (RaHV-2)	frog virus 4
Cyprinivirus	*Anguillid herpesvirus 1*	anguillid herpesvirus 1 (AngHV-1)	European eel herpesvirus
	Cyprinid herpesvirus 1	cyprinid herpesvirus 1 (CyHV-1)	carp pox herpesvirus
	Cyprinid herpesvirus 2	cyprinid herpesvirus 2 (CyHV-2)	goldfish hematopoietic necrosis virus
	Cyprinid herpesvirus 3	cyprinid herpesvirus 3 (CyHV-3)	koi herpesvirus
Ictalurivirus	*Acipenserid herpesvirus 2*	acipenserid herpesvirus 2 (AciHV-2)	white sturgeon herpesvirus 2
	Ictalurid herpesvirus 1	ictalurid herpesvirus 1 (IcHV-1)	channel catfish virus
	Ictalurid herpesvirus 2	ictalurid herpesvirus 2 (IcHV-2)	Ictalurus melas herpesvirus
Salmonivirus	*Salmonid herpesvirus 1*	salmonid herpesvirus 1 (SalHV-1)	herpesvirus salmonis
	Salmonid herpesvirus 2	salmonid herpesvirus 2 (SalHV-2)	Oncorhynchus masou herpesvirus
	Salmonid herpesvirus 3	salmonid herpesvirus 3 (SalHV-3)	epizootic epitheliotropic disease virus

[a]From Waltzek et al. (2009). In instances in which a virus is known by several alternative names, a single example is given.

tree of nine of the classified viruses, based on the complete sequence of the viral DNA polymerase, is shown in Fig. 1B. A tree of all 12 viruses, plus 3 others not yet classified (cyprinid herpesvirus 4 [CyHV-4, sichel herpesvirus], acipenserid herpesvirus 1 [AciHV-1, white sturgeon herpesvirus 1], and gadid herpesvirus 1 [GadHV-1, Atlantic cod herpesvirus]), based on a short segment of the same gene, is shown in Fig. 1C. As indicated by the bootstrap values, the robustness of the former tree is greater than that of the latter.

Table 2 Data on Complete Genome Sequences of Members of the Family *Alloherpesviridae*

Species Name	Virus Name and Abbreviation	Genome Size (bp)	Genome G+C (%)	ORFs (No.)[a]	GenBank Accession	References
Anguillid herpesvirus 1	anguillid herpesvirus 1 (AngHV-1)	248,526	53	134	FJ940765.3	van Beurden et al. (2010) van Beurden, Gatherer, et al. (2012)
Cyprinid herpesvirus 1	cyprinid herpesvirus 1 (CyHV-1)	291,144	51	143	JQ815363.1	Davison et al. (2013)
Cyprinid herpesvirus 2	cyprinid herpesvirus 2 (CyHV-2)	290,304	52	154	JQ815364.1	Davison et al. (2013)
Cyprinid herpesvirus 3	cyprinid herpesvirus 3 (CyHV-3)	295,146	59	163	DQ657948.1[b]	Aoki et al. (2007) Davison et al. (2013)
Ictalurid herpesvirus 1	ictalurid herpesvirus 1 (IcHV-1)	134,226	56	90	M75136.2	Davison (1992)
Ranid herpesvirus 1	ranid herpesvirus 1 (RaHV-1)	220,859	55	132	DQ665917.1	Davison, Cunningham, Sauerbier, and McKinnell (2006)
Ranid herpesvirus 2	ranid herpesvirus 2 (RaHV-2)	231,801	53	147	DQ665652.1	Davison et al. (2006)

[a]Predicted to encode functional proteins. Includes ORFs duplicated in repeated sequences.
[b]Additional genome sequences: DQ177346.1 (Aoki et al., 2007), AP008984.1 (Aoki et al., 2007), and KJ627438.1 (Li, Lee, Weng, He, & Dong, 2015).

Nonetheless, the trees are similar in overall shape, and they support the arrangement of the family into the four genera. The phylogeny of two of the unclassified viruses (AciHV-1 and GadHV-1) is not clear from the limited data used in Fig. 1C. However, the positions of these viruses, and others not included in Fig. 1C, have been examined with greater discrimination using sequences from other genes (Doszpoly, Benko, Bovo, Lapatra, & Harrach, 2011; Doszpoly et al., 2008, 2015; Doszpoly, Somogyi, et al., 2011; Kelley et al., 2005; Kurobe, Kelley, Waltzek, & Hedrick, 2008; Marcos-Lopez et al., 2012).

There has been some consideration of establishing subfamilies in the family *Alloherpesviridae*, as has taken place in the family *Herpesviridae*. These could number two (genus *Cyprinivirus* in one subfamily and the other three genera in another (Waltzek et al., 2009)) or three (genus *Cyprinivirus* in one subfamily, genus *Batrachovirus* in another, and the other two genera in the third (Doszpoly, Somogyi, et al., 2011)). For various reasons, this would seem premature at present.

The overall genetic coherence of the family *Alloherpesviridae* is evident from the presence of 12 convincingly conserved genes in fully sequenced members (Davison et al., 2013). This modest number suggests a last common ancestor that is considerably older than that of the family *Herpesviridae*. Patterns of coevolution between virus and host are apparent only toward the tips of phylogenetic trees and therefore are relevant to a more recent evolutionary period (Waltzek et al., 2009). For example, in Fig. 1B and C, the cyprinid herpesviruses 1 and 2 (CyHV-1 and CyHV-2) cluster with CyHV-3, and salmonid herpesviruses 1, 2, and 3 (SalHV-1, SalHV-2, and SalHV-3) cluster together. However, one of the sturgeon herpesviruses (AciHV-1) is deeply separated from the other viruses, whereas the other (AciHV-2) is most closely related to the ictalurid herpesviruses. Also, the branch point of the frog viruses falls within the fish herpesviruses rather than outside. The apparently smaller degree of coevolution of the family *Alloherpesviridae* compared with the family *Herpesviridae* may be due to several factors, not least those relating to the respective environments and the lengths of time the two families have been evolving.

2.2 Main Biological Properties

All members of the order *Herpesvirales* seem to share common biological properties (Ackermann, 2004; Pellett et al., 2011b): (i) they produce virions with the structure described above (Fig. 2A); (ii) they encode their own

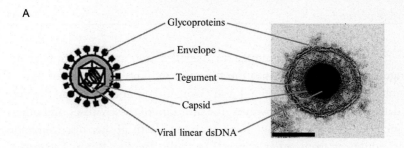

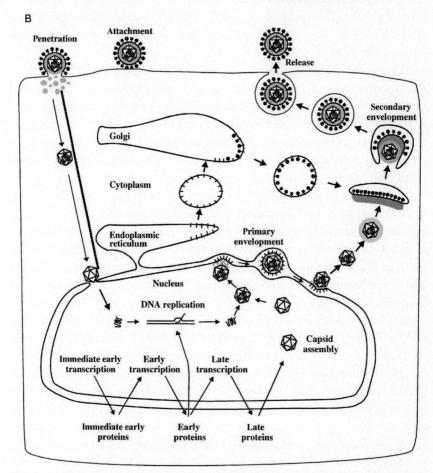

Figure 2 Virion structure and replication cycle of herpesviruses. (A) Schematic representation (left) and electron microscopy examination (right) of CyHV-3 virion. Bar represents 100 nm. (B) Replication cycle of CyHV-3. Diagrammatic representation of the herpesvirus replication cycle, including virus entry and dissociation of the tegument, transport of incoming capsids to the nuclear pore, and release of viral DNA into the nucleus where transcription occurs in a cascade-like fashion, and DNA replication ensues. Capsid assembly, DNA packaging, and primary and secondary envelopment are also illustrated. *Panel (A): Adapted with permission from Mettenleiter (2004) and Mettenleiter, Klupp, and Granzow (2009). Copyright © Elsevier. Panel (B): Reproduced with permission from Mettenleiter (2004). Copyright © Elsevier.*

DNA synthesis machinery, with viral replication as well as nucleocapsid assembly taking place in the nucleus (Fig. 2B); (iii) production of progeny virions is usually associated with lysis of the host cell; (iv) they are able to establish lifelong latent infection, which is characterized by the absence of regular viral transcription and replication and the lack of production of infectious virus particles, but presence of intact viral genomic DNA and the transcription of latency-associated genes. Latency can eventually be interrupted by reactivation that leads to lytic replication and the excretion of infectious particles by infected subjects despite the adaptive immune response developed against the virus; and (v) their ability to establish persistent infection in immunocompetent hosts (Pellett et al., 2011b) is the consequence of immune evasion mechanisms targeting major components of the immune system.

In addition to these properties that are considered to be common to all members of the order *Herpesvirales*, fish alloherpesviruses seem to share several biological properties that differentiate them from *Herpesviridae* (herpesviruses infecting mammals, birds, and reptiles). First, while herpesviruses generally show only modest pathogenicity in their natural immunocompetent hosts, fish herpesviruses can cause outbreaks associated with mortality reaching 100%. The markedly higher virulence of fish herpesviruses could reveal a lower adaptation level of these viruses to their hosts (see Section 2.1.2). However, it could also be explained by other factors such as the high-density rearing conditions and inbreeding promoted by intensive aquaculture. Second, the tropism of members of the family *Herpesviridae* is generally restricted to their natural host species or closely related species. In contrast, whereas some alloherpesviruses induce severe disease in only one or few closely related members of the same genus, others are able to establish subclinical infections in a broader range of hosts. Thus, although CyHV-3 causes a disease only in common and koi carp, its genome has been detected in a wide range of fish species (see Section 3.2.1.1). Third, an age-dependent pathogenesis has been described for several fish herpesviruses, in that AciHV-1, AciHV-2, CyHV-1, CyHV-2, SalHV-2, SalHV-3, and ictalurid herpesvirus 2 (IcHV-2) are particularly pathogenic for young fry (Hanson et al., 2011; van Beurden & Engelsma, 2012). Fourth, a marked difference in the outcome of herpesvirus infection in poikilothermic hosts is related to their temperature dependency, both *in vitro* and *in vivo*. For example, anguillid herpesvirus 1 (AngHV-1), infecting Japanese eel (*Anguilla japonica*) and European eel (*Anguilla anguilla*), only propagates in eel kidney 1 cells between 15 and 30 °C, with an optimum around

20–25 °C (Sano, Fukuda, & Sano, 1990; van Beurden, Engelsma, et al., 2012). *In vivo*, replication of ranid herpesvirus 1 (RaHV-1) is promoted by low temperature, whereas induction of tumor metastasis is promoted by high temperature (McKinnell & Tarin, 1984). In general, fish herpesvirus-induced infection is less severe or even asymptomatic if the ambient water temperature is suboptimal for virus replication, which explains the seasonal occurrence of certain fish herpesviruses, including CyHV-3 (Gilad et al., 2003). In practice, these biological properties have been utilized successfully to immunize naturally carp against CyHV-3 (Ronen et al., 2003) and to reduce the clinical signs and mortality rates of AngHV-1 infections in eel culture systems (Haenen et al., 2002). In addition, temperature could play a role in the induction of latency and reactivation of fish herpesviruses (see Sections 3.2.3.2 and 3.2.3.3).

2.3 Herpesviruses Infecting Fish

The first description of lesions caused by a fish herpesvirus dates from the sixteenth century, when the Swiss naturalist Conrad Gessner described a pox disease of carp. Four hundred years later, the pox-like lesions were found to be associated with herpesvirus-like particles (Schubert, 1966), later designated as CyHV-1 (Sano, Fukuda, & Furukawa, 1985). However, the alloherpesviruses that were first studied in detail originated from the North American leopard frog (*Rana pipiens*). Lucké tumor herpesvirus or RaHV-1 was identified as the etiological agent of renal adenocarcinoma or Lucké tumor (Fawcett, 1956), and frog virus 4 or ranid herpesvirus 2 (RaHV-2) was isolated subsequently from the pooled urine of tumor-bearing frogs (Gravell, Granoff, & Darlington, 1968).

Alloherpesviruses infect a wide range of fish species worldwide, including several of the most important aquaculture species such as catfish, salmon, carp, sturgeon, and eel. As a result of host specificity, the prevalence of specific fish herpesviruses may be restricted to certain parts of the world. For example, pilchard herpesvirus 1 has been described only in wild Australasian pilchards (*Sardinops sagax neopilchardus*) in Australia and New Zealand (Whittington, Crockford, Jordan, & Jones, 2008), whereas CyHV-2 has a worldwide prevalence due to the international trade in goldfish (Goodwin, Sadler, Merry, & Marecaux, 2009).

Currently, 10 herpesviruses infecting fish are included in the family *Alloherpesviridae* (Table 1). At least a dozen of other fish herpesviruses have been described, but many of these viruses have not been isolated yet, and the availability of limited sequence data hampers their official classification (Hanson et al., 2011; Waltzek et al., 2009). Interestingly, all but one of these viruses occur in bony fish, the exception having been found in a shark. Based on the number of different herpesvirus species recognized in humans (i.e., nine) and domestic animals, it is probable that each of the numerous fish species hosts multiple herpesviruses. It is likely that the alloherpesvirus species currently known are biased toward commercially relevant hosts, and the species that cause significant disease.

Channel catfish virus (IcHV-1) has been the prototypic fish herpesvirus for decades (Hanson et al., 2011; Kucuktas & Brady, 1999). In the late 1960s, the extensive catfish (*Ictalurus punctatus*) industry in the United States experienced high mortality rates among fry and fingerlings (Wolf, 1988). The causative virus was isolated and shown by electron microscopy to possess the distinctive morphological features of a herpesvirus (Wolf & Darlington, 1971). The genome sequence of IcHV-1 revealed that fish herpesviruses have evolved separately from herpesviruses infecting mammals, birds, and reptiles (Davison, 1992; see Section 2.1.2).

In the late 1990s, mass mortalities associated with epidermal lesions, gill necrosis, and nephritis occurred worldwide in koi and common carp aquaculture (Haenen, Way, Bergmann, & Ariel, 2004). This highly contagious and virulent disease was called KHV disease (KHVD) and was shown to be caused by a herpesvirus, which was later designated CyHV-3 (Bretzinger et al., 1999; Hedrick et al., 2000; Waltzek et al., 2005). Due to its economic impact on carp culture and its rapid spread across the world, CyHV-3 was listed as a notifiable disease by the World Organization for Animal Health OIE (Michel, Fournier, Lieffrig, Costes, & Vanderplasschen, 2010). Although IcHV-1 had been the model of fish herpesviruses for more than three decades, the associated problems mainly affected the catfish industry in the USA and could be limited by management practices (Hanson et al., 2011; Kucuktas & Brady, 1999). Meanwhile, the desire to protect common and koi carp from the negative impact of CyHV-3 infection prompted an increased interest to study this virus. In addition, the natural host of CyHV-3, the common carp, has been a traditional species for fundamental research on fish immunology, making it a perfect model to study host–virus interactions (Adamek, Steinhagen, et al., 2014; Rakus et al., 2013). As a consequence, advancement in our understanding of CyHV-3 now far exceeds that of any other alloherpesvirus.

3. CYPRINID HERPESVIRUS 3

3.1 General Description

3.1.1 Morphology and Morphogenesis

Like all members of the order *Herpesvirales*, CyHV-3 virions are composed of an icosahedral capsid containing a single copy of a large, linear, double-stranded DNA genome, a host-derived lipid envelope bearing viral glycoproteins and an amorphous proteinaceous layer termed the tegument, which resides between the capsid and the envelope (Fig. 2A; Mettenleiter, 2004; Mettenleiter et al., 2009). The diameter of CyHV-3 virions varies somewhat according to the infected cell type both *in vitro* (180–230 nm in koi fin cells KF-1 (Hedrick et al., 2000) and 170–200 nm in koi fin derived cells (KF-1, NGF-2, and NGF-3) (Miwa, Ito, & Sano, 2007)) and *in vivo* (167–200 nm in various organs (Miyazaki, Kuzuya, Yasumoto, Yasuda, & Kobayashi, 2008)). Despite the very limited sequence conservation in proteins involved in morphogenesis, members of the families *Herpesviridae*, *Alloherpesviridae*, and *Malacoherpesviridae* exhibit a common structure, suggesting that the mechanisms used are similar (Mettenleiter et al., 2009). Indeed, the structure of the CyHV-3 virion and its morphogenesis are entirely typical of herpesviruses (Figs. 2B and 3). Assembly of the nucleocapsids (size 100 nm) takes place in the nucleus (Miwa et al., 2007; Miyazaki et al., 2008), where marginalization of chromatin occurs at the inner nuclear membrane (Miwa et al., 2007; Miyazaki et al., 2008). Mature nucleocapsids with an electron-dense core composed of the complete viral genome bud at the inner nuclear membrane into the perinuclear space and are then released into the cytoplasm according to the envelopment/de-envelopment model (Miwa et al., 2007; Miyazaki et al., 2008). Viral nucleocapsids in the cytoplasm, prior to envelopment, are surrounded by a layer of electron-dense material composed of tegument proteins (Fig. 3). A similar feature is found in members of the subfamily *Betaherpesvirinae* but not in the subfamilies *Alpha-* and *Gammaherpesvirinae*, where they appear to be naked (Mettenleiter et al., 2009). Finally, the lipid envelope bearing viral glycoproteins is acquired by budding into vesicle membranes derived from the Golgi apparatus (Mettenleiter et al., 2009; Miwa et al., 2007; Miyazaki et al., 2008).

3.1.2 Genome

The complete DNA sequences of four CyHV-3 strains derived from different geographical locations have been determined (Aoki et al., 2007; Li et al.,

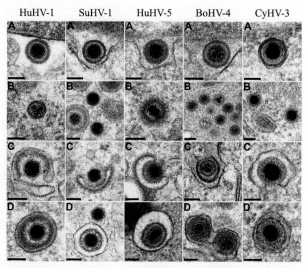

Figure 3 Primary and secondary envelopment of some herpesviruses. (A) Primary-enveloped virions in the perinuclear space. In comparison with Fig. 2, the electron-dense sharply bordered layer of tegument underlying the envelope and the conspicuous absence of envelope glycoprotein spikes are noteworthy. (B) After translocation into the cytosol, capsids of HuHV-1, SuHV-1, and BoHV-4 appear "naked," whereas those of HuHV-5 and CyHV-3 are covered with a visible layer of "inner" tegument. (C) Secondary envelopment and (D) presence of enveloped virions within a cellular vesicle during transport to the plasma membrane. The same stages can be observed for the members of the *Herpesviridae* and *Alloherpesviridae* families. Bars represent 100 nm. HuHV-1, *Human herpesvirus 1* (herpesvirus simplex 1, HSV-1); SuHV-1, *Suid herpesvirus 1* (pseudorabies virus, PrV); HuHV-5, *Human herpesvirus 5* (human cytomegalovirus, HCMV); BoHV-4, *Bovine herpesvirus 4*; CyHV-3, *Cyprinid herpesvirus 3*. *Adapted with permission from Mettenleiter et al. (2009). Copyright © Elsevier.*

2015). CyHV-3 is notable for having the largest known genome among the herpesviruses, at 295 kbp. It is followed by its two closest relatives, CyHV-1 (291 kbp) and CyHV-2 (290 kbp) (Aoki et al., 2007; Davison et al., 2013). Like all other fully sequenced alloherpesvirus genomes, the CyHV-3 genome contains two copies of a terminal direct repeat (TR), which, in the case of CyHV-3, are 22 kbp in size. The arrangement of open reading frames (ORFs) in the CyHV-3 genome that are predicted to encode functional proteins was first described by Aoki et al. (2007), and later refined on the basis of a full comparison with the genomes of other viruses in the genus *Cyprinivirus*, as well as members of the other genera (Davison et al., 2013). A map of the predicted CyHV-3 genes is shown in Fig. 4; the central part of the genome and the TR encode 148 (ORF9–ORF156) and

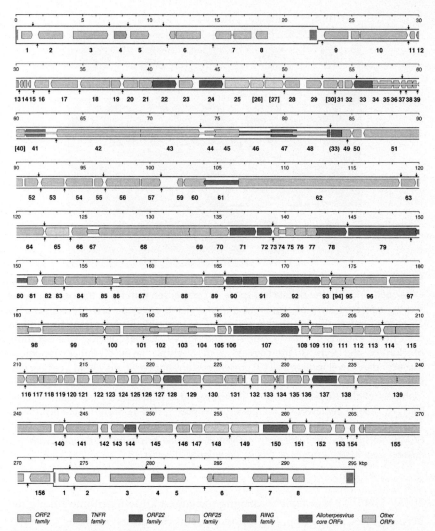

Figure 4 Map of the CyHV-3 genome. The terminal direct repeat (TR) is shown in a thicker format than the rest of the genome. ORFs predicted to encode functional proteins are indicated by arrows (see the key at the foot), with nomenclature lacking the ORF prefix given below. Introns are shown as narrow white bars. The colors (gray shades in the print version) of protein-coding regions indicate core ORFs that are convincingly conserved among members of the family *Alloherpesviridae*, families of related ORFs, and other ORFs. Telomere-like repeats at the ends of TR are shown by gray-shaded blocks. Predicted poly(A) sites are indicated by vertical arrows above and below the genome for rightward- and leftward-oriented ORFs, respectively. *Reproduced with permission from Davison et al. (2013). Copyright © American Society for Microbiology.*

8 (ORF1–ORF8) ORFs, respectively. The latter are therefore duplicated in the copies of TR. One of the unusual features in the sequenced CyHV-3 genomes is the presence of fragmented, and therefore probably non-functional, ORFs. The precise set of such ORFs varies from strain to strain, and there is evidence that at least some originated *in vivo* rather than during viral isolation in cell culture. It is possible that loss of gene functions may have contributed to emergence of disease in carp populations.

Consistent with their close relationships, the 3 cyprinid herpesviruses share 120 conserved genes, of which up to 55 have counterparts in the more distantly related AngHV-1, which is also a member of the genus *Cyprinivirus*. However, as mentioned above, only 12 genes are conserved across the family *Alloherpesviridae* (see Section 2.1.2). The relevant ORFs are marked in Fig. 4, and their characteristics are listed in the upper part of Table 3. There are perhaps two additional genes in this core class (ORF66 and ORF99; not listed in Table 3), but the evidence for their conservation is minimal. Comments may be made on the features or functions of a sizeable number of the remaining gene products, as shown in the lower part of Table 3. This list omits genes that are members of related families and lack other clearly identifiable characteristics, such as incorporation into virions or similarity to other genes. It also excludes genes encoding proteins of which the only identifiable features are those indicating that they might be associated with membranes (e.g., the presence of potential signal peptides or hydrophobic transmembrane regions), which are numerous in CyHV-3. Also, the ancestors of CyHV-3 have evidently captured several genes from the host cell (e.g., the deoxyuridine triphosphatase and interleukin-10 genes) or other viruses (e.g., genes of which the closest relatives are found in iridoviruses or poxviruses) (Ilouze, Dishon, Kahan, & Kotler, 2006).

The CyHV-3 genome also contains five gene families that have presumably arisen by gene duplication, a mechanism for generating diversity that has been used commonly by herpesviruses in all three families. They are shaded in distinguishing colors (gray shades) in Fig. 4. These are the ORF2 family (ORF2, ORF3, ORF9, ORF129, ORF130, and ORF135), the TNFR family (ORF4 and ORF12, encoding proteins related to tumor necrosis factor receptor), the ORF22 family (ORF22, ORF24, and ORF137), the ORF25 family (ORF25, ORF26, ORF27, ORF65, ORF148, and ORF149, encoding potential membrane proteins containing an immunoglobulin domain), and the RING family (ORF41, ORF128, ORF144, and ORF150). Some of the proteins encoded by these genes are virion components (ORF137, ORF25, ORF27, ORF65,

Table 3 Information on Selected CyHV-3 ORFs[a]

ORF Name	Function or Features of Encoded Protein
Conserved among all sequenced members of the family *Alloherpesviridae*	
ORF33	DNA packaging terminase subunit 1
ORF46	Putative helicase–primase primase subunit
ORF47	Putative DNA packaging terminase subunit 2
ORF61	
ORF71	Putative helicase–primase helicase subunit
ORF72	Capsid triplex subunit 2; virion protein
ORF78	Capsid maturation protease; virion protein
ORF79	DNA polymerase catalytic subunit
ORF80	
ORF90	Virion protein
ORF92	Major capsid protein
ORF107	
Additional ORFs with recognizable features	
ORF4	Tumor necrosis factor receptor; member of TNFR gene family
ORF11	*Virion protein*
ORF12	Tumor necrosis factor receptor; member of TNFR gene family
ORF16	Predicted membrane protein; similar to G protein-coupled receptors
ORF19	Deoxyguanosine kinase
ORF23	Ribonucleotide reductase subunit 2
ORF25	Predicted membrane protein; contains an immunoglobulin domain; virion protein; member of ORF25 gene family
ORF27	Predicted membrane protein; contains an immunoglobulin domain; *virion protein*; member of ORF25 gene family
ORF28	Contains an NAD(P)-binding Rossmann-fold domain; similar to bacterial NAD-dependent epimerase/dehydratase
ORF31	Similar to eukaryotic PLAC8 proteins; virion protein
ORF32	Similar to a family of Singapore grouper iridovirus proteins; predicted membrane protein; virion protein
ORF34	*Virion protein*

Table 3 Information on Selected CyHV-3 ORFs—cont'd

ORF Name	Function or Features of Encoded Protein
ORF35	*Virion protein*
ORF36	*Virion protein*
ORF41	Contains a RING-type C3HC4 zinc finger domain; member of RING gene family
ORF42	Virion protein
ORF43	Virion protein
ORF44	*Virion protein*
ORF45	Virion protein
ORF48	Similar to protein kinases
ORF51	Virion protein
ORF54	Contains a putative zinc-binding domain
ORF55	Thymidine kinase
ORF57	Similar to crocodile poxvirus protein CRV155; virion protein
ORF59	Predicted membrane protein; virion protein
ORF60	Virion protein
ORF62	Contains an OTU-like cysteine protease domain; virion protein
ORF64	Predicted membrane protein; similar to equilibrative nucleoside transporter ENT1
ORF65	Predicted membrane protein; contains an immunoglobulin domain; virion protein; member of ORF25 gene family
ORF66	Capsid triplex subunit 1; virion protein
ORF68	Similar to myosin and related proteins; virion protein
ORF69	Virion protein
ORF70	Virion protein
ORF81	Multiple transmembrane protein; virion protein
ORF83	Predicted membrane protein; *virion protein*
ORF84	Virion protein
ORF89	Virion protein
ORF91	*Virion protein*
ORF94	Predicted membrane protein; similar to trypsin-like serine proteases

Continued

Table 3 Information on Selected CyHV-3 ORFs—cont'd

ORF Name	Function or Features of Encoded Protein
ORF95	Virion protein
ORF97	Virion protein
ORF98	Uracil-DNA glycosylase
ORF99	Predicted membrane protein; virion protein
ORF104	Similar to protein kinases
ORF106	*Virion protein*
ORF108	Predicted membrane protein; virion protein
ORF112	Contains a double-stranded nucleic acid-binding domain (helix–turn–helix); virion protein
ORF114	Predicted membrane protein; similar to *Danio rerio* LOC569866
ORF115	Predicted membrane protein; virion protein
ORF116	Predicted membrane protein; *virion protein*
ORF123	Deoxyuridine triphosphatase; *virion protein*
ORF128	Contains a RING-type C3HC4 zinc finger domain; similar to SPRY and TRIM proteins; member of RING gene family
ORF131	Predicted membrane protein; virion protein
ORF132	Predicted membrane protein; virion protein
ORF134	Interleukin-10
ORF136	Predicted membrane protein; virion protein
ORF137	*Virion protein*; member of ORF22 gene family
ORF139	Predicted membrane protein; similar to poxvirus B22R proteins
ORF140	Thymidylate kinase
ORF141	Ribonucleotide reductase subunit 1
ORF144	Contains a RING-type C3HC4 zinc finger domain; member of RING gene family
ORF148	Predicted membrane protein; contains an immunoglobulin domain; virion protein; member of ORF25 gene family
ORF149	Predicted membrane protein; contains an immunoglobulin domain; virion protein; member of ORF25 gene family
ORF150	Contains a RING-type C3HC4 zinc finger domain; member of RING gene family

[a]Data derived from Aoki et al. (2007), Michel, Leroy, et al. (2010), Yi et al. (2014), and Davison et al. (2013).

Italic-type indicates virion proteins detected in only some of the strains tested (Michel, Leroy, et al., 2010; Yi et al., 2014).

ORF148, and ORF149). Members of each of these gene families are also present in CyHV-1 and CyHV-2, whereas AngHV-1 lacks all but the TNFR family, having instead several other families that are absent from the cyprinid herpesviruses (Davison et al., 2013).

Herpesvirus genomes are described as infectious because their transfection into permissive cells is sufficient to initiate replication and the production of progeny virions. This property has been exploited to produce recombinant viruses by using bacterial artificial chromosome (BAC) cloning of the entire viral genome and prokaryotic recombination technologies. Such an approach has been used extensively for members of the *Herpesviridae* family (Tischer & Kaufer, 2012) and has been demonstrated to be also applicable to CyHV-3 (Costes et al., 2008).

3.1.3 Genotypes

Early investigations on CyHV-3 genetic diversity comparing partial DNA polymerase gene and partial major envelope protein gene sequences of CyHV-3 isolates from Japan, the USA, and Israel showed a high degree of nucleotide sequence identity (Ishioka et al., 2005). Similar sequence identities were also found among isolates from Poland and Germany (Antychowicz, Reichert, Matras, Bergmann, & Haenen, 2005; El-Matbouli, Saleh, & Soliman, 2007), suggesting that the virus causing disease in carp worldwide represented a single virus entity. Comparison of the complete genome sequences of three isolates from Japan (CyHV-3 J), the USA (CyHV-3 U), and Israel (CyHV-3 I) also revealed more than 99% identity (Aoki et al., 2007) which was consistent with this scenario.

Despite this close genetic relationship between isolates, the alignment of three complete CyHV-3 sequences revealed numerous minor deletions/insertions and single-nucleotide substitutions. These variations enabled a distinction between the CyHV-3 J lineage and the lineage represented by CyHV-3 U and CyHV-3 I isolates (Aoki et al., 2007; Bercovier et al., 2005). Recently, the full-length sequencing of a fourth strain, CyHV-3 GZ11 (isolated from a mass mortality outbreak in adult koi in China), revealed a closer relationship of this isolate with the CyHV-3 U/I lineage (Li et al., 2015). The existence of two lineages was confirmed on a larger set of European and Asian isolates using a PCR-based approach targeting two distinct regions of the genome (Bigarré et al., 2009). Marker I, located between ORF29 and ORF31 of CyHV-3 (Aoki et al., 2007), was 168 bp in length (designated I^{++}) for CyHV-3 J and only 130 bp (I^{--}) for the CyHV-3 U/I. Marker II, located upstream of ORF133, was 352 bp in length in the CyHV-3 J sequence (II$^+$) compared to 278 bp (II$^-$) in the other two sequences. These studies also provided the first evidence of other potential genotypes, describing a unique genotype of CyHV-3 in koi carp from Poland that was identical to the CyHV-3 U/I viruses in marker II (II$^-$) but shared features of both the CyHV-3 J and CyHV-3 U/I genotypes in marker I (I^{-+}) (Bigarré et al., 2009). A similar profile was observed for a CyHV-3 strain from Korea and the GZ11 strain from China (Kim & Kwon, 2013; Li et al., 2015). The same markers were used to identify another novel "intermediate" genotype of CyHV-3 in Indonesia that resembled the CyHV-3 J genotype in marker I (I^{++}) but was identical to the CyHV-3 U/I genotype in marker II (II$^-$) (Sunarto et al., 2011). Sunarto et al. (2011) speculated that genotype I^{--}II$^-$ has evolved through genetic intermediates, I^{-+}II$^-$ and I^{++}II$^-$, to give rise to I^{++}II$^+$, and that the first genotype I^{--}II$^-$ (corresponding to E1 genotype based on the thymidine kinase (TK) gene sequence, see below) may be the origin of CyHV-3. Alternatively, it is suggested that an ancestral form diverged to give rise to two lineages, CyHV-3 J and CyHV-3 U/I (Aoki et al., 2007).

Analysis of the TK gene sequence (Bercovier et al., 2005), particularly the region immediately downstream of the stop codon, provided significantly more resolution (Table 4). In combination with sequence data for the SphI-5 (coordinates 93604–93895, NCBI: DQ657948) and the 9/5 (coordinates 165399–165882, NCBI: DQ657948) regions (Gilad et al., 2002; Gray et al., 2002), nine different genotypes were identified (Kurita et al., 2009). The CyHV-3 from Asia showed a high degree of sequence homology, although two variants were differentiated based on a single-nucleotide polymorphism in the TK gene (A1 and A2). In contrast, seven genotypes were identified in CyHV-3 from outside of Asia (E1–E7).

Interestingly, a study by Han et al. (2013) identified a sequence insertion in a glycoprotein gene (ORF125) of a Korean isolate (CyHV-3 K) compared with the viruses from Japan (CyHV-3 J), the USA (CyHV-3 U), and Israel (CyHV-3 I). This suggests that the CyHV-3 K is distinct from the CyHV-3 A1 and A2 genotypes. However, in the absence of comparable data from the TK gene, marker I or II regions, it is not possible to confirm this hypothesis (Han et al., 2013). In addition, some recent CyHV-3 isolates from Korea, Malaysia, and China were shown to belong to the E4 genotype which suggests the emergence of European lineages in Asia (Chen et al., 2014; Dong et al., 2013; Kim & Kwon, 2013; Li et al., 2015).

Besides the nucleotide mismatches, insertions, or deletions, much of the sequence differences between CyHV-3 isolates occurred at the level of

Table 4 Genotyping Scheme for CyHV-3 Based on Three Distinct Regions of the Genome: the 9/5 Region (Gray, Mullis, LaPatra, Groff, & Goodwin, 2002), the SphI-5 Region (Gilad et al., 2002), and the TK Gene (Bercovier et al., 2005)

		9/5 Region		SphI-5 Region							TK Gene		
Genotype	Country of Origin	184–187	212–218	209	586–588	94	778	813–814	849–850	877–885	945–956	957–958	961–967
A1	Japan[a,b], Indonesia[a], Taiwan[a], Philippines[a], South Korea[c], Malaysia[b], China[b]	TTTT	AAAAAA	C	–	C	A	–	AA	TTTTTT	CTTTAAAAAAAA	–	AGATATT
A2	Indonesia[a], Taiwan[a]	TTTT	AAAAAA	C	–	C	A	–	AA	TTTTTTTT	CTTTAAAAAAAA	–	AGATATT
E1	USA[a], Netherlands[a]	TTTT	AAAAAAA	C	AAC	C	G	AT	–	TTTTTTTT	CTTTAAAAAAAA	CA	AGATATT
E2	Netherlands[a]	TTTT	AAAAAAA	T	AAC	C	G	AT	–	TTTTTTTT	CTTTAAAAAAAA	CA	AGATATT
E3	Netherlands[a]	–	AAAAAAA	C	AAC	C	G	AT	–	TTTTTTTT	CTTTAAAAAAAA	CA	AGATATT
E4	Netherlands[a], South Korea[c], Malaysia[b], China (TK)[d]	TTTT	AAAAAAA	C	AAC	C	G	AT	AA	TTTTTTTT	CTTTAAAAAAAA	CA	AGATATT
E5	Netherlands[a]	TTTT	AAAAAAA	C	AAC	C	G	AT	–	TTTTTTTT	–	–	–
E6	Israel[a]	TTTT	AAAAAAA	C	AAC	T	G	AT	–	TTTTTTTT	CTTTAAAAAAAA	CA	AGATATT
E7	UK[a]	TTTT	AAAAAAA	C	AAC	C	G	AT	–	TTTTTTTT	CTTTAAAAAAAA	CA	AGATATT

[a]From Kurita et al. (2009).
[b]From Chen et al. (2014).
[c]From Kim and Kwon (2013).
[d]From Dong, Li, Weng, Xie, and He (2013).
Adapted from Kurita et al. (2009).

variable number of tandem repeat (VNTR) sequences (Avarre et al., 2011). In agreement with other genetic studies (Bigarré et al., 2009; Kim & Kwon, 2013), analyses using multiple VNTR loci identified two lineages which were equivalent to the Asian and European viruses, but, with the increased discriminatory power of VNTR analysis, allowed the identification of up to 87 haplotypes (Avarre et al., 2011, 2012). As expected, several of the isolates from the Netherlands showed a close relationship to CyHV-3 J and were assigned to the same lineage, but the isolates from France and the Netherlands generally showed a closer relationship to CyHV-3 U/I and were assigned to the European lineage (Bigarré et al., 2009). Surprisingly, the Indonesian isolates, with a $I^{++}II^-$ haplotype (Sunarto et al., 2011), are closely related to CyHV-3 J and were assigned to the same lineage. No VNTR data were available for CyHV-3 K and GZ11 strains.

VNTR polymorphism has shown great potential for differentiating isolates of large DNA viruses such as human herpesvirus 1 (Deback et al., 2009). However, since the mechanism of VNTR evolution in CyHV-3 is not fully understood, it remains possible for the different phylogeographic types to share some VNTR features but have acquired them through separate evolutionary routes. Therefore, in future epidemiological studies on CyHV-3, it may be necessary to consider undertaking an initial phylogeographic analysis using the non-VNTR polymorphisms (insertions, deletions, and point mutations) observed throughout the genome and, only after, exploit the power of the VNTR variability to provide resolution to the isolate level.

3.1.4 Transcriptome

Herpesvirus gene expression follows a coordinated temporal pattern upon infection of permissive cells as shown in Fig. 2B (Pellett et al., 2011b). Immediate-early (IE) genes are first transcribed in the absence of *de novo* protein synthesis and regulate the subsequent expression of other genes. Expression of early (E) genes is dependent on IE-gene expression, and they encode enzymes and proteins involved in the modification of host cell metabolism and the viral DNA replication complex. The late (L) genes form the third and last set to be expressed, dependent on viral DNA synthesis, and primarily encode the viral structural proteins. The first indication that fish herpesvirus gene expression follows a similar temporal pattern came from *in vitro* studies on IcHV-1 transcription (Hanson et al., 2011).

More recently, two extensive genome-wide gene expression analyses of CyHV-3 (Ilouze, Dishon, & Kotler, 2012a) and AngHV-1 (van Beurden, Peeters, Rottier, Davison, & Engelsma, 2013) explored the kinetic class of each annotated ORF following two approaches. First, gene expression was studied by RT-PCR or RT-qPCR during the first hours post-infection (hpi). Second, cycloheximide (CHX), and either cytosine-β-D-arabinofuranoside (Ara-C) or phosphonoacetic acid (PAA), were used to block *de novo* protein synthesis and viral DNA replication, respectively. In the presence of CHX, only IE genes are expressed, whereas in the presence of Ara-C or PAA, the IE and E genes but not the L genes are expressed. For CyHV-3, viral RNA synthesis was evident as early as 1 hpi, and viral DNA synthesis initiated between 4 and 8 hpi (Ilouze et al., 2012a). Transcription of 59 ORFs was detectable from 2 hpi, 63 ORFs from 4 hpi and 28 ORFs from 8 hpi. Transcription of six ORFs was only evident at 24 hpi (Table 5). Expression kinetics for related AngHV-1 genes were analyzed differently, thus hampering direct comparison, but in general followed the same pattern (van Beurden et al., 2013). RNAs from all 156 predicted ORFs of CyHV-3 were detected (including ORF58 which was initially predicted based on a marginal prediction but recently removed from the predicted genome map (Fig. 4; Davison et al., 2013)), and based on the observation that antisense transcription for related AngHV-1 was very low, it is expected that all annotated ORFs indeed code for viral RNAs (Aoki et al., 2007; Ilouze et al., 2012a; van Beurden, Gatherer, et al., 2012).

By blocking protein synthesis or viral DNA replication, 15 IE, 112 E, and 22 L genes were identified for CyHV-3, whereas for 7 ORFs no classification was possible (Ilouze et al., 2012a; Table 5). In general, this classification followed the expression kinetics determined for each ORF, with most IE genes being expressed at 1 or 2 hpi, most E genes between 2 and 4 hpi and most L genes at 8 hpi. For AngHV-1, 4 IE genes, 54 E or E-L genes, and 68 L genes were found (van Beurden et al., 2013). As there is no clear boundary between the E-L (or leaky-late) and L genes, these differences may be explained by sensitivity of the method used to determine the onset of gene expression and data analysis. Similar to mammalian herpesviruses, gene transcripts known to be involved in DNA replication were expressed early, while proteases and enzymes involved in virion assembly and maturation were expressed late (Ilouze et al., 2012a; van Beurden et al., 2013). Inhibition of some E genes involved in DNA replication (e.g., TK and DNA polymerase) by specific siRNA decreased viral release from infected cells (Gotesman, Soliman, Besch, & El-Matbouli, 2014).

Interestingly, in IcHV-1, CyHV-3, and AngHV-1, the IE genes show a clear clustering in or near the TRs, suggesting positional conservation of these regulatory genes (Ilouze et al., 2012a; Stingley & Gray, 2000; van Beurden

Table 5 Transcriptomic Classification of CyHV-3 ORFs

ORF	Putative Function[a]	Kinetic Class[b]	22 °C[c] (hpi)	30 °C[d] (dpi)
1L/R		IE	2	1–8
2L/R		E	2	1
3L/R		IE	2	1
4L/R	Immune regulation	E	2	1–8
5L/R		E	4	1
6L/R		IE	2	–
7L/R		IE	2	1–8
8L/R		IE	2	1
9		IE	2	1
10		IE	2	1–8
11	Virion protein	IE	2	1
12	Immune regulation	L	8	1
13		E	2	1
14		E	2	1
15		E	2	1
16	Intracellular signaling	E	2	1
17		L	2	1
18		E	2	1
19	Nucleotide metabolism	E	4	1
20		E	4	1
21		E	4	1
22		E	4	–
23	Nucleotide metabolism	E	2	1
24		IE	2	1
25	Virion protein	E	4	1
26		E	2	1
27	Virion protein	E	2	1
28		E	4	–

Table 5 Transcriptomic Classification of CyHV-3 ORFs—cont'd

ORF	Putative Function	Kinetic Class	22 °C (hpi)	30 °C (dpi)
29		E	2	1
30		E	4	1–8
31	Virion protein	E	4	1
32	Virion protein	E	2	1–8
33	DNA encapsidation	E	4	1
34	Virion protein	UN	24	–
35	Virion protein	E	4	1
36	Virion protein	E	4	1
37		E	2	1–8
38		E	2	1–8
39		E	2	1–8
40		E	2	1
41		E	2	1
42	Virion protein	E	4	1–8
43	Virion protein	E	4	1
44	Virion protein	L	8	–
45	Virion protein	E	4	1
46	DNA replication	E	4	1
47	DNA encapsidation	L	8	1
48	Protein phosphorylation	E	2	1
49		UN	24	–
50		E	2	1
51	Virion protein	E	4	1
52		E	4	1
53		E	2	1
54		IE	2	1
55	Nucleotide metabolism	E	2	1
56		E	2	1–8

Continued

Table 5 Transcriptomic Classification of CyHV-3 ORFs—cont'd

ORF	Putative Function	Kinetic Class	22 °C (hpi)	30 °C (dpi)
57	Virion protein	UN	8	1
58		E	4	–
59	Virion protein	E	4	1
60	Virion protein	E	4	1
61		E	4	1
62	Virion protein	L	8	–
63		E	4	–
64		E	2	–
65	Virion protein	L	8	–
66	Virion protein/capsid morphogenesis	E	4	1
67		E	4	–
68	Virion protein	L	8	1
69	Virion protein	UN	24	–
70	Virion protein	UN	24	1–8
71	DNA replication	E	4	1
72	Virion protein/capsid morphogenesis	E	4	–
73		E	8	1
74		L	8	–
75		L	8	1
76		L	8	1
77		E	4	–
78	Virion protein/capsid morphogenesis	L	8	–
79	DNA replication	E	4	–
80		E	4	1
81	Virion protein	E	4	1
82		E	4	–

Table 5 Transcriptomic Classification of CyHV-3 ORFs—cont'd

ORF	Putative Function	Kinetic Class	22 °C (hpi)	30 °C (dpi)
83	Virion protein	E	8	1–8
84	Virion protein	E	4	1
85		L	8	1
86		L	8	1
87		E	4	1
88		IE	4	–
89	Virion protein	L	8	–
90	Virion protein	E	8	–
91	Virion protein	E	4	–
92	Virion protein/major capsid protein	E	4	1
93		E	4	1
94		E	4	–
95	Virion protein	L	8	1
96		E	4	1
97	Virion protein	L	8	1
98	DNA repair	E	4	1
99	Virion protein	E	8	–
100		E	4	1
101		E	4	–
102		E	4	–
103		E	4	–
104	Protein phosphorylation	E	4	1
105		UN	24	–
106	Virion protein	L	8	–
107		E	4	–
108	Virion protein	E	4	–
109		E	4	–

Continued

Table 5 Transcriptomic Classification of CyHV-3 ORFs—cont'd

ORF	Putative Function	Kinetic Class	22 °C (hpi)	30 °C (dpi)
110		L	8	–
111		E	2	1
112	Virion protein/immune regulation	IE	2	–
113		E	8	–
114		L	8	1–18
115	Virion protein	L	8	1–18
116	Virion protein	E	4	1
117		E	2	1
118		E	2	1
119		UN	24	–
120		E	2	1
121		E	2	1
122		E	4	1
123	Virion protein/nucleotide metabolism	E	4	1
124		E	4	1–8
125		L	8	1
126		L	8	–
127		E	2	1
128		E	4	1
129		E	4	1
130		E	2	1
131	Virion protein	E	4	1
132	Virion protein	E	2	1
133		E	4	1
134	Immune regulation	E	2	1
135		E	4	–
136	Virion protein	E	4	1

Table 5 Transcriptomic Classification of CyHV-3 ORFs—cont'd

ORF	Putative Function	Kinetic Class	22 °C (hpi)	30 °C (dpi)
137	Virion protein	E	2	1
138		E	2	1
139	Immune regulation	E	2	1–8
140	Nucleotide metabolism	E	4	–
141	Nucleotide metabolism	E	8	–
142		E	2	1
143		E	2	1–8
144		E	4	–
145		E	4	–
146		IE	2	1
147		E	2	–
148	Virion protein	E	4	1
149	Virion protein	IE	2	–
150		E	2	–
151		E	2	1
152		E	2	1
153		E	2	1
154		E	2	1
155		IE	2	1–8
156		E	2	1

[a]Putative gene functions were adapted from Davison et al. (2013).
[b]Kinetic class as determined by transcription analysis in the presence of CHX or Ara-C (adapted from Ilouze et al., 2012a). Light grey: immediate early (IE) gene; intermediate grey: early (E) gene; dark grey: late (L) gene; white: unknown (UN).
[c]Initiation of viral mRNA transcription at permissive temperature (adapted from Ilouze et al., 2012a).
[d]Presence of CyHV-3 transcripts at restrictive temperature (adapted from Ilouze, Dishon, & Kotler, 2012b).
dpi, days post-infection; hpi, hours post-infection.

et al., 2013). The E and L genes are mainly located in the unique long region of the genome, with almost half of the CyHV-3 E genes clustered and transcribed simultaneously (Ilouze et al., 2012a). This observation may be biased, however, by 3′-coterminality of transcripts, which was shown to be abundant in the AngHV-1 genome (van Beurden, Gatherer, et al., 2012).

3.1.5 Structural Proteome and Secretome

Initial predictions of the structural proteome of CyHV-3 were based on comparison with experimental findings obtained for IcHV-1 and bioinformatically predicted properties of the putative CyHV-3 encoded proteins (Aoki et al., 2007; Davison & Davison, 1995). More recently, two independent studies explored the structural proteome of one European and two Chinese CyHV-3 isolates by a combination of virus particle purification, gel electrophoresis, and mass spectrometry-based proteomic approaches (Michel, Leroy, et al., 2010; Yi et al., 2014). A total of 34 structural proteins were identified for all 3 CyHV-3 isolates, and another 12 proteins were found in only 1 or 2 of the 3 studied isolates (Table 6 and Fig. 5). The latter were generally of low abundance, suggesting that these small differences in protein constitution indicate either strain-specific variation or interstudy variation. Overall, the total number of structural proteins of viral origin reported for CyHV-3 (46) corresponds with the number reported for closely related AngHV-1 (40) and is in line with numbers reported for members of the *Herpesviridae* family, e.g., 44 for herpes simplex virus 1 (Loret, Guay, & Lippe, 2008).

Comparisons of homologous genes with similar studies for related alloherpesviruses IcHV-1 and AngHV-1, as well as bioinformatical predictions of protein properties, enabled putative localization of the proteins within the virion (Table 6). Based on these predictions, five capsid proteins were identified, including the highly conserved major capsid protein, capsid triplex subunit 1 and 2, and the capsid maturation protease. Indeed, the architecture and protein composition of fish herpesvirus capsids generally mirror that of mammalian herpesviruses, with the exception of the small protein which forms the hexon tips in mammalian herpesviruses (Booy et al., 1996; Davison & Davison, 1995). Comparison with the closely related AngHV-1 resulted in the identification of 11 tegument or tegument-associated proteins, including the large tegument protein ORF62 (Michel, Leroy, et al., 2010; van Beurden, Leroy, et al., 2011; Yi et al., 2014). Bioinformatical predictions for signal peptides, transmembrane domains, and glycosylation allowed the identification of a total of 16 putative membrane proteins (Aoki et al., 2007; Michel, Leroy, et al., 2010; Yi et al., 2014).

In addition, several studies dedicated to specific virion proteins have been carried out (Aoki et al., 2011; Dong et al., 2011; Fuchs, Granzow, Dauber, Fichtner, & Mettenleiter, 2014; Rosenkranz et al., 2008; Tu et al., 2014; Vrancken et al., 2013; Yi et al., 2014). Some of these proteins have been studied in more detail, notably ORF81, which is a type 3 membrane protein and is thought to be one of the most immunogenic (major) membrane

Table 6 Structural Proteome of CyHV-3

ORF	NCBI ID	Predicted MM (kDa)	Predicted Localization	Protein Description[a]	No. of Peptides[b]		
					FL	GZ11	GZ10
11	131840041	13.1	Unknown	–	–	1	2
25	131840055	67.1	Envelope[c]	Predicted membrane protein; ORF25 gene family	7	6	8
27	380708459	47.9	Envelope[c]	Predicted membrane protein; ORF25 gene family	–	1	1
31	131840058	13.9	Unknown	Similar to eukaryotic PLAC8 proteins	2	3	7
32	131840059	22.3	Envelope[c]	Predicted membrane protein; similar to a family of Singapore grouper iridovirus proteins	3	2	3
34	131840061	17	Unknown	–	2	3	–
35	131840062	36.3	Unknown	–	1	–	1
36	131840063	30.3	Unknown	–	1	–	–
42	131840068	53.5	Tegument[d]	Related to AngHV-1 ORF18	13	18	24
43	131840069	159.4	Unknown	–	48	51	59
44	131840070	97.5	Unknown	–	4	–	–
45	131840045	97.5	Tegument[d]	Related to AngHV-1 ORF20	5	4	6
51	131840077	165.9	Tegument-associated[d]	Related to AngHV-1 ORF34	41	38	48

Continued

Table 6 Structural Proteome of CyHV-3—cont'd

ORF	NCBI ID	Predicted MM (kDa)	Predicted Localization	Protein Description	No. of Peptides		
					FL	GZ11	GZ10
57	131840083	54	Tegument-associated[d]	Similar to crocodile poxvirus CRV155; related to AngHV-1 ORF35	17	11	20
59	131840085	14.6	Envelope[c]	Predicted membrane protein	2	1	2
60	131840086	59.9	Tegument-associated[d]	Related to AngHV-1 ORF81	10	4	12
62	131840088	442.2	Tegument-(associated)[d,e]	Contains an OTU-like cysteine protease domain; related to AngHV-1 ORF83 and IcHV-1 ORF65	76	83	92
65	131840091	63.5	Envelope[c]	Predicted membrane protein; member of ORF25 gene family	10	6	10
66	131840092	45.4	Capsid[d]	Capsid triplex subunit 1; related to AngHV-1 ORF42	13	10	21
68	131840094	253	Unknown	Similar to myosin-related proteins; related to IcHV-1 ORF22, RaHV-1 ORF56 and ORF89, and RaHV-2 ORF126	59	77	75
69	131840095	58.9	Tegument[d]	Related to AngHV-1 ORF39	1	1	3
70	131840096	51.1	Tegument[d]	Related to AngHV-1 ORF38	2	4	3
72	131840098	40.7	Capsid[d,e]	Capsid triplex subunit 2; related to AngHV-1 ORF36, IcHV-1 ORF27, RaHV-1 ORF95, and RaHV-2 ORF131	10	11	13

78	131840104	76.9	Capsid[d,e]	Capsid maturation protease; related to AngHV-1 ORF57, IcHV-1 ORF28, RaHV-1 ORF63, and RaHV-2 ORF88	5	2	5
81	131840107	28.6	Envelope[c–e]	Multiple transmembrane protein; related to AngHV-1 ORF51, positionally similar to IcHV-1 ORF59, RaHV-1 ORF83, and RaHV-2 ORF117	3	5	3
83	131840109	26.9	Envelope[c,d]	Predicted multiple transmembrane protein; related to AngHV-1 ORF49	–	2	3
84	131840110	85.6	Unknown	–	25	21	32
89	131840115	53.5	Unknown	–	7	5	10
90	131840116	86.1	Capsid[d]	Related to AngHV-1 ORF100, IcHV-1 ORF37, RaHV-1 ORF52, and RaHV-2 ORF78	9	11	14
91	131840117	26.4	Tegument[d]	Related to AngHV-1 ORF103	–	–	1
92	131840118	140.4	Capsid[d,e]	Major capsid protein; related to AngHV-1 ORF104, IcHV-1 ORF39, RaHV-1 ORF54, and RaHV-2 ORF80	45	32	45
95	131840121	24.2	Unknown	–	3	1	5

Continued

Table 6 Structural Proteome of CyHV-3—cont'd

ORF	NCBI ID	Predicted MM (kDa)	Predicted Localization	Protein Description	No. of Peptides		
					FL	GZ11	GZ10
97	131840123	117.5	Tegument-associated[d]	Related to AngHV-1 ORF30	19	20	22
99	131840125	170.7	Envelope[c,d]	Predicted membrane protein; related to AngHV-1 ORF67, IcHV-1 ORF46, RaHV-1 ORF46, RaHV-2 ORF72	34	14	16
106	131840132	7.5	Unknown	–	–	1	–
108	131840134	21	Envelope[c]	Predicted membrane protein	2	1	3
112	131840138	31	Unknown	Contains a double-stranded nucleic acid-binding domain (helix–turn–helix)	1	1	1
115	131840141	86.2	Envelope[c]	Predicted membrane protein	14	12	17
116	131840142	30.4	Envelope[c]	Predicted membrane protein	–	–	1
123	131840149	29.5	Tegument[c]	Deoxyuridine triphosphatase; related to AngHV-1 ORF5, IcHV-1 ORF49, and RaHV-2 ORF142; also encoded by some iridoviruses and poxviruses	2	–	4
131	131840157	30.6	Envelope[c]	Predicted membrane protein	5	3	4
132	131840158	19	Envelope[c]	Predicted membrane protein	2	4	1

		MM		Description[a]			
136	131840162	17	Envelope[c]	Predicted membrane protein	2	3	3
137	131840163	69.7	Unknown	Member of ORF22 gene family	4	–	–
148	131840174	64.8	Envelope[c]	Predicted membrane protein; member of ORF25 gene family	7	4	6
149	131840175	72.8	Envelope[c]	Predicted membrane protein; member of ORF25 gene family	7	9	9

[a]Protein descriptions adapted from Michel, Leroy, et al. (2010).
[b]Number of peptides detected as determined by Michel, Leroy, et al. (2010) (FL strain) and Yi et al. (2014) (GZ11 and GZ10 strains).
[c]Predicted based on bioinformatical predictions, adapted from Aoki et al. (2007).
[d]Predicted based on sequence homology with AngHV-1 as determined by van Beurden, Leroy, et al. (2011).
[e]Predicted based on sequence homology with IcHV-1 as determined by Davison and Davison (1995).
MM, molecular mass.

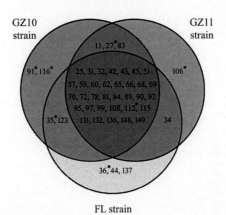

Figure 5 Structural proteome of CyHV-3. Schematic representation of virion-associated proteins from two CyHV-3 Chinese isolates (GZ10 and GZ11) (Yi et al., 2014) and one European isolate (FL) (Michel, Leroy, et al., 2010). Numbers indicate CyHV-3 ORFs. A total of 46 viral proteins were identified from which 34 were consistently identified in the three CyHV-3 isolates. Asterisks indicate viral proteins in which only one matched peptide was detected. *Adapted with permission from Yi et al. (2014). Copyright © Elsevier.*

proteins of CyHV-3 (Rosenkranz et al., 2008). Based on their high abundance and unique locations upon SDS–PAGE of purified proteins, Yi et al. (2014) marked ORF43, ORF51, ORF62, ORF68, ORF72, ORF84, and ORF92 as the major structural proteins of CyHV-3 (Yi et al., 2014). Two of these proteins, namely the large tegument protein encoded by ORF62 and ORF68, had previously been identified as major antigenic CyHV-3 proteins by immunoscreening (Aoki et al., 2011). Moreover, sera from infected carp reacted also against cells transfected with plasmids encoding for ORF25, ORF65, ORF148, ORF149 (four members of the ORF25 family; envelope proteins), ORF99 (envelope protein), and ORF92 (major capsid protein) (Fuchs et al., 2014).

The degree of conservation of the tegument and envelope proteins among fish herpesviruses is limited, with only one large tegument protein and potentially two envelope proteins being conserved between CyHV-3, AngHV-1, and IcHV-1 (van Beurden, Leroy, et al., 2011). For AngHV-1, the distribution of the structural proteins across the different viral compartments resembles that of other herpesviruses, with decreasing numbers for the different proteins from tegument to envelope to capsid (van Beurden, Leroy, et al., 2011). Although the localization of the CyHV-3 structural proteins remains to be demonstrated experimentally, a similar ratio may be expected, implying that most of the yet unclassified proteins could be located in the tegument.

Both studies on the CyHV-3 structural proteome also identified 18–27 cellular proteins associated with extracellular CyHV-3 virions (Michel, Leroy, et al., 2010; Yi et al., 2014). Similar to mammalian herpesviruses, these include proteins involved in stress response, signal transduction, vesicular trafficking, metabolism, cytoskeleton organization, translational control, immunosuppression, and cell-signaling regulation. Except for the so-called virus-induced stress protein identified by Yi et al. (2014), host cellular proteins were generally low in abundance suggesting them as minor components of the virions (Michel, Leroy, et al., 2010; Yi et al., 2014).

The viral secretome of CyHV-3 was examined by analyzing concentrated supernatants of infected cell cultures by mass spectrometry (Ouyang et al., 2013). Five viral proteins were identified, of which the two most abundant were ORF12 encoding a soluble TNF receptor homolog, and ORF134 encoding an IL-10 homolog. Three additional viral proteins (encoded by ORF52, ORF116, and ORF119) had previously been predicted to be potential membrane proteins, but were not convincingly identified as such. Overall, the identification of the viral and cellular protein composition of the virions and viral secretome represents a milestone in fundamental CyHV-3 research and may facilitate the development of diagnostic and prophylactic applications (see, for example, Fuchs et al., 2014; Vrancken et al., 2013).

3.1.6 Viral Replication in Cell Culture
3.1.6.1 Cell Lines Permissive to CyHV-3
CyHV-3 can be cultivated in cell lines derived from common carp brain (CCB) (Davidovich, Dishon, Ilouze, & Kotler, 2007; Neukirch, Böttcher, & Bunnajrakul, 1999), gills (CCG) (Neukirch et al., 1999), and fin (CaF-2, CCF-K104, MFC) (Imajoh et al., 2014; Neukirch & Kunz, 2001; Zhou et al., 2013). Permissive cell lines have also been derived from koi fin: KF-1 (Hedrick et al., 2000), KFC (Hutoran et al., 2005; Ronen et al., 2003), KCF-1 (Dong et al., 2011), NGF-2 and NGF-3 (Miwa et al., 2007), and KF-101 (Lin, Cheng, Wen, & Chen, 2013; Table 7). Other permissive cell lines were developed from snout tissues (MSC, KS) (Wang et al., 2015; Zhou et al., 2013). Non-carp cell lines, such as silver carp fin (Tol/FL) and goldfish fin (Au), were also described as permissive to CyHV-3 (Davidovich et al., 2007). One report showed cytopathic effect (CPE) in a cell line from fathead minnow (FHM cell line) after inoculation with CyHV-3 (Grimmett et al., 2006), but this observation was not confirmed by others (Davidovich et al., 2007; Hedrick et al., 2000; Neukirch et al., 1999). Similarly, Neukirch et al. (1999) and Neukirch and Kunz

Table 7 Cell Lines Susceptible to CyHV-3 Infection

| Origin | Name | Cytopathic Effect | |
		Yes	No
Cyprinus carpio			
Brain			
Common carp brain	CCB	Davidovich et al. (2007) Neukirch et al. (1999)	
Gills			
Common carp gill	CCG	Neukirch et al. (1999)	
Fins/skin			
Common carp fin	CaF-2	Neukirch and Kunz (2001)	
	CCF-K104	Imajoh et al. (2014)	
	MFC	Zhou et al. (2013)	
Common carp skin tumor	EPC	Neukirch et al. (1999) Neukirch and Kunz (2001)	Hedrick et al. (2000)[a] Hutoran et al. (2005) Davidovich et al. (2007)
Koi carp fin	KF-1	Hedrick et al. (2000)	
	KFC	Hutoran et al. (2005)	
	KCF-1	Dong et al. (2011)	
	NGF-2(-3)	Miwa et al. (2007)	
	KF-101	Lin et al. (2013)	
Common carp snout	MSC	Zhou et al. (2013)	
Koi carp snout	KS	Wang et al. (2015)	
Other species			
Silver carp fin (*Hypophthalmichthys molitrix*)	Tol/FL	Davidovich et al. (2007)	
Goldfish fin (*Carassius auratus*)	Au	Davidovich et al. (2007)	
Fathead minnow connective tissue and muscle (*Pimephales promelas*)	FHM	Grimmett, Warg, Getchell, Johnson, and Bowser (2006)	Neukirch et al. (1999) Hedrick et al. (2000) Davidovich et al. (2007)
Chinook salmon embryo (*Oncorhynchus tshawytscha*)	CHSE-214		Neukirch et al. (1999)
Channel catfish ovary (*Ictalurus punctatus*)	CCO		Davidovich et al. (2007)

[a]Only transient CPE.

(2001) reported CPE in EPC (*epithelioma papulosum cyprini*) cells but this observation was also not confirmed (Davidovich et al., 2007; Hedrick et al., 2000; Hutoran et al., 2005). This discrepancy could be partially explained by the controversial origin of the EPC cell line. This cell line was initially described as originating from common carp but was recently found to derive from fathead minnow (Winton et al., 2010). Other commonly used cell lines such as CHSE-214 (chinook salmon embryo) (Neukirch et al., 1999) and CCO (channel catfish ovary) (Davidovich et al., 2007) are not permissive to CyHV-3 infection. Typical CPE induced by CyHV-3 includes vacuolization and increased cell volume. Infected cells form characteristic plaques that grow according to time post-infection, frequently leading to the formation of syncytia (Ilouze, Dishon, & Kotler, 2006). Finally, infected cells become rounded before they detach from the substrate. Infectious virions are mostly retrieved from the infected cell supernatant (cell-free fraction) (Gilad et al., 2003). Isolation and adaptability of CyHV-3 *in vitro* seem to vary according to the field strain and cell line used. However, well-adapted laboratory strains usually reach titers up to 10^6–10^7 pfu/ml (Ilouze, Dishon, & Kotler, 2006).

3.1.6.2 Temperature Restriction

CyHV-3 replication is restricted by temperature *in vitro* and *in vivo*. *In vitro*, optimal viral growth was observed in KF-1 cell line at temperatures between 15 and 25 °C (Gilad et al., 2003); however, within this range, temperature affected the time at which viral production peaked (e.g., peak of viral titer observed at 7 days post-infection (dpi) and 13 dpi after incubation at 20–25 and 15 °C, respectively) (Gilad et al., 2003).

Virus production, virus gene transcription, and genome replication are gradually turned off when cells are moved from permissive temperature to the nonpermissive temperature of 30 °C (Dishon, Davidovich, Ilouze, & Kotler, 2007; Ilouze et al., 2012b; Imajoh et al., 2014). Although most of the 110 ORFs still transcribed 24 h after the temperature shift are gradually shut off (Table 5), few ORFs such as ORF114 and 115 were still expressed 18 days after temperature shift. However, infected cells maintained for 30 days at 30 °C preserve the potential to reinitiate a productive infection when returned to permissive temperatures (Dishon et al., 2007). This state of abortive infection with the potential to reactivate resembles latency as described for *Herpesviridae*. Putatively, the viral membrane protein encoded by ORF115 may represent an Epstein–Barr virus-like membrane-bound antigen associated with latency.

3.2 CyHV-3 Disease

3.2.1 Epidemiology

3.2.1.1 Fish Species Susceptible to CyHV-3 Infection

There is evidence that CyHV-3 can infect a wide range of species but that it only induces disease in common and koi carp. Hybrids of koi × goldfish and koi × crucian carp can be infected by CyHV-3, with mortality rates of 35% and 91%, respectively (Bergmann, Sadowski, et al., 2010). Common carp × goldfish hybrids have also been reported to show some susceptibility to CyHV-3 infection; however, the mortality rate observed was rather limited (5%) (Hedrick, Waltzek, & McDowell, 2006). PCR detection of CyHV-3 performed on cyprinid and non-cyprinid fish species, but also on freshwater mussels and crustaceans, suggested that these species could act as a reservoir of the virus (Table 8; El-Matbouli et al., 2007; El-Matbouli & Soliman, 2011; Fabian et al., 2013; Kempter & Bergmann, 2007; Kempter et al., 2009, 2012; Kielpinski et al., 2010; Radosavljevic et al., 2012). Cohabitation experiments suggest that some of these fish species (goldfish, tench, vimba, common bream, common roach, European perch, ruffe, gudgeon, rudd, northern pike, Prussian carp, silver carp, and grass carp) can carry CyHV-3 asymptomatically and transmit it to naive carp (Bergmann, Lutze, et al., 2010; El-Matbouli & Soliman, 2011; Fabian et al., 2013; Kempter et al., 2012; Radosavljevic et al., 2012). Recent studies provided increasing evidence that CyHV-3 can infect goldfish asymptomatically (Bergmann, Lutze, et al., 2010; El-Matbouli & Soliman, 2011; Sadler, Marecaux, & Goodwin, 2008), although some discrepancies exist in the literature (Yuasa et al., 2013). Consistent with this observation, *in vitro* studies showed that CyHV-3 can replicate and cause CPE in cell cultures derived not only from common and koi carp but also from silver carp and goldfish (Davidovich et al., 2007). Finally, the World Organization for Animal Health (OIE) lists one KHVD susceptible species (*C. carpio* and its hybrids) and several suspected carrier fish species (goldfish, grass carp, ide, catfish, Russian sturgeon, and Atlantic sturgeon) (OIE, 2012).

3.2.1.2 Geographical Distribution and Prevalence

The geographical range of the disease caused by CyHV-3 has become extensive since the first outbreaks in Germany in 1997 and in the USA and Israel in 1998 (Bretzinger et al., 1999; Hedrick et al., 2000; Perelberg et al., 2003). Worldwide trade in common and koi carp is generally held responsible for the spread of the virus before methods of detection were available and

Table 8 Organisms Tested for CyHV-3 Infection

Common Name (Species)	Detection of CyHV-3			Detection of CyHV-3 Genome in Carp After Cohabitation
	DNA	Transcript	Antigen	
Vertebrates				
Cyprinidae				
Goldfish (*Carassius auratus*)	Yes$^{a-d,k}$/no^e	Yes^b	Yes^c	Yes^{b-d}/no^e
Ide (*Leuciscus idus*)	Yes^{a,g}	nt	nt	nt
Grass carp (*Ctenopharyngodon idella*)	Yes^{a,d,g}	nt	nt	Yes^{d,g}
Silver carp (*Hypophthalmichthys molitrix*)	Yes^{d,g}	nt	nt	Yes^{d,g}
Prussian carp (*Carassius gibelio*)	Yes^{d,g}/no^h	nt	nt	Yes^d/no^h
Crucian carp (*Carassius carassius*)	Yes^g	nt	nt	nt
Tench (*Tinca tinca*)	Yes^{d,g,h}	nt	nt	Yes^{d,g,h}
Vimba (*Vimba vimba*)	Yes^{f,g}	nt	nt	Yes^g
Common bream (*Abramis brama*)	Yes^{g,h}	nt	nt	Yes^g
Common roach (*Rutilus rutilus*)	Yes^{g,h}	nt	nt	Yes^g/no^h
Common dace (*Leuciscus leuciscus*)	Yes^{f,g,h}	nt	nt	No^h
Gudgeon (*Gobio gobio*)	Yes^{g,h}	nt	nt	Yes^h
Rudd (*Scardinius erythrophthalmus*)	Yes^h	nt	nt	Yes^h
European chub (*Squalius cephalus*)	Yes^g/no^h	nt	nt	nt
Common barbel (*Barbus barbus*)	Yes^g	nt	nt	nt
Belica (*Leucaspius delineatus*)	Yes^g	nt	nt	nt

Continued

Table 8 Organisms Tested for CyHV-3 Infection—cont'd

Common Name (Species)	Detection of CyHV-3			Detection of CyHV-3 Genome in Carp After Cohabitation
	DNA	Transcript	Antigen	
Common nase (*Chondrostoma nasus*)	Yes[g]	nt	nt	nt
Acipenseridae				
Russian sturgeon (*Acipenser gueldenstaedtii*)	Yes[i]	nt	nt	nt
Atlantic sturgeon (*Acipenser oxyrinchus*)	Yes[i]	nt	nt	nt
Cobitidae				
Spined loach (*Cobitis taenia*)	Yes[g]	nt	nt	nt
Cottidae				
European bullhead (*Cottus gobio*)	Yes[g]	nt	nt	nt
Esocidae				
Northern pike (*Esox lucius*)	Yes[g,h]	nt	nt	Yes[h]
Gasterosteidae				
Three-spined stickleback (*Gasterosteus aculeatus*)	Yes[h]	nt	nt	No[h]
Ictaluridae				
Brown bullhead (*Ameiurus nebulosus*)	Yes[h]	nt	nt	No[h]
Loricariidae				
Ornamental catfish (*Ancistrus sp.*)	Yes[a]	nt	nt	nt
Percidae				
European perch (*Perca fluviatilis*)	Yes[g,h]	nt	nt	Yes[g]/no[h]
Ruffe (*Gymnocephalus cernua*)	Yes[g]/no[h]	nt	nt	Yes[g,h]

Table 8 Organisms Tested for CyHV-3 Infection—cont'd

| Common Name (Species) | Detection of CyHV-3 | | | Detection of CyHV-3 Genome in Carp After Cohabitation |
	DNA	Transcript	Antigen	
Invertebrates				
Swan mussels (*Anodonta cygnea*)	Yes[j]	nt	nt	nt
Scud (*Gammarus pulex*)	Yes[j]	nt	nt	nt

[a]Bergmann et al. (2009).
[b]El-Matbouli and Soliman (2011).
[c]Bergmann, Lutze, et al. (2010).
[d]Radosavljevic et al. (2012).
[e]Yuasa, Sano, and Oseko (2012).
[f]Kempter and Bergmann (2007).
[g]Kempter et al. (2012).
[h]Fabian, Baumer, and Steinhagen (2013).
[i]Kempter et al. (2009).
[j]Kielpinski et al. (2010).
[k]El-Matbouli et al. (2007).
nt, not tested.
Adapted with permission from Rakus et al. (2013); Original publisher BioMed Central.

implemented (OIE, 2012). The disease is now known to occur in, or has been reported in fish imported into, at least 28 different countries (OIE, 2012).

In Europe, reports of widespread mass mortality have been notified in carp farms and fisheries in Germany, Poland, and the UK (Bergmann, Kempter, Sadowski, & Fichtner, 2006; Gotesman, Kattlun, Bergmann, & El-Matbouli, 2013; Taylor, Dixon, et al., 2010). The disease is also known to occur in, or has been recorded in fish imported into, Austria, Belgium, Czech Republic, Denmark, France, Hungary, Italy, Luxembourg, The Netherlands, Republic of Ireland, and Switzerland (Haenen et al., 2004; McCleary et al., 2011; Pokorova et al., 2010; Pretto et al., 2013). Most recently, KHVD outbreaks have been reported to the OIE from Romania, Slovenia, Spain, and Sweden (OIE, 2012). Three novel CyHV-3-like viruses were also identified by PCR in The Netherlands, UK, Austria, and Italy, sharing only 95–98% nucleotide identity with the CyHV-3 J, CyHV-3 I, and CyHV-3 U strains. Carp carrying the CyHV-3 variants did not show clinical signs consistent with CyHV-3 infection and originated from locations with no actual CyHV-3 outbreaks. These strains might represent low- or nonpathogenic variants of CyHV-3 (Engelsma et al., 2013).

In Asia, in the Middle East, the first disease outbreaks with mass mortalities were seen in Israel in 1998 and in the following 3 years, the virus had spread to 90% of all carp farms (Perelberg et al., 2003). In southeastern Asia, the first outbreaks of KHVD, with mass mortalities of cultured koi carp, occurred in Indonesia in 2002 and were associated with an importation of koi from Hong Kong (Haenen et al., 2004; Sunarto et al., 2011). Later in 2002, the first occurrence of CyHV-3 infection was reported in koi carp in Taiwan (Tu, Weng, Shiau, & Lin, 2004). In 2003, detection of CyHV-3 was first reported in Japan following mass mortalities of cage-cultured common carp in the Ibaraki prefecture (Sano et al., 2004). Since then, the virus has been confirmed in 90% of the 109 class A natural rivers and in 45 of the 47 prefectures (Lio-Po, 2011; Minamoto, Honjo, Yamanaka, Uchii, & Kawabata, 2012). Similarly, CyHV-3 spread rapidly in Indonesia with disease outbreaks reported on most of the major islands by 2006 (Lio-Po, 2011). CyHV-3 has also been detected in China (Dong et al., 2011), South Korea (Gomez et al., 2011), Singapore (Lio-Po, 2011), Malaysia (Musa, Leong, & Sunarto, 2005), and Thailand (Lio-Po, 2011; Pikulkaew, Meeyam, & Banlunara, 2009).

In North America, the first reports of CyHV-3 infection were from disease outbreaks at koi dealers (Gray et al., 2002; Hedrick et al., 2000). Then, in 2004, CyHV-3 was confirmed from mass mortalities of wild common carp in South Carolina and New York states (Grimmett et al., 2006; Terhune et al., 2004). In Canada, CyHV-3 was first detected during disease outbreaks in wild common carp in Ontario in 2007 and further outbreaks were reported in Ontario and Manitoba in 2008 (Garver et al., 2010). More recently, mass mortalities of common carp have been reported along the US/Canada border in Michigan and Wisconsin (Gotesman et al., 2013) (S. Marcquenski, personal communication).

There are no reports of KHVD or CyHV-3 detections from South America or Australasia, and the only reports from the African continent are from South Africa (OIE, 2012).

Horizontal transmission of the disease is very rapid (see Section 3.2.3.1.3). Several hypotheses were suggested to explain the swift spread of the virus: (i) The practice of mixing koi carp in the same tanks at koi shows has been held responsible for spreading the disease, particularly within a country (Gilad et al., 2002). (ii) In Israel, piscivorous birds are suspected to be responsible for the rapid spread of CyHV-3 from farm to farm (Ilouze, Davidovich, Diamant, Kotler, & Dishon, 2010). (iii) Disposal of infected fish by selling them below the market price was one suspected route of dissemination of the virus in Indonesia (Sunarto, Rukyani, & Itami, 2005). (iv) It was suggested that the outbreaks of disease in public parks and ponds in Taiwan without recent introduction of fish were the result of members of the public releasing infected fish into the ponds (Tu et al., 2004). (v) Additionally, the virus has also been spread nationally and internationally before regulators were aware of the disease and methods to detect CyHV-3 were available. This is evidenced by the detection of CyHV-3 DNA in archive histological specimens collected during unexplained mass mortalities of koi and common carp in the UK in 1996 and in cultured common carp in South Korea in 1998 (Haenen et al., 2004; Lee, Jung, Park, & Do, 2012).

There are limited published observations of virus prevalence in wild or farmed populations of carp. A PCR survey, performed 2 years after the KHVD outbreaks in Lake Biwa, Japan, found a higher prevalence of CyHV-3 in larger common carp (>3 cm, 54% of seropositive fish) compared to smaller ones (<3 cm, 0% seropositive fish) (Uchii, Matsui, Iida, & Kawabata, 2009). Again in Japan, CyHV-3 DNA was detected in 3.9% (3/76), 5.1% (4/79), and 16.7% (12/72) of brain samples in three rivers of the Kochi prefecture (Fujioka et al., 2015).

In England, three sites experiencing clinical outbreaks of disease in 2006 and having no introduction of fish since that time were revisited in 2007, and found to have detectable serum anti-CyHV-3 antibodies in the surviving carp with a seroprevalence of 85–93% (Taylor, Dixon, et al., 2010). Similarly, studies to determine the prevalence of CyHV-3 in a country's carp farms or natural water bodies have been few. In the UK, common carp positive for CyHV-3 antibodies were found to be widely distributed in fisheries (angling waters) but the majority of carp farms remained negative. The main route of spread of CyHV-3 was determined to be live fish movements but alternative routes, including the stocking of imported ornamental fish, were also suggested (Taylor, Norman, Way, & Peeler, 2011; Taylor, Way, Jeffery, & Peeler, 2010).

Further evidence of widespread dissemination of CyHV-3 is provided by molecular epidemiology studies using the approaches described in Section 3.1.3. Two major lineages, CyHV-3 J and CyHV-3 U/I, have been identified with lineage J representing the major lineage in eastern Asia (Aoki et al., 2007; Kurita et al., 2009). Further studies have identified potential subgenotypes within the European (CyHV-3 U/I) and Asian (CyHV-3 J) lineages, with the European viruses showing the most variation (Kurita et al., 2009). The CyHV-3 J lineage has been detected in samples of infected koi and common carp from France and The Netherlands (Bigarré et al.,

2009), and the same study also identified a unique genotype of CyHV-3, intermediate between J and U/I, in koi carp from Poland. In Austria, the sequence analysis that was undertaken indicates that the CyHV-3 J was the only lineage detected in infected tissues from 15 koi carp from different locations in 2007 and suggests that the presence of the CyHV-3 J lineage in Europe may be linked to imports of Asian koi. In the UK, VNTR analysis similar to that described by Avarre et al. (2011) identified 41 distinct virus VNTR profiles for 68 disease cases studied between 2000 and 2010, and since these were distributed throughout three main clusters, CyHV-3 J, CyHV-3 I, and CyHV-3 U, and an intermediate lineage (D. Stone, personal communication), it suggests multiple incursions of CyHV-3 into the UK during that period.

In eastern and southeastern Asia, the U/I or European lineage has been detected but only at low frequency. In Indonesia, analysis of infected tissues from 10 disease outbreaks, from 2002 to 2007, identified two Asian genotypes and also another intermediate genotype (Sunarto et al., 2011). A study in South Korea identified from disease outbreaks in 2008, both a European genotype in samples of infected common carp and the expected Asian genotype in koi carp (Kim & Kwon, 2013). More recently, a European genotype of CyHV-3 was detected from a disease outbreak in 2011 in China (Dong et al., 2013), and in imported carp from Malaysia in Singapore (Chen et al., 2014).

3.2.1.3 Persistence of CyHV-3 in the Natural Environment

CyHV-3 remains infectious in water for at least 4 h, but not for 21 h, at water temperatures of 23–25 °C (Perelberg et al., 2003). Other studies in Japan have displayed a significant reduction in the infectious titer of CyHV-3 within 3 days in environmental water or sediment samples at 15 °C, while the infectivity remained for more than 7 days when CyHV-3 was exposed to sterilized water samples, thus suggesting the roles of microorganisms in the inactivation of CyHV-3 (Shimizu, Yoshida, Kasai, & Yoshimizu, 2006). Supporting this hypothesis, a recent report showed that bacteria isolated from carp habitat waters and carp intestine contents possessed some anti-CyHV-3 activity (Yoshida, Sasaki, Kasai, & Yoshimizu, 2013). These studies suggest that, in the absence of hosts, CyHV-3 can be rapidly inactivated in environmental water.

In Japan, the detection of CyHV-3 DNA in river water samples at temperatures of 9–11 °C has been reported 4 months before an outbreak of KHVD in a river (Haramoto, Kitajima, Katayama, & Ohgaki, 2007).

Japanese researchers have quantified CyHV-3 in environmental samples by cation-coated filter concentration of virus linked to a quantitative PCR (qPCR) (Haramoto, Kitajima, Katayama, Ito, & Ohgaki, 2009; Honjo et al., 2010). Using this technique, CyHV-3 was detected at high levels in water samples collected at eight sites along the Yura river system during, and 3 months after an episode of mass mortality caused by KHVD, and at water temperatures ranging from 28.4 down to 14.5 °C (Minamoto, Honjo, Uchii, et al., 2009). The seasonal distribution of CyHV-3 in Lake Biwa, Japan, was investigated using qPCR, which found the virus to be distributed all over the lake 5 years after the first KHVD outbreak in 2004. Mean concentrations of CyHV-3 in the lake water showed annual variation, with a peak in the summer and a trough in winter, and also indicated that the virus is more prevalent in reductive environments such as the turbid, eutrophic water found in reed zones (Minamoto, Honjo, & Kawabata, 2009). These areas are the main spawning sites of carp in Lake Biwa and support the hypothesis of increased prevalence of CyHV-3 during spawning (Uchii et al., 2011). The researchers suggested that, in highly turbid water, viruses may escape degradation by attaching to organic or nonorganic particles (Minamoto, Honjo, & Kawabata, 2009). Further studies of carp spawning areas in Lake Biwa reported the detection of CyHV-3 DNA in plankton samples and in particular the *Rotifera* species (Minamoto et al., 2011).

Finally, as explained earlier (see Section 3.2.1.1), other vertebrate and invertebrate species could play a significant role in CyHV-3 persistence in aquatic environments and should be considered as an epidemiological risk for carp farms (Fabian et al., 2013).

3.2.1.4 Use of CyHV-3 for Biological Control of Common Carp

In Australia, common carp is considered as an important invasive pest species. Its population and geographical range drastically expanded after an accidental escape from isolated farms in southeastern Australia due to flooding in the 1970s. In the early 2000s, an integrated pest management plan was developed to counteract common carp invasion. CyHV-3 was proposed as a biological control agent to reduce common carp populations (McColl, Cooke, & Sunarto, 2014). With regard to this goal, CyHV-3 possesses some interesting characteristics such as inducing high morbidity/mortality, high contagiosity, and a narrow host range for induction of the disease (not for asymptomatic carriers). These viral characteristics coupled with some epidemiological conditions specific to Australia, such as the absence of CyHV-3 and other cyprinid fish species together with the relatively low abundance of common and koi carp aquaculture, suggest that CyHV-3 could be a successful biocontrol agent. However, as stated by the authors, the use of exotic viruses as biocontrol agents is not trivial and studies addressing the safety and the efficacy of this measure are essential before applying it to the field (McColl et al., 2014).

3.2.2 Clinical Aspects

KHVD is seasonal, occurring mainly at water temperatures between 18 and 28 °C (Gotesman et al., 2013; Rakus et al., 2013). It is a highly contagious and extremely virulent disease with mortality rates up to 80–100% (Ilouze, Dishon, & Kotler, 2006). The disease can be reproduced experimentally by immersion of fish in water containing the virus, by ingestion of contaminated food, by cohabitation with freshly infected fish, and, more artificially, by injection of infectious material (Fournier et al., 2012; Perelberg et al., 2003). Fish infected with CyHV-3 using these various routes, and kept at permissive temperature, die between 5 and 22 dpi with a peak of mortality between 8 and 12 dpi (Fournier et al., 2012; Hedrick et al., 2000; Perelberg et al., 2003; Rakus, Wiegertjes, Adamek, et al., 2009). Furthermore, CyHV-3-infected fish are more susceptible to secondary infections by bacterial, parasitic, or fungal pathogens, which may contribute to the mortalities observed in the infected population (McDermott & Palmeiro, 2013).

3.2.2.1 Clinical Signs

The first clinical signs usually appear 2–3 dpi, while the first mortalities are frequently delayed to 6–8 dpi (McDermott & Palmeiro, 2013). The course of infection and the clinical signs observed are variable between individual fish, even after simultaneous and controlled experimental CyHV-3 inoculation. Fish can express the following clinical signs: folding of the dorsal fin; increased respiratory frequency; gathering near well-aerated areas; skin changes including gradual hyperemia at the base of fins, increased (sometimes decreased) mucus secretion, hemorrhages and ulcers on the skin, sloughing of scales and fin erosion, sandpaper-like texture of the skin, and skin herpetic lesions; gasping at the water surface; lethargy (lying at the bottom of the tank, hanging in head-down position in the water column) associated with anorexia; sunken eyes; and neurological symptoms with erratic swimming and loss of equilibrium (Hedrick et al., 2000; McDermott & Palmeiro, 2013; Rakus et al., 2013; Walster, 1999). None of these clinical signs are pathognomonic of KHVD.

3.2.2.2 Anatomopathology

The external post-mortem lesions that can be observed on the skin include pale and irregular patches, hemorrhages, fin erosions, ulcers, and peeling away of the epithelium. The main lesion in the gills is a mild to severe necrosis with multifocal or diffuse discoloration, sometimes associated with extensive erosions of the primary lamellae. Some of these anatomopathological lesions are illustrated in Fig. 6a. Other inconsistent necropsy changes include enlarging, darkening, and/or mottling of some internal organs associated with petechial hemorrhages, accumulation of abdominal fluid, and abdominal adhesions (Bretzinger et al., 1999; Hedrick et al., 2000; McDermott & Palmeiro, 2013; Walster, 1999). None of the lesions listed above are pathognomonic of KHVD.

3.2.2.3 Histopathology

Histopathological alterations are observed in the gills, skin, kidneys, heart, spleen, liver, gut, and brain of CyHV-3-infected fish (Hedrick et al., 2000; Miyazaki et al., 2008). In the skin, the lesions can appear as soon as 2 dpi and worsen with time (Fig. 6b; Miwa et al., 2014). The cells exhibiting degeneration and necrosis show various stages of nuclear degeneration (e.g., pale coloration, karyorrhexis, pyknosis), frequently associated with characteristic intranuclear inclusion bodies (Fig. 6b, D). These cells, shown to be infected by CyHV-3 using EM, are characterized by a basophilic material within the nucleus associated with marginal hyperchromatosis (Miyazaki et al., 2008). The number of goblet cells is reduced by 50% in infected fish, and furthermore, they appear mostly slim and slender, suggesting that mucus has been released and not replenished (Adamek et al., 2013). At later stages, erosion of skin epidermis is frequently observed (Adamek et al., 2013; Miwa et al., 2014). A recent report revealed that the damages caused to the skin of the body and fins were the most pronounced lesions (Miwa et al., 2014).

During the course of CyHV-3 infection, important histopathological changes are observed in the two compartments of the gills, the gill lamellae and gill rakers (Fig. 6c; Miyazaki et al., 2008; Pikarsky et al., 2004). The lesions observed in the gill lamellae involve infiltration of inflammatory cells, hyperplasia, hypertrophy, degeneration and necrosis of epithelial cells, congestion, and edema (Miyazaki et al., 2008; Ouyang et al., 2013; Pikarsky et al., 2004). As a consequence of the pronounced hyperplasia, the secondary lamellae interspace is progressively filled by cells. At later stages, the gill lamellae architecture can be completely lost by necrosis, erosion, and fusion of the primary lamellae (Pikarsky et al., 2004). These lesions can be

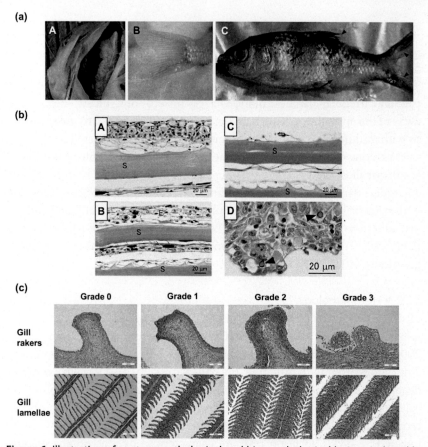

Figure 6 Illustration of anatomopathological and histopathological lesions induced by CyHV-3. (a) Anatomopathological lesions. (A) Severe gill necrosis. (B) Hyperemia at the base of the caudal fin. (C) Extensive necrosis of the skin covering the body (arrows indicate circular herpetic lesion) and fin erosion (arrowheads). (b) Histopathological lesions in the skin. Sections of the skin of carp stained with hematoxylin and eosin. S, scale; E, epidermis. (A) The skin of a mock-infected fish. (B) The skin of a moribund specimen sampled 6 dpi. Most of the cells exhibit degenerescence and necrosis as well as marginalization of the chromatin. (C) The skin of a moribund fish sampled 5 dpi. The epidermis has detached from the underlying dermis probably as a consequence of extensive necrosis. (D) High magnification of the skin of an infected fish 2 dpi. Note the characteristic chromatin marginalization observed in some epithelial cells (arrowheads). (c) Histopathological lesions in the gills. Five-micrometer sections were stained with hematoxylin and eosin. A grading system was developed to characterize the lesions observed in gill rakers and gill lamellae. The grading system evaluates the degree of epithelial hyperplasia, the presence of intranuclear viral inclusions, and cell degeneration. Briefly, grade 0 = physiological state; grade 1 = mild hyperplasia without evidence of degenerated cells and viral inclusions; grade 2 = severe hyperplasia and presence of few degenerated cells and viral inclusions; and grade 3 = presence of abundant degenerated cells and viral inclusions (gill lamellae and gill rakers), massive epithelial hyperplasia filling the entire secondary lamellae interspace (gill lamellae), and ulcerative erosion of the epithelium (gill rakers). Scale bars = 100 μm. *Panel (a): Adapted with permission from Michel, Fournier, et al. (2010). Panel (b): Adapted from with permission from Miwa, Kiryu, Yuasa, Ito, and Kaneko (2014). Copyright © Wiley & Sons, Inc. Panel (c): Adapted with permission from Boutier et al. (2015).*

visualized macroscopically and are frequently associated with secondary infections (Pikarsky et al., 2004). In the gill rakers, the changes are even more recognizable (Pikarsky et al., 2004). These include subepithelial inflammation, infiltration of inflammatory cells, and congestion at early stages (Pikarsky et al., 2004), followed by hyperplasia, degeneration, and necrosis of cells presenting intranuclear inclusion bodies. At ulterior stages, complete erosion of the epithelium can be observed. Based on these histo-pathological observations, a grading system (Fig. 6c) has been proposed by Boutier et al. (2015). This grading system classifies the lesions according to three criteria, i.e., (i) hyperplasia of epithelial cells, (ii) presence and extent of degeneration and necrosis, and (iii) presence and abundance of intranuclear inclusion bodies. As the number of presumed infected cells does not always correlate with the severity of the lesions, the combination of these criteria is necessary to obtain a reliable histopathological grading system (Boutier et al., 2015; Miwa et al., 2014).

In the kidney, a weak peritubular inflammatory infiltrate is evident as early as 2 dpi and increases with time. It is accompanied by blood vessel congestion and degeneration of the tubular epithelium in many nephrons (Pikarsky et al., 2004). Intranuclear inclusion bodies are mainly found in hematopoietic cells (Miwa et al., 2014; Miyazaki et al., 2008). In the spleen, the main susceptible cells are the splenocytes. In extreme cases, the lesions include large numbers of necrotic splenocytes accompanied by hemorrhages (Miyazaki et al., 2008). In the heart, many myocardial cells exhibit nuclear degeneration and alteration of the myofibril bundles with disappearance of the cross-striation (Miyazaki et al., 2008).

In the intestine and stomach, the lesions induced are mainly the consequence of the hyperplasia of the epithelium, forming projections inside the lumen. Cells of the epithelium expressing intranuclear inclusion bodies and necrosis detach from the mucosa and locate in the lumen of the organ (El-Din, 2011). In the liver, hepatocytes are the most affected cell type (Miyazaki et al., 2008) and mild inflammatory infiltrates can be observed in the parenchyma (Pikarsky et al., 2004).

In the brain, focal meningeal and parameningeal inflammation is observed (Pikarsky et al., 2004). Analysis of brains from fish that showed clear neurologic signs revealed congestion of capillaries and small veins associated with edematous dissociation of nerve fibers in the valvula cerebelli and medulla oblongata (Miyazaki et al., 2008). Infected cells were detected at 12 dpi in all compartments of the brain. These cells were mainly ependymal cells and, to a lesser extent, neurons (Fig. 7; Miwa et al., 2014). At 20 dpi, the

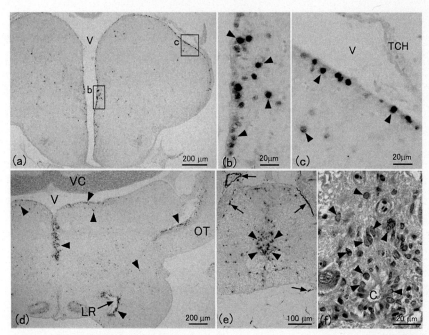

Figure 7 Illustration of histopathological lesions induced in the central nervous system of carp by CyHV-3. (A), (D), and (E) show sections of telencephalon, mesencephalon, and spinal cord hybridized for the viral genome, respectively. Fish were sampled 12 dpi. The hybridization signals (arrowheads) are observed along the ependyma as well as in some neurons in the neuropil and around the central canal. The rectangles in (A) are shown enlarged in (B) and (C). Arrows indicate melanin. (F) A section of the spinal cord stained with hematoxylin and eosin. Arrowheads indicate nuclei of cells presumably infected with CyHV-3. V, ventricle; TCH, tela choroidea; VC, valvula cerebelli; OT, optic tectum; LR, lateral recess; C, central canal. *Reproduced with permission from Miwa et al. (2014). Copyright © Wiley & Sons, Inc.*

lesions are accompanied by perivascular lymphocyte infiltration and gliosis. The peak of nervous lesions coincides in time with the peak of neurological clinical signs (Miwa et al., 2014).

3.2.3 *Pathogenesis*

All members of the family *Herpesviridae* exhibit two distinct phases in their infection cycle: lytic replication and latency. While lytic replication is associated with production of viral particles, latency entails the maintenance of the viral genome as a nonintegrated episome and the expression of very few viral genes and microRNAs. Upon reactivation, lytic replication ensues. Studies on a few members of the *Alloherpesviridae* family also suggest the existence of these two types of infection. Most of these studies are on CyHV-3 and suggest that the temperature of the water could regulate the switch between latency and lytic replication and *vice versa*, allowing the virus to persist in the host population throughout the seasons even when the temperature is nonpermissive (Uchii, Minamoto, Honjo, & Kawabata, 2014). Below, we have summarized the data available for CyHV-3 for the two types of infection.

3.2.3.1 Productive Infection

3.2.3.1.1 Portals of Entry In early reports, it has been suggested that CyHV-3 may enter the host through infection of the gills (Hedrick et al., 2000; Ilouze, Dishon, & Kotler, 2006; Miyazaki et al., 2008; Miyazaki, Yasumoto, Kuzuya, & Yoshimura, 2005; Pikarsky et al., 2004; Pokorova, Vesely, Piackova, Reschova, & Hulova, 2005) and the intestine (Dishon et al., 2005; Ilouze, Dishon, & Kotler, 2006). These hypotheses rely on several observations: (i) the gills undergo histopathological lesions early after inoculation by immersion in infectious water (Hedrick et al., 2000; Pikarsky et al., 2004), (ii) viral DNA can be detected in the gills and the gut as early as 1 dpi (as in virtually all organs including skin mucus) (Gilad et al., 2004), and (iii) the gills are an important portal of entry for many fish pathogens. More recent studies using *in vivo* bioluminescent imaging system (IVIS) demonstrated that the skin is the major portal of entry of CyHV-3 after immersion in virus-containing water (Fig. 8; Costes et al., 2009; Fournier et al., 2012). The epidermis of teleost fish is a living stratified squamous epithelium that is capable of mitotic division at all levels (even the outermost squamous layer). The scales are dermal structures and, consequently, are covered by the epidermis (Costes et al., 2009). A discrete luciferase signal was detected as early as 12 hpi in most of the fish, while all fish were clearly positive at 24 hpi with the positive signal preferentially localized on the fins (Costes et al., 2009). This finding is supported by independent reports that show early CyHV-3 RNA expression in the skin as early as 12 hpi (Adamek et al., 2013) and detection of viral DNA in infected cells by *in situ* hybridization (ISH) in the fin epithelium as early as 2 dpi (the earliest positive organ) (Miwa et al., 2014). Fish epidermis has also been shown to support early infection of a Novirhabdovirus (IHNV, infectious hematopoietic necrosis virus) in trout, suggesting that the skin is an important portal of entry of viruses in fish (Harmache, LeBerre, Droineau, Giovannini, & Bremont, 2006).

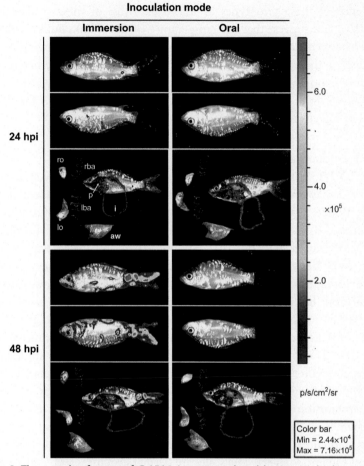

Figure 8 The portals of entry of CyHV-3 in carp analyzed by *in vivo* bioluminescent imaging. Two groups of fish (mean weight 10 g) were infected with a recombinant CyHV-3 strain expressing luciferase as a reporter gene either by bathing them in water containing the virus (immersion, left column) or by feeding them with food pellets contaminated with the virus (oral, right column). At the indicated times post-infection, six fish per group were analyzed by IVIS. Each fish was analyzed lying on its right and left side. The internal signal was analyzed after euthanasia and dissection. Dissected fish and isolated organs were analyzed for *ex vivo* bioluminescence using IVIS. One representative fish is shown for each time point and inoculation mode. Images collected over the course of the experiment were normalized using an identical pseudocolor scale ranging from violet (black in the print version; least intense) to red (dark gray in the print version; most intense) using Living Image 3.2 software. rba, right branchial arches; lba, left branchial arches; ro, right operculum; lo, left operculum; p, pharynx; aw, abdominal wall; i, intestine. *Reproduced with permission from Fournier et al. (2012). Original publisher BioMed Central.*

The data listed above demonstrated that the skin is the major portal of entry after inoculation of carp by immersion in water containing CyHV- 3. While this mode of infection mimics natural conditions in which infection takes place, other epidemiological conditions could favor entry of virus through the digestive tract. To test this hypothesis, carp were fed with material contaminated with a CyHV-3 recombinant strain expressing luciferase as a reporter gene, and bioluminescence imaging analyses were performed at different times post-infection (Fig. 8; Fournier et al., 2012). These experiments demonstrated that the pharyngeal periodontal mucosa is the major portal of entry after oral contamination. This mode of inoculation led to the dissemination of the infection to the various organs tested, inducing clinical signs and mortality rates comparable to the infection by immersion (Fournier et al., 2012). More recently, Monaghan, Thompson, Adams, Kempter, and Bergmann (2015) claimed that the gills and gut represent additional portals of entry by using ISH analysis. In this report, several organs were tested after infection by immersion and positive signal was detected as early as 1–2 hpi in gills, gut, and blood vessels of internal organs. Surprisingly, this early detection occurs far before viral DNA replication, which starts 4–8 hpi *in vitro* (Ilouze et al., 2012a). Moreover, this report is in contradiction with another study that detected positive cells only after 2 days of infection in the fins by using the very same technique (Miwa et al., 2014). Further evidence that the skin, and not the gills, is the major portal of entry after inoculation by immersion in infectious water was recently provided by a study aiming to develop an attenuated recombinant vaccine (Boutier et al., 2015). The study of the tropism of a recombinant strain deleted for ORF56 and 57 (Δ56-57) demonstrated that it also spreads from the skin to all tested organs. However, compared to the wild-type strain, its systemic spread to the other organs was much slower, and its replication was reduced in intensity and duration (Boutier et al., 2015). The slower spread of the Δ56-57 vaccine strain within infected fish allowed better discrimination of the portal(s) of entry from secondary sites of infection. Though the skin of all fish was positive as early as 2 dpi, all of the other tested organs (including gills and gut) were positive in the majority of fish after 6 dpi. These data further demonstrate that the skin is the major portal of entry of CyHV-3 after infection by immersion and suggest that the other organs (including gills and gut) represent secondary sites of replication.

3.2.3.1.2 Secondary Sites of Infection
After infection at the portals of entry, CyHV-3 rapidly spreads in infected fish as demonstrated by the

detection of CyHV-3 DNA in almost all tissues as early as 1–2 dpi (Boutier et al., 2015; Gilad et al., 2003; Ouyang et al., 2013; Pikarsky et al., 2004). The tropism of CyHV-3 for white blood cells most probably explains such a rapid spread of the virus within the body (Eide, Miller-Morgan, Heidel, Bildfell, & Jin, 2011). CyHV-3 DNA can be isolated from blood as early as 1 dpi (Pikarsky et al., 2004). During the first days post-infection, most of the organs (including those that act as portals of entry) support increasing viral replication according to time post-infection (Boutier et al., 2015). The cause of death is more controversial. The severe CyHV-3 infection observed in gills and kidneys, together with the associated histopathological alterations, could be responsible for acute death (Gilad et al., 2004; Hedrick et al., 2000). It has also been proposed that the severe skin alterations could lead to hypo-osmotic shock (Miwa et al., 2014).

3.2.3.1.3 Excretion and Transmission Horizontal transmission of CyHV-3 could occur either by direct contact between fish or by indirect transmission. Study of the CyHV-3 portals of entry demonstrated that, according to specific epidemiological conditions, CyHV-3 can enter carp through either infection of the skin or infection of the pharyngeal periodontal mucosa (Fig. 8). Therefore, direct transmission could result from skin to skin contact between acutely infected or carrier fish with naive ones, or from cannibalistic and necrophagous behaviors of carp (Fournier et al., 2012; Raj et al., 2011). Interestingly, horizontal transmission in natural ponds seems accentuated in hot spots of carp breeding behavior and mating (Uchii et al., 2011), which could favor this skin-to-skin mode of transmission (Raj et al., 2011). Several potential vectors could be involved in the indirect transmission of CyHV-3 including fish droppings (Dishon et al., 2005), plankton (Minamoto et al., 2011), sediments (Honjo, Minamoto, & Kawabata, 2012), aquatic invertebrates feeding by water filtration (Kielpinski et al., 2010), and finally the water as the major abiotic vector (Minamoto, Honjo, Uchii et al., 2009). Indeed, virus replication in organs such as the gills, skin, and gut probably represents a source of viral excretion into the water and the ability of CyHV-3 to remain infective in water has been extensively studied experimentally (see Section 3.2.1.3; Adamek et al., 2013; Costes et al., 2009; Dishon et al., 2005; Pikarsky et al., 2004).

The spread of CyHV-3 was recently studied using two experimental settings designed to allow transmission of the virus through infectious water (water sharing) or through infectious water and physical contact between infected and naive sentinel fish (tank sharing) (Boutier et al., 2015). The difference in transmission kinetics observed between the two systems demonstrated that direct contact between subjects promotes transmission of CyHV-3 as postulated. Nevertheless, transmission through infectious water was still highly efficient (Boutier et al., 2015). To date, there is no evidence of CyHV-3 vertical transmission.

3.2.3.2 Latent Infection

Although latency has not been demonstrated conclusively in members of the *Alloherpesviridae* family as it has been for *Herpesviridae*, increasing evidence supports the existence of a latent phase. The evidence related to CyHV-3 is discussed in this section.

Low amounts of CyHV-3 DNA have been detected 2 months post-infection in the gills, kidneys, and brain of fish that survived primary infection and no longer showed clinical signs (Gilad et al., 2004). Independent studies confirmed the presence of CyHV-3 DNA in the brain of fish as late as 1 year post-infection (Miwa et al., 2014; Yuasa & Sano, 2009). In addition, CyHV-3 DNA, but no infectious particles, has been detected in several organs of fish after CyHV-3 infection (Eide, Miller-Morgan, Heidel, Bildfell, et al., 2011). Finally, CyHV-3 DNA can be routinely detected in apparently healthy fish (Cho et al., 2014).

CyHV-3 can persist in farmed (Baumer, Fabian, Wilkens, Steinhagen, & Runge, 2013) or wild carp populations (Uchii et al., 2009, 2014). At least 2 years after an initial outbreak, CyHV-3 DNA was detected in the brain of both large-sized seropositive fish and small-sized seronegative fish from a wild population of common carp (Uchii et al., 2009). These data suggest that transmission occurred between latently infected fish that survived previous outbreaks and the new naive generation (Uchii et al., 2009). In a more recent report, Uchii et al. (2014) suggests that it is the seasonal reactivation that enables CyHV-3 to persist in a wild population. Indeed, they were able to detect RNA expression of CyHV-3 replicative-related genes in the brain of seropositive fish, suggesting reactivation, while some fish expressed only presumed latency-related genes (Ilouze et al., 2012a; Uchii et al., 2014).

St-Hilaire et al. (2005) described that fish can express symptoms and die from CyHV-3 infection following a temperature stress several months after the initial exposure to the virus. Reactivation of infectious virions was demonstrated by contamination of naive fish. In another report, a netting stress induced viral reactivation without symptoms 81 days after initial infection as detected by qPCR on gill samples (Bergmann & Kempter, 2011).

Recent studies suggested that white blood cells could support CyHV-3 latency (Eide, Miller-Morgan, Heidel, Bildfell, et al., 2011; Eide, Miller-Morgan, Heidel, Kent, et al., 2011; Reed et al., 2014; Xu, Bently, et al., 2013). First, koi carp with previous exposure to the virus displayed CyHV-3 DNA in white blood cells in the absence of any clinical signs or detectable infectious viral particles (Eide, Miller-Morgan, Heidel, Bildfell, et al., 2011). Similar results were found in wild carp collected from ponds in Oregon with no history of CyHV-3 outbreaks (Xu, Bently, et al., 2013). Interestingly, Eide, Miller-Morgan, Heidel, Kent, et al. (2011) detected low amounts of CyHV-3 DNA ranging from 2 to 60 copies per µg of isolated DNA in white blood cells of previously infected koi. These numbers are similar to those reported during the latency of *Herpesviridae*. Notably, similar viral DNA copies were found in all other tissues with no evidence of whether this widespread tissue distribution reflects detection of latently infected circulating white blood cells or latently infected resident cells (Eide, Miller-Morgan, Heidel, Kent, et al., 2011). Among white blood cells, it seems that the IgM$^+$ B cells are the main cell type supporting CyHV-3 latency (Reed et al., 2014). Indeed, the amount of CyHV-3 DNA copies was 20 times higher in IgM$^+$-purified B cells compared to the remaining white blood cells. However, it has to be noted that the latter still contained 10% of IgM$^+$ B cells due to lack of selectivity of the IgM$^+$-sorting method. Therefore, it is still not known whether the low amount of CyHV-3 DNA found in the remaining white blood cells could be explained by the existence of another cell type also supporting latent infection or by the IgM$^+$ B cell contamination. This study also investigated the CyHV-3 transcriptome in latently infected IgM$^+$ B cells (Reed et al., 2014). It demonstrated that CyHV-3 ORF6 transcription was associated with latent infection of IgM$^+$ B cells (ORF1–5 and 7–8 were not transcribed). Interestingly, one domain of ORF6 (aa 342–472) was found to be similar to the consensus sequences of EBNA-3B (EBV nuclear antigen) and the N-terminal regulator domain of ICP4 (infected-cell polypeptide 4). The EBNA-3B is one of the proteins expressed by the gammaherpesvirus EBV during latency and is potentially involved in regulation of cellular gene expression, while ICP4 is found in alphaherpesviruses and acts also as a transcriptional regulator (Reed et al., 2014).

A hallmark of herpesviruses is their capacity to establish a latent infection. Recent studies on CyHV-3 highlighted potential latency in white blood cells and, more precisely, in the B cell fraction as observed for some gammaherpesviruses. On the other hand, CyHV-3 DNA was found in various tissues of long-term infected fish and especially in the brain. Whether the nervous system represents an additional site of latency as observed in alphaherpesviruses requires further investigation.

3.2.3.3 Effect of Water Temperature

KHVD occurs naturally when the water temperature is between 18 and 28 °C (Gotesman et al., 2013; Rakus et al., 2013). Experimentally, KHVD has been reproduced in temperatures ranging from 16 to 28 °C (Gilad et al., 2003, 2004; Yuasa, Ito, & Sano, 2008) and the lowest temperature associated with a CyHV-3 outbreak was 15.5 °C in a field survey in Japan (Hara, Aikawa, Usui, & Nakanishi, 2006). Interestingly, CyHV-2 induces mortalities in goldfish at a slightly enlarged temperature range from 15 to 30 °C (Ito & Maeno, 2014) suggesting a similar but adaptable temperature range in cyprinid herpesviruses. In CyHV-3 infections, the onset of mortality was affected by the water temperature; the first mortalities occurred between 5–8 and 14–21 dpi when the fish were kept between 23–28 and 16–18 °C, respectively (Gilad et al., 2003; Yuasa et al., 2008). Moreover, daily temperature fluctuations of ±3 °C induce important stress in fish, which increases cortisol release in the water and also their susceptibility to CyHV-3 (higher mortality rate and viral excretion) (Takahara et al., 2014).

Several studies demonstrated that transfer of recently infected fish (between 1 and 5 dpi) to nonpermissive low (≤13 °C) (St-Hilaire, Beevers, Joiner, Hedrick, & Way, 2009; St-Hilaire et al., 2005; Sunarto et al., 2014) or high temperatures (30 °C) (Ronen et al., 2003) significantly reduces the mortality. Some observations suggest that the virus can replicate at low temperatures without inducing mortalities. Indeed, relatively high amounts of CyHV-3 DNA, together with the detectable expression of viral genes encoding structural proteins (ORF149 (glycoprotein member of the ORF25 family), ORF72 (Capsid triplex subunit 2)), and nonstructural proteins (ORF55 (TK), ORF134 (vIL-10)) were detected in fish maintained at low temperature (Baumer et al., 2013; Gilad et al., 2004; Sunarto et al., 2012, 2014), while no infectious particles could be isolated (Sunarto et al., 2014). In addition, CyHV-3-infected fish maintained at low temperature (≤13 °C) and then returned to permissive temperature frequently expressed the disease (Eide, Miller-Morgan, Heidel, Kent, et al., 2011; Gilad et al., 2003; St-Hilaire et al., 2005, 2009; Sunarto et al., 2014) and were able to contaminate naive cohabitants (St-Hilaire et al., 2005). Together, these observations suggest that the temperature of the water could regulate the switch between latency and lytic replication and *vice versa*, thus allowing the virus to persist in the host population throughout the seasons even when the temperature is nonpermissive for productive viral replication.

The studies described above suggest that the effect of temperature on the biological cycle of CyHV-3 *in vivo* is twofold. First, it could control the switch from latency to lytic infection and *vice versa*. Second, it clearly regulates the amplitude of viral replication during lytic infection. Further studies are required to clarify the relative importance of these two effects and their putative interactions.

3.2.4 Host–Pathogen Interactions

3.2.4.1 Susceptibility of Common Carp According to the Developmental Stage
Carp of all ages are affected by CyHV-3, but younger fish (1–3 months, 2.5–6 g) seem to be more susceptible to infection than mature fish (1 year, ≈ 230 g) (Perelberg et al., 2003). Ito, Sano, Kurita, Yuasa, and Iida (2007) suggested that carp larvae are not susceptible to CyHV-3 since larvae (3 days post-hatching) infected with the virus showed no mortality, whereas most of the carp juveniles (>13 days post-hatching) died after infection. This conclusion was challenged recently. Using a CyHV-3 recombinant strain expressing luciferase as a reporter gene and IVIS, Ronsmans et al. (2014) demonstrated that carp larvae are sensitive and permissive to CyHV-3 infection immediately after hatching and that their sensitivity increases with the developmental stages (Ronsmans et al., 2014). However, the sensitivity of the two early stages (embryo and larval stages, 1–21 days post-hatching) was limited compared to the older stages (juvenile and fingerling stages; >21 days post-hatching) (Ronsmans et al., 2014).

3.2.4.2 Susceptibility of Common Carp According to Host Genetic Background
Common carp originated from the Eurasian continent and consist of at least two subspecies *C. carpio carpio* (Europe) and *C. carpio haematopterus* (East Asia) (Chistiakov & Voronova, 2009). During the long history of domestication, common carp of multiple origins have been intensively submitted to selective breeding which led to a high variety of breeds, strains, and hybrid fish (Chistiakov & Voronova, 2009). In addition, domesticated common carp were spread worldwide by human activities (Uchii, Okuda, Minamoto, & Kawabata, 2013). Fish from genetically distant populations may differ in their resistance to diseases. Traditional selective breeding methods as well as marker-associated selection proved to be a relevant

approach to reduce the economic losses induced by infectious diseases (Midtlyng, Storset, Michel, Slierendrecht, & Okamoto, 2002).

Differences in resistance to CyHV-3 have been described among different carp strains and crossbreeds. Zak, Perelberg, Magen, Milstein, and Joseph (2007) reported that the crossbreeding of some Hungarian strains (Dinnyes and Szarvas-22 bred at the research Institute for Fisheries, Aquaculture and Irrigation (HAKI) in Szarvas) with the Dor-70 strain (bred in Israel) does not improve the resistance to CyHV-3. On the other hand, independent research groups demonstrated that resistance to CyHV-3 can be significantly increased by crossbreeding domesticated carp strains with wild carp strains. Shapira et al. (2005) reported that crossing the domesticated carp Dor-70 (bred in Israel) and Našice (introduced in Israel from ex-Yugoslavia in the 1970s) with a wild carp strain Sassan (originated from the Amur river) significantly increases the resistance to CyHV-3 (Shapira et al., 2005). Carp genetic resistance to CyHV-3 has also been investigated using 96 carp families derived from diallelic crossbreeding of two wild carp strains (Amur and Duna, native of the Amur and Danube rivers) and two domesticated Hungarian strains (Tat, and Szarvas 15) (Dixon et al., 2009; Ødegård et al., 2010). These studies showed that overall the more resistant families derived from wild-type strains, even if important variations were observed according to the pair of genitors used (Dixon et al., 2009). Similarly, Piackova et al. (2013) demonstrated that most of the Czech strains and crossbreeds which are genetically related to wild Amur carp were significantly more resistant to CyHV-3 infection than strains with no relation to Amur carp.

In Japan, common carp of two different genetic origins inhabit Lake Biwa: an ancient Japanese indigenous type and an introduced domesticated Eurasian type (Mabuchi, Senou, Suzuki, & Nishida, 2005). During the CyHV-3 outbreak in Lake Biwa in 2004, mortalities were mainly recorded in the Japanese indigenous type (Ito, Kurita, & Yuasa, 2014; Uchii et al., 2013). This higher susceptibility of the Japanese indigenous type to CyHV-3 was later confirmed experimentally (Ito et al., 2014) and is supposed to be one factor responsible for the important decline of this ancient lineage in Lake Biwa (Uchii et al., 2013).

Recently, resistance to CyHV-3 among common carp strains has also been linked to the polymorphism of genes involved in the immune response, i.e., the MHC class II B genes (Rakus, Wiegertjes, Jurecka, et al., 2009) and carp IL-10 gene (Kongchum et al., 2011). All together, these findings support the hypothesis that the outcome of the disease can be correlated to some extent to genetic factors of the host, and consequently, that selection of resistant carp breeds is one of the potential ways to reduce the negative impact of CyHV-3 on carp aquaculture.

3.2.4.3 Common Carp Innate Immune Response Against CyHV-3

CyHV-3 enters fish through infection of the skin and/or the pharyngeal periodontal mucosa (Fig. 8; Costes et al., 2009; Fournier et al., 2012). These mucosal epithelia are covered by mucus that acts as a physical, chemical, and immunological barrier against pathogens. The mucus layer contains numerous proteins, such as antimicrobial peptides, mucins, immunoglobulins, enzymes, and lytic agents, capable of neutralizing microorganisms (Ellis, 2001; Shephard, 1994; van der Marel et al., 2012). Interestingly, Raj et al. (2011) demonstrated that skin mucus acts as an innate immune barrier and inhibits CyHV-3 binding to epidermal cells at least partially by neutralization of viral infectivity as shown by *in vitro* assay. Recently, the low sensitivity of carp larvae to CyHV-3 infection was circumvented by a mucus removal treatment suggesting a critical role of skin mucus in protecting larvae against infectious diseases (Fig. 9). Such an innate protection is likely to play a key role in the immune protection of this developmental stage which does not yet benefit from a mature adaptive immune system (Ronsmans et al., 2014). The anti-CyHV-3 immune response has been studied in the skin and the intestine of common carp (Adamek et al., 2013; Syakuri et al., 2013). In the skin, CyHV-3 infection leads to downregulation of genes encoding several important components of the skin mucosal barrier, including antimicrobial peptides (beta defensin 1 and 2), mucin 5B, and tight junction proteins (claudin 23 and 30). This probably contributes to the disintegration of the skin (downregulation of claudins), the decreased amount of mucus, and the sandpaper-like surface of the skin (downregulation of mucins), as well as changes in the cutaneous bacterial flora and subsequent development of secondary bacterial infections (Adamek et al., 2013). These studies also revealed an upregulation of proinflammatory cytokine IL-1β, the inducible nitric oxide synthase, and activation of interferon (IFN) class I pathways (Adamek et al., 2013; Syakuri et al., 2013).

IFNs are secreted mediators that play essential roles in the innate immune response against viruses. *In vitro* studies demonstrated that CCB cells can secrete IFN type I in response to spring viremia of carp virus (SVCV) but not CyHV-3 infection, suggesting that CyHV-3 can inhibit this critical antiviral pathway *in vitro* (Adamek et al., 2012). Poly I:C stimulation of CCB cells prior to CyHV-3 infection activates the IFN type I response and

Time of inoculation (days post hatching)

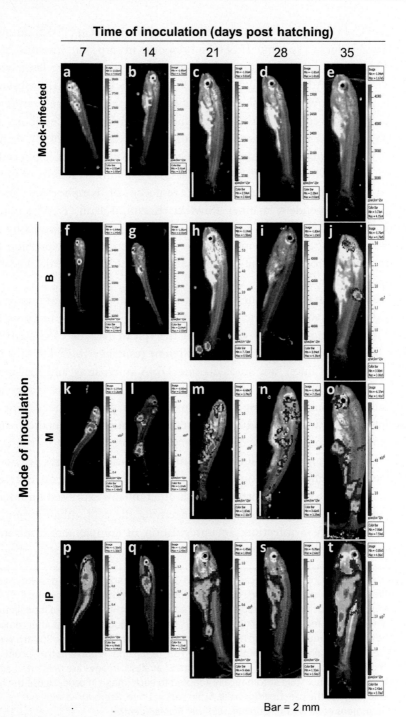

Bar = 2 mm

Figure 9 See legend on next page.

reduces CyHV-3 spreading in the cell culture (Adamek et al., 2012). *In vivo*, CyHV-3 induces a systemic IFN type I response in carp skin, intestine, and head kidney, and the magnitude of IFN type I expression is correlated with the virus load (Adamek, Rakus, et al., 2014; Adamek et al., 2013; Syakuri et al., 2013). However, no significant difference in the IFN type I response could be observed between two carp lines with different susceptibility to CyHV-3 (i.e., R3 and K carp lines) (Adamek, Rakus, et al., 2014). Additional *in vitro* studies demonstrated that CyHV-3 does not induce apoptosis, unlike SVCV (Miest et al., 2015), and that CyHV-3 inhibits activity of stimulated macrophages and proliferative response of lymphocytes, in a temperature-dependent manner (Siwicki, Kazuń, Kazuń, & Majewicz-Zbikowska, 2012). Finally, stimulation of the apoptosis intrinsic pathway *in vivo* following CyHV-3 infection, as determined by the expression of proapoptotic proteins (Apaf-1, p53, and Caspase 9), was delayed to 14 dpi (Miest et al., 2015).

Recently, a transcriptomic study uncovered the wide array of immune-related genes involved in the systemic anti-CyHV-3 immune response of carp by sampling the head kidney and the spleen (Rakus et al., 2012). The response of two carp lines with different resistance to CyHV-3 (i.e., R3 and K carp lines) was studied using DNA microarray and real-time PCR. Significantly higher expression of several immune-related genes including a number of those that are involved in pathogen recognition, complement activation, MHC class I-restricted antigen presentation, and development of adaptive mucosal immunity, was noted in the more resistant carp line. In this same line, further real-time PCR-based analyses provided evidence for higher activation of CD8$^+$ T cells. Thus, differences in resistance to CyHV-3 can be correlated with differentially expressed immune-related genes (Rakus et al., 2012). Concerning the acute-phase response following CyHV-3 infection, an upregulation of complement-associated proteins and C-reactive proteins was also detected by 72 hpi, suggesting a

Figure 9 Sensitivity of common carp to CyHV-3 during the early stages of development. At different times post-hatching, carp were inoculated with a recombinant CyHV-3 strain expressing luciferase as a reporter gene, according to three modes of inoculation: by immersion in infectious water (B), by immersion in infectious water just after removing the epidermal mucus (M), and by IP injection of the virus (IP). At 24 hpi, 30 carp were analyzed individually by IVIS. Mock-infected fish (A–E) and representative positive-infected fish (F–T) are shown for each time point of analysis. Images are presented with a relative photon flux scale automatically adapted to each image in order to use the full dynamic range of the pseudocolor scale. Scale bars = 2 mm. *Reproduced with permission from Ronsmans et al. (2014). Original publisher BioMed Central.*

strong and quick innate immune response (Pionnier et al., 2014). A summary of immune responses of common carp against CyHV-3 is shown in Table 9.

3.2.4.4 Common Carp Adaptive Immune Response Against CyHV-3

The systemic immune response to CyHV-3 has been evaluated by measuring anti-CyHV-3 antibodies in the serum of infected carp (Adkison et al., 2005; Perelberg et al., 2008; Ronen et al., 2003; St-Hilaire et al., 2009). Some studies reported slight cross-reaction by enzyme-linked immunosorbent assay (ELISA) and Western blot of anti-CyHV-3 antibodies to CyHV-1, probably due to shared epitopes between these two closely related viruses (Adkison et al., 2005; Davison et al., 2013; St-Hilaire et al., 2009). Detection of anti-CyHV-3 antibodies starts between 7 and 14 dpi, rises till 20–40 dpi, and finally progressively decreases with significant titers still detected at 150 dpi (Perelberg et al., 2008; Ronen et al., 2003). During these periods, the anti-CyHV-3 antibody response correlates with protection against CyHV-3 disease. On the other hand, at 280 dpi, the titer of anti-CyHV-3 antibodies in previously infected fish is only slightly higher or comparable to that of naive fish. Nevertheless, immunized fish, even those in which antibodies are no longer detectable, are resistant to a lethal challenge, possibly because of the subsequent rapid response of B and T memory cells to antigen restimulation (Perelberg et al., 2008).

Temperature strongly influences the adaptive immune response of fish (Bly & Clem, 1992). The cutoff between permissive and nonpermissive temperature for effective cellular and humoral immune response of carp is 14 °C (Bly & Clem, 1992). Therefore, fish kept below this temperature are supposed to be less immunocompetent than fish kept at higher temperature. This has been shown in CyHV-3 infection with a temperature-dependent expression of anti-CyHV-3 antibodies from 14 (slow antibody response shown at 40 dpi) to 31 °C (quick antibody response at 10 dpi) (Perelberg et al., 2008). In another study, only 40% of CyHV-3 exposed fish were able to seroconvert when kept at 12 °C and experienced mortalities due to CyHV-3 disease when brought back to permissive temperature, suggesting a reduced immunocompetence in low-temperature conditions (St-Hilaire et al., 2009).

Recently, the knowledge on mucosal immune response of teleost fish increased with the discovery of a new immunoglobulin isotype, IgT (or IgZ) (Hansen, Landis, & Phillips, 2005; Ryo et al., 2010), specialized in mucosal immunity (Xu, Parra, et al., 2013; Zhang et al., 2010). This specific

Table 9 Immune Responses of *Cyprinus carpio* to Cyprinid Herpesvirus 3 Infection

Immune Response	Antiviral Action	Organ/Cell Type	Phenotype	References
Antimicrobial peptides	Destroying virus particles	Skin	Downregulated/no response	Adamek et al. (2013)
Mucins	Physical protection	Skin	Downregulated	Adamek et al. (2013)
		Gut	No response	Syakuri et al. (2013)
Claudins	Physical protection, tissue permeability	Skin	Downregulated	Adamek et al. (2013)
		Gut	Upregulated	Syakuri et al. (2013)
Type I IFNs and IFN-stimulated genes	Limiting virus replication, inducing antiviral state of the cell	Fibroblasts	No response	Adamek et al. (2012)
		Head kidney leukocytes	Upregulated	Adamek et al. (2012)
		Head kidney	Upregulated	Adamek, Rakus, et al. (2014) Rakus et al. (2012)
		Skin	Upregulated	Adamek et al. (2013) Adamek, Rakus, et al. (2014)
		Gut	Upregulated	Syakuri et al. (2013)
Apoptosis	Death of infected cell	Gills Head kidney Spleen	Upregulated (delay)	Miest et al. (2015)

Component	Function	Tissue	Response	Reference
Proinflammatory cytokines/ chemokines	Activating the immune response, proinflammatory action	Skin	Upregulated	Adamek et al. (2013)
		Gut	Upregulated	Syakuri et al. (2013)
		Spleen	Upregulated/ downregulated	Rakus et al. (2012) Ouyang et al. (2013)
Anti-inflammatory cytokines	Regulation of inflammatory response	Spleen	Upregulated	Rakus et al. (2012) Ouyang et al. (2013)
Acute-phase response (CRP and complement)	Neutralizing viral particles, lysis of infected cells	Serum	Upregulated/no response	Rakus et al. (2012)
		Serum Liver Head kidney Spleen Gills	Upregulated	Pionnier et al. (2014)
MHC class I	Antigen presentation	Head kidney	Upregulated	Rakus et al. (2012)
Cytotoxic CD8$^+$ T cells	Killing infected cells	Spleen	Upregulated	Rakus et al. (2012)
Antibody response	Coating, neutralizing of virus particles	Serum	Upregulated	Perelberg, Ilouze, Kotler, and Steinitz (2008) Ronen et al. (2003) Adkison, Gilad, and Hedrick (2005)
Genetic markers associated with resistance				
Cyprinus carpio IL-10a				Kongchum et al. (2011)
Cyprinus carpio MHC class II *B*				Rakus, Wiegertjes, Adamek, et al. (2009)

Adapted with Permission from Adamek, Steinhagen, et al. (2014). Copyright © Elsevier.

mucosal adaptive immune response further supports the importance of anti-gen presentation at the pathogen's portal of entry to induce topologically adequate immune protection capable of blocking pathogen entry into the host (Gomez, Sunyer, & Salinas, 2013; Rombout, Yang, & Kiron, 2014). In a recent study, a CyHV-3 recombinant attenuated vaccine candidate used by immersion was shown to infect the skin mucosa and to induce a strong immune response at this CyHV-3 portal of entry. Indeed, the vaccine induced a protective mucosal immune response capable of preventing the entry of wild-type CyHV-3 expressing luciferase as a reporter (Fig. 10; Boutier et al., 2015). Whether this protection is related to the stimulation of the IgZ-secreting B cells associated with a higher concentration of IgZ in the mucus represents an interesting fundamental research question that could be addressed in the future using this CyHV-3 mucosal immunity model.

3.2.4.5 CyHV-3 Genes Involved in Immune Evasion

In silico analyses but also *in vitro* and *in vivo* experiments suggest that CyHV-3 may express immune evasion mechanisms that could explain the acute and dramatic clinical signs associated with KHVD. Members of the *Herpesviridae* family have developed sophisticated immune evasion mechanisms (Horst, Ressing, & Wiertz, 2011). Bioinformatics analysis of the CyHV-3 genome revealed several genes encoding putative homologs of host or viral immune-related genes (Aoki et al., 2007). These genes are ORF4 and ORF12 encoding TNF receptor homologs, ORF16 encoding a G protein-coupled receptor homolog, ORF112 encoding a Zalpha domain-containing protein, ORF134 encoding an IL-10 homolog, and ORF139 encoding a poxvirus B22R protein homolog (Aoki et al., 2007). The potential roles of some of these genes in immune evasion mechanisms have been addressed in a few studies. Their main results are summarized below.

Ouyang et al. (2013) characterized the secretome of CyHV-3 and demonstrated that ORF12 was the most abundant secreted viral protein in the supernatant of infected CCB cells. Recently, it was established that infected carp produce antibodies raised against the ORF12 protein (Kattlun, Menanteau-Ledouble, & El-Matbouli, 2014). These observations are consistent with the hypothesis that ORF12 could act *in vivo* as a soluble TNFα receptor as suggested by bioinformatics analyses.

When exploring the usefulness of a CyHV-3 BAC clone to produce recombinant viruses, a CyHV-3 ORF16 deleted strain was produced (Costes et al., 2008). No significant reduction of virulence was observed, suggesting a minor role of this gene in the pathogenesis of the infection at least in the experimental conditions tested.

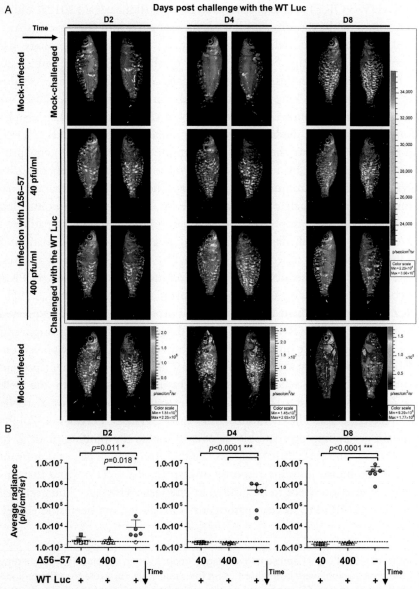

Figure 10 Immune protection conferred by the Δ56–57 attenuated CyHV-3 vaccine revealed by *in vivo* bioluminescent imaging. Common carp (mean±SD weight 13.82±5.00 g, 9 months old) were infected for 2 h by immersion in water containing 40 or 400 pfu/ml of the Δ56–57 attenuated CyHV-3 strain or mock-infected. None of the fish died from primary infection. Forty-two days post-primary infection, fish were challenged by immersion for 2 h in water containing 200 pfu/ml of the WT Luc strain. At the indicated times post-challenge, fish (*n* = 6) were analyzed using the IVIS.

(Continued)

CyHV-3 ORF112 is expressed as an IE gene (Ilouze et al., 2012a) and its 278 amino-acid expression product is incorporated into the virion (structural protein; Fig. 5; Michel, Leroy, et al., 2010). No homology has been detected for the N-terminal part of the protein. In contrast, its C-terminal end encodes a functional Zalpha domain. Zalpha domains are 66 amino acid-long domains which bind to left-handed dsDNA (Z-DNA) or left-handed dsRNA (Z-RNA) (Athanasiadis, 2012). Zalpha domains have been described in three cellular proteins (ADAR1, DAI, and PKZ) belonging to the host innate immune system and in two viral proteins (E3L encoded by most *Chordopoxvirinae* and ORF112 encoded by CyHV-3), acting as immune evasion factors. These data suggest that unusual conformation of nucleic acids detected by Zalpha domain-containing proteins could be interpreted by the innate immune system as pathogen (PAMP) or host cell damage (DAMP). In cells, Z-DNA formation is induced by negative supercoiling generated by moving RNA polymerases. One of the three cellular proteins containing Zalpha domains is PKZ encoded by Cypriniforms and Salmoniforms (Rothenburg et al., 2005). PKZ is a paralog of the dsRNA-dependent protein kinase (PKR) expressed by all vertebrates. PKR is an IFN-induced protein that plays an important role in antiviral innate immunity, mainly (but not exclusively) by phosphorylation of the eukaryotic initiation factor 2 alpha and consequent protein synthesis shutdown when detecting right-handed dsRNA in the cell. PKZ induces the same effects when detecting Z-DNA and/or Z-RNA in infected cells. The demonstration that CyHV-3 encodes a Zalpha domain-containing protein able to over-compete the binding of PKZ to Z-DNA (Tome et al., 2013) suggests that the latter protein plays a significant role in the innate immune response of carp against CyHV-3 and that this immune reaction needs to be evaded by the virus. However, the potential function of ORF112 in virus pathogenesis *in vivo* has not been studied yet.

Figure 10—Cont'd (A) Representative images. Images within the blue frame (light gray in the print version) were normalized using the same scale. (B) Average radiance (individual values, mean + SD) measured on the entire body surface of the fish (individual values represent the mean of the left and right sides obtained for each fish). The discontinuous line represents the cutoff for positivity, which is the mean + 3 SD ($p < 0.00135$) of the values obtained (not presented) for mock-infected and mock-challenged fish (negative control). Positive fish are represented by red filled dots (dark gray in the print version). Significant differences in the mean of the average radiance were identified by post hoc *t*-test after two-way ANOVA analysis taking the treatment and the time post-challenge as variables. *Reproduced with permission from Boutier et al. (2015).*

CyHV-3 ORF134 encodes a viral homolog of cellular IL-10 (Aoki et al., 2007). Cellular IL-10 is a pleiotropic cytokine with both immunostimulating and immunosuppressive properties (Ouyang et al., 2014). Herpesviruses and poxviruses encode orthologs of cellular IL-10, called viral IL-10s, which appear to have been acquired from their host on multiple independent occasions during evolution (Ouyang et al., 2014). Common carp IL-10 was recently shown to possess the prototypical activities described in mammalian IL-10s such as anti-inflammatory activities on macrophages and neutrophils, stimulation of $CD8^+$ memory T cells, stimulation of·the differentiation and antibody secretion by IgM^+ B cells (Piazzon, Savelkoul, Pietretti, Wiegertjes, & Forlenza, 2015). Whether CyHV-3 ORF134 exhibits similar properties to carp IL-10 still needs to be investigated. The CyHV-3 ORF134 expression product is a 179 amino-acid protein (Sunarto et al., 2012) which exhibits 26.9% identity (67.3% similarity) with the common carp IL-10 over 156 amino acids (van Beurden, Forlenza, et al., 2011). Transcriptomic analyses revealed that ORF134 is expressed from a spliced transcript belonging to the early (Ilouze et al., 2012a) or early-late class (Ouyang et al., 2013). Proteomic analyses of CyHV-3-infected cell supernatant demonstrated that the ORF134 expression product is the second most abundant protein of the CyHV-3 secretome (Ouyang et al., 2013). In CyHV-3-infected carp, ORF134 is highly expressed during acute and reactivation phases, while it is expressed at a low level during low-temperature-induced persistent phase (Sunarto et al., 2012). *In vivo* study using a zebrafish embryo model suggested that CyHV-3 ORF134 encodes a functional IL-10 homolog. Indeed, injection of mRNA encoding CyHV-3 IL-10 into zebrafish embryos increased the number of lysozyme-positive cells to a similar degree as observed with zebrafish IL-10. Moreover, this effect was abrogated when downregulation of the IL-10 receptor long chain (IL-10R1) was performed using a specific morpholino (Sunarto et al., 2012). Recently, a CyHV-3 strain deleted for ORF134 and a derived revertant strain were produced using BAC cloning technologies (Ouyang et al., 2013). The recombinant ORF134 deleted strain replicated comparably to the parental and the revertant strains both *in vitro* and *in vivo*, leading to a similar mortality rate. These results demonstrated that the IL-10 homolog encoded by CyHV-3 is essential neither for viral replication *in vitro* nor for virulence *in vivo*. In addition, quantification of carp cytokine expression by RT-qPCR at different times post-infection did not reveal any significant difference between the groups of fish infected with the three virus genotypes (Ouyang et al., 2013).

3.2.5 Diagnosis

Diagnosis of KHVD in clinically affected fish can be achieved by numerous methods. The manual of diagnostic tests for aquatic animals lists gross clinical signs, histopathological alterations, and transmission electron microscopy as suitable for presumptive diagnosis of KHVD and descriptions of these can be found earlier in this review (OIE, 2012). However, final diagnosis must rely on direct detection of viral DNA or virus isolation and identification (OIE, 2012). The manual details virus detection methods that include single-round conventional PCR assays, virus isolation in cell culture, indirect fluorescent antibody tests (FATs) on kidney imprints, and formalin-fixed paraffin wax sections followed by confirmatory identification using PCR and nucleotide sequencing. However, none of the tests are fully validated and the manual suggests that diagnosis of KHVD should not rely on just one test but rather a combination of two or three that include clinical examination as well as virus detection (OIE, 2012).

3.2.5.1 PCR-Based Methods

A number of conventional PCR assays have been published, which have been shown to detect CyHV-3 DNA in cell culture supernatant or directly in fish tissues (Bercovier et al., 2005; Gilad et al., 2002; Gray et al., 2002; Hutoran et al., 2005; Ishioka et al., 2005). A PCR based on amplification of the TK gene of CyHV-3 was reported to be more sensitive than other published PCR assays (Gilad et al., 2002; Gray et al., 2002) and could detect 10 fg of CyHV-3 DNA (Bercovier et al., 2005), while the PCR of Ishioka et al. (2005), based on the DNA polymerase gene, detected 100 fg of CyHV-3 DNA. The PCR developed by Gray et al. (2002) was improved by Yuasa, Sano, Kurita, Ito, and Iida (2005) and has been incorporated in the official Japanese guidelines for the diagnosis of KHVD. The Yuasa et al. (2005) and Bercovier et al. (2005) assay protocols are recommended by, and detailed in, the manual of diagnostic tests for aquatic animals (OIE, 2012).

Alternatively, many diagnostic laboratories favor the use of qPCR assays for detection of CyHV-3. The most commonly used quantitative assay for detection of CyHV-3 is the Gilad Taqman real-time PCR assay (Gilad et al., 2004), which has been shown to detect and quantitatively assess very low copy numbers of target nucleic acid sequences and is widely acknowledged to be the most sensitive published PCR method available (OIE, 2012). There are a small number of studies that have compared the sensitivity of the published PCR assays, and different primer sets, for detection of

CyHV-3 (Bergmann, Riechardt, Fichtner, Lee, & Kempter, 2010; Monaghan, Thompson, Adams, & Bergmann, 2015; Pokorova et al., 2010). Conventional PCR assays that include a second round with nested primers have also been shown to be comparable in sensitivity to real-time assays (Bergmann, Riechardt, et al., 2010).

Loop-mediated isothermal amplification (LAMP) is a rapid single-step assay which does not require a thermal cycler, and is widely favored for pond-side diagnosis. LAMP of the TK gene has been developed for detection of CyHV-3 and shown to be more or equally sensitive as conventional PCR assays (Gunimaladevi, Kono, Venugopal, & Sakai, 2004; Yoshino, Watari, Kojima, & Ikedo, 2006; Yoshino, Watari, Kojima, Ikedo, & Kurita, 2009). An assay incorporating DNA hybridization technology and antigen–antibody reactions in combination with LAMP has also been developed and reported to have improved sensitivity and specificity (Soliman & El-Matbouli, 2010).

3.2.5.2 Virus Isolation in Cell Culture

Cell lines permissive to CyHV-3 replication have been described earlier in this review (see Section 3.1.6.1). The CCB and KF-1 cell lines are recommended for isolation of CyHV-3, but cell culture isolation is not considered to be as sensitive as the published PCR-based methods for detecting CyHV-3 DNA. Consequently, virus isolation in cell culture is not a reliable diagnostic method for KHVD (OIE, 2012). Furthermore, viruses isolated in cell culture must be definitively identified, as a number of different viruses have been isolated from carp exhibiting clinical signs resembling those of KHVD (Haenen et al., 2004; Neukirch et al., 1999; Neukirch & Kunz, 2001). The most reliable method for confirmatory identification of a CPE is by PCR and nucleotide sequence analysis (OIE, 2012). A variety of tissues in different combinations have been used for inoculation of cell cultures, such as gill, kidney, spleen, liver, skin, and encephalon (Gilad et al., 2003; Gilad et al., 2002; Hedrick et al., 2000; Neukirch & Kunz, 2001; Sano et al., 2004; Yuasa, Sano, et al., 2012). There is no definitive study that has demonstrated the advantages of certain tissues over others but in the early stages of clinical infection, before clinical signs are observed, virus levels are higher in gill tissue than in kidney tissue (Yuasa, Sano, et al., 2012).

3.2.5.3 Immunodiagnostic Methods

Immunodiagnostic (antibody-based) assays have been little used for the diagnosis of KHVD. Pikarsky et al. (2004) identified the virus in touch

imprints of liver, kidney, and brain of infected fish by FAT; positive immunofluorescence (IF) was the highest in the kidney. The same FAT method
was subsequently used by Shapira et al. (2005) who followed the course of
KHVD in different strains of fish and detected virus on a kidney imprint
1 dpi. Pikarsky et al. (2004) also detected virus antigen in infected tissues
by an immunoperoxidase-staining method. The virus antigen was detected
at 2 dpi in the kidney and also observed in the gills and liver. However, the
results of antibody-based identification methods must be interpreted with
care, as positive cells were seen in a small number of control fish which
could have originated from a serologically related virus, or a cross-reaction
with nonviral proteins (Pikarsky et al., 2004). ELISA-based methods have
not been widely favored by diagnostic laboratories. Currently, one published ELISA method is available to detect CyHV-3 in fish droppings
(Dishon et al., 2005). Recently, a CyHV-3-detection kit (The FASTest
Koi HV kit) adapted to field conditions has been developed and proved
to detect 100% of animals which died from CyHV-3. This lateral flow
device relies on the detection of the ORF65 glycoprotein of CyHV-3. It
is recommended to be performed on gill swabs and takes 15 min
(Vrancken et al., 2013).

3.2.5.4 Other Diagnostic Assays

Assays developed for research applications include a primer probe designed
against an exonic mRNA-coding sequence that allows the detection of
replicating CyHV-3 (Yuasa, Kurita, et al., 2012). IF and ISH methods, performed on separated fish leucocytes obtained by nondestructive (nonlethal)
techniques, have also been used in research applications for detection or
identification of CyHV-3 (Bergmann, Lutze, et al., 2010; Bergmann
et al., 2009). ISH has also been applied to successfully detect CyHV-3
DNA in archive paraffin-embedded tissue specimens collected during
unexplained mass mortalities of koi and common carp in the UK in
1996, and in cultured common carp in South Korea in 1998 (Haenen
et al., 2004; Lee et al., 2012).

3.2.6 Vaccination

The economic losses induced by CyHV-3 stimulated the development of
prophylactic measures. Passive immunization by administration of pooled sera
from immunized fish (Adkison et al., 2005) and addition of anti-CyHV-3 IgY
antibodies to fish food (Liu et al., 2014) showed partial effect on the onset
of clinical signs but did not significantly reduce mortalities. In contrast,

several vaccine candidates conferring efficient protection were developed. They are reviewed in this section.

3.2.6.1 Natural Immunization

Soon after the identification of CyHV-3 as the causative agent of KHVD, an original protocol was developed to induce a protective adaptive immune response in carp (Ronen et al., 2003). This approach relied on the fact that CyHV-3 replication is drastically altered at temperatures above 30 °C (Dishon et al., 2007). According to this protocol, healthy fingerlings are exposed to the virus by cohabitation with sick fish for 3–5 days at permissive temperature (22–23 °C). After that, the fish are transferred to ponds for 25–30 days at nonpermissive water temperature ($\approx$30 °C). Despite its ingenuity, this protocol has several disadvantages. (i) Fish that are infected with this protocol become latently infected carriers of a fully virulent strain and are therefore likely to represent a potential source of CyHV-3 outbreaks if they later cohabit with naive carp. (ii) The increase of water temperature to nonpermissive is costly and correlates with increasing susceptibility of the fish to secondary infections. (iii) Finally, after this procedure, only 60% of infected fish were sufficiently immunized to be resistant to a CyHV-3 challenge (Ronen et al., 2003).

3.2.6.2 Vaccine Candidates

In addition to the safety/efficacy issues that apply to all vaccines independent of the target species (humans or animals), vaccines for fish and production animals in general are under additional constraints (Boutier et al., 2015). First, the vaccine must be compatible with mass vaccination and administered via a single dose as early as possible in life. Second, the cost–benefit ratio should be as low as possible, implying the lowest cost for vaccine production and administration (Sommerset, Krossoy, Biering, & Frost, 2005). Ideally, cost-effective mass vaccination of young fish is performed by immersion vaccination, meaning that the fish are bathed in water containing the vaccine. This procedure allows vaccination of a large number of subjects when their individual value is still low and their susceptibility to the disease is the highest (Brudeseth et al., 2013). Immersion vaccination is particularly adapted to common carp culture that is a low-cost and low industrial scale production compared to other sectors (Brudeseth et al., 2013). The use of injectable vaccines for mass vaccination of fish is restricted to limited circumstances, i.e., when the value of individual subject is relatively high and when vaccination can be delayed until an age when the size of the fish is compatible with their manipulation (Plant & Lapatra, 2011).

Various anti-CyHV-3 vaccine candidates have been developed. An inactivated vaccine candidate was described which consists of formalin-inactivated CyHV-3 trapped within a liposomal compartment. This vaccine could be used for oral immunization by addition to fish food. It reduced by 70% the mortality induced by a challenge (Yasumoto, Kuzuya, Yasuda, Yoshimura, & Miyazaki, 2006). Injectable DNA vaccines consisting of plasmids encoding envelope glycoproteins ORF25 and ORF81 were shown efficacious under experimental conditions (Zhou, Wang, et al., 2014; Zhou, Xue, et al., 2014) but are unfortunately incompatible with most of the field constraints described above. Nevertheless, they could represent a solution for individual vaccination of koi carp.

Attenuated vaccines could meet the constraints of mass vaccination listed above. However, they raise safety concerns, such as residual virulence, reversion to virulence, and spread from vaccinated to naive subjects (Boutier et al., 2015). A conventional anti-CyHV-3 attenuated vaccine has been developed by serial passages in cell culture and UV irradiation (O'Connor et al., 2014; Perelberg et al., 2008; Perelberg, Ronen, Hutoran, Smith, & Kotler, 2005; Ronen et al., 2003; Weber et al., 2014). This vaccine is commercialized in Israel for the vaccination of koi and common carp by immersion in water containing the attenuated strain. Recently launched in the US market, it was withdrawn from sale after just a year. This vaccine has two major disadvantages. First, the attenuated strain has residual virulence for fish weighing less than 50 g (Weber et al., 2014; Zak et al., 2007), which restricts the use of this vaccine. Second, the determinism of the attenuation is unknown, and consequently, reversions to a pathogenic phenotype cannot be excluded (Meeusen, Walker, Peters, Pastoret, & Jungersen, 2007).

Due to scientific advances in molecular biology and molecular virology, the development of attenuated vaccines is evolving from empirical to rational design (Rueckert & Guzman, 2012). A viral genome can be edited to delete genes encoding virulence factors in such a way that reversion to virulence can be excluded. This approach has been tested for CyHV-3 by targeting different genes thought to encode virulence factors, such as ORF16, ORF55, ORF123, and ORF134, which encode a G protein-coupled receptor, TK, deoxyuridine triphosphatase, and an IL-10 homolog, respectively. Unfortunately, none of the recombinants expressed a safety/efficacy profile compatible with its use as an attenuated recombinant vaccine (Costes et al., 2008; Fuchs, Fichtner, Bergmann, & Mettenleiter, 2011; Ouyang et al., 2014).

Recently, a vaccine candidate based on the double deletion of ORF56 and ORF57 was produced using BAC cloning technology (Boutier et al., 2015). This strain exhibited properties compatible with its use as an attenuated recombinant vaccine for mass vaccination of carp by immersion in water containing the virus: (i) it replicates efficiently *in vitro* (essential for vaccine production); (ii) the deletion performed makes reversion impossible; (iii) it expresses a safe attenuated phenotype as demonstrated by the absence of residual virulence even for young subjects and by its limited spreading from vaccinated to naive subjects; (iv) it induces a protective mucosal immune response against a lethal challenge by blocking viral infection at the portal of entry (Fig. 10). Although the two ORFs deleted in this vaccine candidate are of unknown function, they are both conserved in cyprinid herpesviruses (CyHV-1 and CyHV-2) and ORF57 is additionally conserved in AngHV-1 and crocodile poxvirus (Davison et al., 2013). These homologs represent evident targets for further development of attenuated recombinants for these pathogenic viruses (Boutier et al., 2015).

4. CONCLUSIONS

It is generally accepted that fundamental research precedes and stimulates applied research. Work on CyHV-3 has demonstrated that events can take place in the reverse order. Since the first description of CyHV-3 in the late 1990s, this virus has been inducing important economic losses in the common and koi carp industries worldwide. It is also producing negative environmental implications by affecting wild carp populations. These negative impacts and the importance of the host species have stimulated studies aimed directly or indirectly at developing diagnostic and prophylactic tools to monitor and treat CyHV-3 disease. Unexpectedly, the data generated by these applied studies have created and highlighted interest in CyHV-3 as a fundamental research model. The CyHV-3/carp model has the advantages that large amounts of information and reagents are available for both the virus and its host, and that it permits the study of the entire biological cycle (including transmission) of an alloherpesvirus during infection of its natural host (i.e., a virus/host homologous model). As highlighted throughout this review, there are many fascinating topics that can be addressed by using the CyHV-3/carp model as the archetype for studying the family *Alloherpesviridae*. These include, for example, how viruses in this family express key biological properties that are shared with members of the family *Herpesviridae* while having relatively few genes in common with them, and how the temperature of the poikiloterm host affects and possibly regulates the switch between lytic and latent infection.

ACKNOWLEDGMENTS

This work was supported by Belgian Science Policy Grant R.SSTC.0454-BELVIR IAP7/45 to A.V. and by the Medical Research Council Grant Number MC_UU_12014/3 to A.D.

REFERENCES

Ackermann, M. (2004). Herpesviruses: A brief overview. *Methods in Molecular Biology, 256*, 199–219. http://dx.doi.org/10.1385/1-59259-753-X:199.

Adamek, M., Rakus, K. L., Brogden, G., Matras, M., Chyb, J., Hirono, I., et al. (2014). Interaction between type I interferon and Cyprinid herpesvirus 3 in two genetic lines of common carp *Cyprinus carpio*. *Diseases of Aquatic Organisms, 111*(2), 107–118. http://dx.doi.org/10.3354/dao02773.

Adamek, M., Rakus, K. L., Chyb, J., Brogden, G., Huebner, A., Irnazarow, I., et al. (2012). Interferon type I responses to virus infections in carp cells: *In vitro* studies on Cyprinid herpesvirus 3 and Rhabdovirus carpio infections. *Fish & Shellfish Immunology, 33*(3), 482–493. http://dx.doi.org/10.1016/j.fsi.2012.05.031.

Adamek, M., Steinhagen, D., Irnazarow, I., Hikima, J., Jung, T. S., & Aoki, T. (2014). Biology and host response to Cyprinid herpesvirus 3 infection in common carp. *Developmental and Comparative Immunology, 43*(2), 151–159. http://dx.doi.org/10.1016/j.dci.2013.08.015.

Adamek, M., Syakuri, H., Harris, S., Rakus, K. L., Brogden, G., Matras, M., et al. (2013). Cyprinid herpesvirus 3 infection disrupts the skin barrier of common carp (*Cyprinus carpio* L.). *Veterinary Microbiology, 162*(2–4), 456–470. http://dx.doi.org/10.1016/j.vetmic.2012.10.033.

Adkison, M. A., Gilad, O., & Hedrick, R. P. (2005). An enzyme linked immunosorbent assay (ELISA) for detection of antibodies to the koi herpesvirus (KHV) in the serum of koi *Cyprinus carpio*. *Fish Pathology, 40*, 53–62.

Antychowicz, J., Reichert, M., Matras, M., Bergmann, S. M., & Haenen, O. L. (2005). Epidemiology, pathogenicity and molecular biology of koi herpesvirus isolated in Poland. *Bulletin of the Veterinary Institute in Pulawy, 49*, 367–373.

Aoki, T., Hirono, I., Kurokawa, K., Fukuda, H., Nahary, R., Eldar, A., et al. (2007). Genome sequences of three koi herpesvirus isolates representing the expanding distribution of an emerging disease threatening koi and common carp worldwide. *Journal of Virology, 81*(10), 5058–5065. http://dx.doi.org/10.1128/JVI.00146-07.

Aoki, T., Takano, T., Unajak, S., Takagi, M., Kim, Y. R., Park, S. B., et al. (2011). Generation of monoclonal antibodies specific for ORF68 of koi herpesvirus. *Comparative Immunology, Microbiology and Infectious Diseases, 34*(3), 209–216. http://dx.doi.org/10.1016/j.cimid.2010.11.004.

Ariav, R., Tinman, S., Paperna, I., & Bejerano, I. (1999). In *First report of newly emerging viral disease of Cyprinus carpio species in Israel. Paper presented at the EAFP 9th international conference, Rhodes, Greece*.

Athanasiadis, A. (2012). Zalpha-domains: At the intersection between RNA editing and innate immunity. *Seminars in Cell and Developmental Biology, 23*(3), 275–280. http://dx.doi.org/10.1016/j.semcdb.2011.11.001.

Avarre, J. C., Madeira, J. P., Santika, A., Zainun, Z., Baud, M., Cabon, J., et al. (2011). Investigation of Cyprinid herpesvirus-3 genetic diversity by a multi-locus variable number of tandem repeats analysis. *Journal of Virological Methods*, *173*(2), 320–327. http://dx.doi.org/10.1016/j.jviromet.2011.03.002.

Avarre, J. C., Santika, A., Bentenni, A., Zainun, Z., Madeira, J. P., Maskur, M., et al. (2012). Spatio-temporal analysis of cyprinid herpesvirus 3 genetic diversity at a local scale. *Journal of Fish Diseases*, *35*(10), 767–774. http://dx.doi.org/10.1111/j.1365-2761.2012.01404.x.

Balon, E. K. (1995). Origin and domestication of the wild carp, *Cyprinus carpio*: From Roman gourmets to the swimming flowers. *Aquaculture*, *129*, 3–48.

Baumer, A., Fabian, M., Wilkens, M. R., Steinhagen, D., & Runge, M. (2013). Epidemiology of cyprinid herpesvirus-3 infection in latently infected carp from aquaculture. *Diseases of Aquatic Organisms*, *105*(2), 101–108. http://dx.doi.org/10.3354/dao02604.

Bercovier, H., Fishman, Y., Nahary, R., Sinai, S., Zlotkin, A., Eyngor, M., et al. (2005). Cloning of the koi herpesvirus (KHV) gene encoding thymidine kinase and its use for a highly sensitive PCR based diagnosis. *BMC Microbiology*, *5*, 13. http://dx.doi.org/10.1186/1471-2180-5-13.

Bergmann, S. M., & Kempter, J. (2011). Detection of koi herpesvirus (KHV) after re-activation in persistently infected common carp (*Cyprinus carpio* L.) using non-lethal sampling methods. *Bulletin of the European Association of Fish Pathologists*, *31*(3), 92–100.

Bergmann, S. M., Kempter, J., Sadowski, J., & Fichtner, D. (2006). First detection, confirmation and isolation of koi herpesvirus (KHV) in cultured common carp (*Cyprinus carpio* L.) in Poland. *Bulletin of the European Association of Fish Pathologists*, *26*(2), 97–104.

Bergmann, S. M., Lutze, P., Schütze, H., Fischer, U., Dauber, M., Fichtner, D., et al. (2010). Goldfish (*Carassius auratus auratus*) is a susceptible species for koi herpesvirus (KHV) but not for KHV disease (KHVD). *Bulletin of the European Association of Fish Pathologists*, *30*(2), 74–84.

Bergmann, S. M., Riechardt, M., Fichtner, D., Lee, P., & Kempter, J. (2010). Investigation on the diagnostic sensitivity of molecular tools used for detection of koi herpesvirus. *Journal of Virological Methods*, *163*(2), 229–233. http://dx.doi.org/10.1016/j.jviromet.2009.09.025.

Bergmann, S. M., Sadowski, J., Kielpinski, M., Bartlomiejczyk, M., Fichtner, D., Riebe, R., et al. (2010). Susceptibility of koi x crucian carp and koi x goldfish hybrids to koi herpesvirus (KHV) and the development of KHV disease (KHVD). *Journal of Fish Diseases*, *33*(3), 267–272. http://dx.doi.org/10.1111/j.1365-2761.2009.01127.x.

Bergmann, S. M., Schütze, H., Fischer, U., Fichtner, D., Riechardt, M., Meyer, K., et al. (2009). Detection KHV genome in apparently health fish. *Bulletin of the European Association of Fish Pathologists*, *29*(5), 145–152.

Bigarré, L., Baud, M., Cabon, J., Antychowicz, J., Bergmann, S. M., Engelsma, M., et al. (2009). Differentiation between Cyprinid herpesvirus type-3 lineages using duplex PCR. *Journal of Virological Methods*, *158*(1–2), 51–57. http://dx.doi.org/10.1016/j.jviromet.2009.01.023.

Bly, J. E., & Clem, W. (1992). Temperature and teleost immune functions. *Fish & Shellfish Immunology*, *2*, 159–171.

Bondad-Reantaso, M. G., Subasinghe, R. P., Arthur, J. R., Ogawa, K., Chinabut, S., Adlard, R., et al. (2005). Disease and health management in Asian aquaculture. *Veterinary Parasitology*, *132*(3–4), 249–272. http://dx.doi.org/10.1016/j.vetpar.2005.07.005.

Booy, F. P., Trus, B. L., Davison, A. J., & Steven, A. C. (1996). The capsid architecture of channel catfish virus, an evolutionarily distant herpesvirus, is largely conserved in the absence of discernible sequence homology with herpes simplex virus. *Virology*, *215*(2), 134–141. http://dx.doi.org/10.1006/viro.1996.0016.

Boutier, M., Ronsmans, M., Ouyang, P., Fournier, G., Reschner, A., Rakus, K., et al. (2015). Rational development of an attenuated recombinant cyprinid herpesvirus 3 vaccine using prokaryotic mutagenesis and in vivo bioluminescent imaging. *PLoS Pathogens*, *11*(2), e1004690. http://dx.doi.org/10.1371/journal.ppat.1004690.

Bretzinger, A., Fischer-Scherl, T., Oumouma, R., Hoffmann, R., & Truyen, U. (1999). Mass mortalities in koi, *Cyprinus carpio*, associated with gill and skin disease. *Bulletin of the European Association of Fish Pathologists*, *19*, 182–185.

Brudeseth, B. E., Wiulsrod, R., Fredriksen, B. N., Lindmo, K., Lokling, K. E., Bordevik, M., et al. (2013). Status and future perspectives of vaccines for industrialised fin-fish farming. *Fish & Shellfish Immunology*, *35*(6), 1759–1768. http://dx.doi.org/10.1016/j.fsi.2013.05.029.

Chen, J., Chee, D., Wang, Y., Lim, G. Y., Chong, S. M., Lin, Y. N., et al. (2014). Identification of a novel cyprinid herpesvirus 3 genotype detected in koi from the East Asian and South-East Asian Regions. *Journal of Fish Diseases*. http://onlinelibrary.wiley.com/doi/10.1111/jfd.12305/abstract.

Chistiakov, D. A., & Voronova, N. V. (2009). Genetic evolution and diversity of common carp *Cyprinus carpio* L. *Central European Journal of Biology*, *4*(3), 304–312. http://dx.doi.org/10.2478/s11535-009-0024-2.

Cho, M. Y., Won, K. M., Kim, J. W., Jee, B. Y., Park, M. A., & Hong, S. (2014). Detection of koi herpesvirus (KHV) in healthy cyprinid seed stock. *Diseases of Aquatic Organisms*, *112*(1), 29–36. http://dx.doi.org/10.3354/dao02784.

Costes, B., Fournier, G., Michel, B., Delforge, C., Raj, V. S., Dewals, B., et al. (2008). Cloning of the koi herpesvirus genome as an infectious bacterial artificial chromosome demonstrates that disruption of the thymidine kinase locus induces partial attenuation in *Cyprinus carpio koi*. *Journal of Virology*, *82*(10), 4955–4964. http://dx.doi.org/10.1128/JVI.00211-08.

Costes, B., Raj, V. S., Michel, B., Fournier, G., Thirion, M., Gillet, L., et al. (2009). The major portal of entry of koi herpesvirus in *Cyprinus carpio* is the skin. *Journal of Virology*, *83*(7), 2819–2830. http://dx.doi.org/10.1128/JVI.02305-08.

Davidovich, M., Dishon, A., Ilouze, M., & Kotler, M. (2007). Susceptibility of cyprinid cultured cells to cyprinid herpesvirus 3. *Archives of Virology*, *152*(8), 1541–1546. http://dx.doi.org/10.1007/s00705-007-0975-4.

Davison, A. J. (1992). Channel catfish virus: A new type of herpesvirus. *Virology*, *186*(1), 9–14.

Davison, A. J. (1998). The genome of salmonid herpesvirus 1. *Journal of Virology*, *72*(3), 1974–1982.

Davison, A. J., Cunningham, C., Sauerbier, W., & McKinnell, R. G. (2006). Genome sequences of two frog herpesviruses. *Journal of General Virology*, *87*(Pt 12), 3509–3514. http://dx.doi.org/10.1099/vir.0.82291-0.

Davison, A. J., & Davison, M. D. (1995). Identification of structural proteins of channel catfish virus by mass spectrometry. *Virology*, *206*(2), 1035–1043. http://dx.doi.org/10.1006/viro.1995.1026.

Davison, A. J., Eberle, R., Ehlers, B., Hayward, G. S., McGeoch, D. J., Minson, A. C., et al. (2009). The order Herpesvirales. *Archives of Virology*, *154*(1), 171–177. http://dx.doi.org/10.1007/s00705-008-0278-4.

Davison, A. J., Kurobe, T., Gatherer, D., Cunningham, C., Korf, I., Fukuda, H., et al. (2013). Comparative genomics of carp herpesviruses. *Journal of Virology*, *87*(5), 2908–2922. http://dx.doi.org/10.1128/JVI.03206-12.

Davison, A. J., Trus, B. L., Cheng, N., Steven, A. C., Watson, M. S., Cunningham, C., et al. (2005). A novel class of herpesvirus with bivalve hosts. *Journal of General Virology*, *86*(Pt 1), 41–53. http://dx.doi.org/10.1099/vir.0.80382-0.

Deback, C., Boutolleau, D., Depienne, C., Luyt, C. E., Bonnafous, P., Gautheret-Dejean, A., et al. (2009). Utilization of microsatellite polymorphism for differentiating herpes simplex virus type 1 strains. *Journal of Clinical Microbiology*, *47*(3), 533–540. http://dx.doi.org/10.1128/JCM.01565-08.

Dishon, A., Davidovich, M., Ilouze, M., & Kotler, M. (2007). Persistence of cyprinid herpesvirus 3 in infected cultured carp cells. *Journal of Virology*, *81*(9), 4828–4836. http://dx.doi.org/10.1128/JVI.02188-06.

Dishon, A., Perelberg, A., Bishara-Shieban, J., Ilouze, M., Davidovich, M., Werker, S., et al. (2005). Detection of carp interstitial nephritis and gill necrosis virus in fish droppings. *Applied and Environmental Microbiology*, *71*(11), 7285–7291. http://dx.doi.org/10.1128/AEM.71.11.7285-7291.2005.

Dixon, P. F., Joiner, C. L., Way, K., Reese, R. A., Jeney, G., & Jeney, Z. (2009). Comparison of the resistance of selected families of common carp, *Cyprinus carpio* L., to koi herpesvirus: Preliminary study. *Journal of Fish Diseases*, *32*(12), 1035–1039. http://dx.doi.org/10.1111/j.1365-2761.2009.01081.x.

Dong, C., Li, X., Weng, S., Xie, S., & He, J. (2013). Emergence of fatal European genotype CyHV-3/KHV in mainland China. *Veterinary Microbiology*, *162*(1), 239–244. http://dx.doi.org/10.1016/j.vetmic.2012.10.024.

Dong, C., Weng, S., Li, W., Li, X., Yi, Y., Liang, Q., et al. (2011). Characterization of a new cell line from caudal fin of koi, *Cyprinus carpio koi*, and first isolation of cyprinid herpesvirus 3 in China. *Virus Research*, *161*(2), 140–149. http://dx.doi.org/10.1016/j.virusres.2011.07.016.

Doszpoly, A., Benko, M., Bovo, G., Lapatra, S. E., & Harrach, B. (2011). Comparative analysis of a conserved gene block from the genome of the members of the genus ictalurivirus. *Intervirology*, *54*(5), 282–289. http://dx.doi.org/10.1159/000319430.

Doszpoly, A., Kovacs, E. R., Bovo, G., LaPatra, S. E., Harrach, B., & Benko, M. (2008). Molecular confirmation of a new herpesvirus from catfish (*Ameiurus melas*) by testing the performance of a novel PCR method, designed to target the DNA polymerase gene of alloherpesviruses. *Archives of Virology*, *153*(11), 2123–2127. http://dx.doi.org/10.1007/s00705-008-0230-7.

Doszpoly, A., Papp, M., Deakne, P. P., Glavits, R., Ursu, K., & Dan, A. (2015). Molecular detection of a putatively novel cyprinid herpesvirus in sichel (*Pelecus cultratus*) during a mass mortality event in Hungary. *Archives of Virology*, *160*(5), 1279–1283. http://dx.doi.org/10.1007/s00705-015-2348-8.

Doszpoly, A., Somogyi, V., LaPatra, S. E., & Benko, M. (2011). Partial genome characterization of acipenserid herpesvirus 2: Taxonomical proposal for the demarcation of three subfamilies in Alloherpesviridae. *Archives of Virology*, *156*(12), 2291–2296. http://dx.doi.org/10.1007/s00705-011-1108-7.

Eide, K. E., Miller-Morgan, T., Heidel, J. R., Bildfell, R. J., & Jin, L. (2011). Results of total DNA measurement in koi tissue by Koi Herpes Virus real-time PCR. *Journal of Virological Methods*, *172*(1–2), 81–84. http://dx.doi.org/10.1016/j.jviromet.2010.12.012.

Eide, K. E., Miller-Morgan, T., Heidel, J. R., Kent, M. L., Bildfell, R. J., Lapatra, S., et al. (2011). Investigation of koi herpesvirus latency in koi. *Journal of Virology*, *85*(10), 4954–4962. http://dx.doi.org/10.1128/JVI.01384-10.

El-Din, M. M. M. (2011). Histopathological studies in experimentally infected koi carp (*Cyprinus carpio koi*) with koi herpesvirus in Japan. *World Journal of Fish and Marine Science*, *3*, 252–259.

Ellis, A. E. (2001). Innate host defense mechanisms of fish against viruses and bacteria. *Developmental and Comparative Immunology*, *25*(8–9), 827–839. http://dx.doi.org/10.1016/s0145-305x(01)00038-6.

El-Matbouli, M., Saleh, M., & Soliman, H. (2007). Detection of cyprinid herpesvirus type 3 in goldfish cohabiting with CyHV-3-infected koi carp (*Cyprinus carpio koi*). *Veterinary Record*, *161*(23), 792–793.

El-Matbouli, M., & Soliman, H. (2011). Transmission of cyprinid herpesvirus-3 (CyHV-3) from goldfish to naive common carp by cohabitation. *Research in Veterinary Science*, *90*(3), 536–539. http://dx.doi.org/10.1016/j.rvsc.2010.07.008.

Engelsma, M. Y., Way, K., Dodge, M. J., Voorbergen-Laarman, M., Panzarin, V., Abbadi, M., et al. (2013). Detection of novel strains of cyprinid herpesvirus closely related to koi herpesvirus. *Diseases of Aquatic Organisms, 107*(2), 113–120. http://dx.doi.org/10.3354/dao02666.

Fabian, M., Baumer, A., & Steinhagen, D. (2013). Do wild fish species contribute to the transmission of koi herpesvirus to carp in hatchery ponds? *Journal of Fish Diseases, 36*(5), 505–514. http://dx.doi.org/10.1111/jfd.12016.

FAO. (2012). Global Aquaculture Production 1950–2012. Retrieved 15/02/2015, from http://www.fao.org/fishery/statistics/global-aquaculture-production/query/fr.

Fawcett, D. W. (1956). Electron microscope observations on intracellular virus-like particles associated with the cells of the Lucke renal adenocarcinoma. *The Journal of Biophysical and Biochemical Cytology, 2*(6), 725–741.

Fournier, G., Boutier, M., Stalin Raj, V., Mast, J., Parmentier, E., Vanderwalle, P., et al. (2012). Feeding *Cyprinus carpio* with infectious materials mediates cyprinid herpesvirus 3 entry through infection of pharyngeal periodontal mucosa. *Veterinary Research, 43*, 6. http://dx.doi.org/10.1186/1297-9716-43-6.

Fuchs, W., Fichtner, D., Bergmann, S. M., & Mettenleiter, T. C. (2011). Generation and characterization of koi herpesvirus recombinants lacking viral enzymes of nucleotide metabolism. *Archives of Virology, 156*(6), 1059–1063. http://dx.doi.org/10.1007/s00705-011-0953-8.

Fuchs, W., Granzow, H., Dauber, M., Fichtner, D., & Mettenleiter, T. C. (2014). Identification of structural proteins of koi herpesvirus. *Archives of Virology, 159*(12), 3257–3268. http://dx.doi.org/10.1007/s00705-014-2190-4.

Fujioka, H., Yamasaki, K., Furusawa, K., Tamura, K., Oguro, K., Kurihara, S., et al. (2015). Prevalence and characteristics of Cyprinid herpesvirus 3 (CyHV-3) infection in common carp (*Cyprinus carpio* L.) inhabiting three rivers in Kochi Prefecture, Japan. *Veterinary Microbiology, 175*(2–4), 362–368. http://dx.doi.org/10.1016/j.vetmic.2014.12.002.

Garver, K. A., Al-Hussinee, L., Hawley, L. M., Schroeder, T., Edes, S., LePage, V., et al. (2010). Mass mortality associated with koi herpesvirus in wild common carp in Canada. *Journal of Wildlife Diseases, 46*(4), 1242–1251. http://dx.doi.org/10.7589/0090-3558-46.4.1242.

Gilad, O., Yun, S., Adkison, M. A., Way, K., Willits, N. H., Bercovier, H., et al. (2003). Molecular comparison of isolates of an emerging fish pathogen, koi herpesvirus, and the effect of water temperature on mortality of experimentally infected koi. *Journal of General Virology, 84*(Pt 10), 2661–2667.

Gilad, O., Yun, S., Andree, K. B., Adkison, M. A., Zlotkin, A., Bercovier, H., et al. (2002). Initial characteristics of koi herpesvirus and development of a polymerase chain reaction assay to detect the virus in koi, *Cyprinus carpio koi*. *Diseases of Aquatic Organisms, 48*(2), 101–108. http://dx.doi.org/10.3354/dao048101.

Gilad, O., Yun, S., Zagmutt-Vergara, F. J., Leutenegger, C. M., Bercovier, H., & Hedrick, R. P. (2004). Concentrations of a Koi herpesvirus (KHV) in tissues of experimentally infected *Cyprinus carpio koi* as assessed by real-time TaqMan PCR. *Diseases of Aquatic Organisms, 60*(3), 179–187. http://dx.doi.org/10.3354/dao060179.

Gomez, D. K., Joh, S. J., Jang, H., Shin, S. P., Choresca, C. H., Han, J. E., et al. (2011). Detection of koi herpesvirus (KHV) from koi (*Cyprinus carpio koi*) broodstock in South Korea. *Aquaculture, 311*(1–4), 42–47. http://dx.doi.org/10.1016/j.aquaculture.2010.11.021.

Gomez, D., Sunyer, J. O., & Salinas, I. (2013). The mucosal immune system of fish: The evolution of tolerating commensals while fighting pathogens. *Fish & Shellfish Immunology, 35*(6), 1729–1739. http://dx.doi.org/10.1016/j.fsi.2013.09.032.

Goodwin, A. E., Sadler, J., Merry, G. E., & Marecaux, E. N. (2009). Herpesviral haematopoietic necrosis virus (CyHV-2) infection: Case studies from commercial goldfish farms. *Journal of Fish Diseases*, *32*(3), 271–278. http://dx.doi.org/10.1111/j.1365-2761.2008.00988.x.

Gotesman, M., Kattlun, J., Bergmann, S. M., & El-Matbouli, M. (2013). CyHV-3: The third cyprinid herpesvirus. *Diseases of Aquatic Organisms*, *105*(2), 163–174. http://dx.doi.org/10.3354/dao02614.

Gotesman, M., Soliman, H., Besch, R., & El-Matbouli, M. (2014). *In vitro* inhibition of cyprinid herpesvirus-3 replication by RNAi. *Journal of Virological Methods*, *206*, 63–66. http://dx.doi.org/10.1016/j.jviromet.2014.05.022.

Gravell, M., Granoff, A., & Darlington, R. W. (1968). Viruses and renal carcinoma of *Rana pipiens*. VII. Propagation of a herpes-type frog virus. *Virology*, *36*(3), 467–475. http://dx.doi.org/10.1016/0042-6822(68)90172-4.

Gray, W. L., Mullis, L., LaPatra, S. E., Groff, J. M., & Goodwin, A. (2002). Detection of koi herpesvirus DNA in tissues of infected fish. *Journal of Fish Diseases*, *25*, 171–178.

Grimmett, S. G., Warg, J. V., Getchell, R. G., Johnson, D. J., & Bowser, P. R. (2006). An unusual koi herpesvirus associated with a mortality event of common carp *Cyprinus carpio* in New York State, USA. *Journal of Wildlife Diseases*, *42*(3), 658–662. http://dx.doi.org/10.7589/0090-3558-42.3.658.

Gunimaladevi, I., Kono, T., Venugopal, M. N., & Sakai, M. (2004). Detection of koi herpesvirus in common carp, *Cyprinus carpio* L., by loop-mediated isothermal amplification. *Journal of Fish Diseases*, *27*(10), 583–589. http://dx.doi.org/10.1111/j.1365-2761.2004.00578.x.

Haenen, O. L. M., Dijkstra, S. G., Van Tulden, P. W., Davidse, A., Van Nieuwstadt, A. P., Wagenaar, F., et al. (2002). Herpesvirus anguillae (HVA) isolations from disease outbreaks in cultured European eel, *Anguilla anguilla* in the Netherlands since 1996. *Bulletin of the European Association of Fish Pathologists*, *22*(4), 247–257.

Haenen, O. L. M., Way, K., Bergmann, S. M., & Ariel, E. (2004). The emergence of koi herpesvirus and its significance to European aquaculture. *Bulletin of the European Association of Fish Pathologists*, *24*(6), 293–307.

Han, J. E., Kim, J. H., Renault, T., Jr., Choresca, C., Shin, S. P., Jun, J. W., et al. (2013). Identifying the viral genes encoding envelope glycoproteins for differentiation of cyprinid herpesvirus 3 isolates. *Viruses*, *5*(2), 568–576. http://dx.doi.org/10.3390/v5020568.

Hansen, J. D., Landis, E. D., & Phillips, R. B. (2005). Discovery of a unique Ig heavy-chain isotype (IgT) in rainbow trout: Implications for a distinctive B cell developmental pathway in teleost fish. *Proceedings of the National Academy of Sciences of the United States of America*, *102*(19), 6919–6924. http://dx.doi.org/10.1073/pnas.0500027102.

Hanson, L., Dishon, A., & Kotler, M. (2011). Herpesviruses that infect fish. *Viruses*, *3*(11), 2160–2191. http://dx.doi.org/10.3390/v3112160.

Hara, H., Aikawa, H., Usui, K., & Nakanishi, T. (2006). Outbreaks of koi herpesvirus disease in rivers of Kanagawa prefecture. *Fish Pathology*, *41*(2), 81–83. http://dx.doi.org/10.3147/jsfp.41.81.

Haramoto, E., Kitajima, M., Katayama, H., Ito, T., & Ohgaki, S. (2009). Development of virus concentration methods for detection of koi herpesvirus in water. *Journal of Fish Diseases*, *32*(3), 297–300. http://dx.doi.org/10.1111/j.1365-2761.2008.00977.x.

Haramoto, E., Kitajima, M., Katayama, H., & Ohgaki, S. (2007). Detection of koi herpesvirus DNA in river water in Japan. *Journal of Fish Diseases*, *30*(1), 59–61. http://dx.doi.org/10.1111/j.1365-2761.2007.00778.x.

Harmache, A., LeBerre, M., Droineau, S., Giovannini, M., & Bremont, M. (2006). Bioluminescence imaging of live infected salmonids reveals that the fin bases are the major portal of entry for Novirhabdovirus. *Journal of Virology*, *80*(7), 3655–3659. http://dx.doi.org/10.1128/JVI.80.7.3655-3659.2006.

Hedrick, R. P., Gilad, O., Yun, S., Spangenberg, J., Marty, R., Nordhausen, M., et al. (2000). A herpesvirus associated with mass mortality of juvenile and adult koi, a strain of common carp. *Journal of Aquatic Animal Health*, *12*, 44–55.

Hedrick, R. P., Waltzek, T. B., & McDowell, T. S. (2006). Susceptibility of koi carp, common carp, goldfish, and goldfish × common carp hybrids to cyprinid herpesvirus-2 and herpesvirus-3. *Journal of Aquatic Animal Health*, *18*(1), 26–34. http://dx.doi.org/10.1577/h05-028.1.

Honjo, M. N., Minamoto, T., & Kawabata, Z. (2012). Reservoirs of Cyprinid herpesvirus 3 (CyHV-3) DNA in sediments of natural lakes and ponds. *Veterinary Microbiology*, *155*(2–4), 183–190. http://dx.doi.org/10.1016/j.vetmic.2011.09.005.

Honjo, M. N., Minamoto, T., Matsui, K., Uchii, K., Yamanaka, H., Suzuki, A. A., et al. (2010). Quantification of cyprinid herpesvirus 3 in environmental water by using an external standard virus. *Applied and Environmental Microbiology*, *76*(1), 161–168. http://dx.doi.org/10.1128/AEM.02011-09.

Horst, D., Ressing, M. E., & Wiertz, E. J. (2011). Exploiting human herpesvirus immune evasion for therapeutic gain: Potential and pitfalls. *Immunology and Cell Biology*, *89*((3), 359–366. http://dx.doi.org/10.1038/icb.2010.129.

Hutoran, M., Ronen, A., Perelberg, A., Ilouze, M., Dishon, A., Bejerano, I., et al. (2005). Description of an as yet unclassified DNA virus from diseased *Cyprinus carpio* species. *Journal of Virology*, *79*(4), 1983–1991. http://dx.doi.org/10.1128/JVI.79.4.1983-1991.2005.

Ilouze, M., Davidovich, M., Diamant, A., Kotler, M., & Dishon, A. (2010). The outbreak of carp disease caused by CyHV-3 as a model for new emerging viral diseases in aquaculture: A review. *Ecological Research*, *26*(5), 885–892. http://dx.doi.org/10.1007/s11284-010-0694-2.

Ilouze, M., Dishon, A., Kahan, T., & Kotler, M. (2006). Cyprinid herpes virus-3 (CyHV-3) bears genes of genetically distant large DNA viruses. *FEBS Letters*, *580*(18), 4473–4478. http://dx.doi.org/10.1016/j.febslet.2006.07.013.

Ilouze, M., Dishon, A., & Kotler, M. (2006). Characterization of a novel virus causing a lethal disease in carp and koi. *Microbiology and Molecular Biology Reviews*, *70*(1), 147–156. http://dx.doi.org/10.1128/MMBR.70.1.147-156.2006.

Ilouze, M., Dishon, A., & Kotler, M. (2012a). Coordinated and sequential transcription of the cyprinid herpesvirus-3 annotated genes. *Virus Research*, *169*(1), 98–106. http://dx.doi.org/10.1016/j.virusres.2012.07.015.

Ilouze, M., Dishon, A., & Kotler, M. (2012b). Down-regulation of the cyprinid herpesvirus-3 annotated genes in cultured cells maintained at restrictive high temperature. *Virus Research*, *169*(1), 289–295. http://dx.doi.org/10.1016/j.virusres.2012.07.013.

Imajoh, M., Fujioka, H., Furusawa, K., Tamura, K., Yamasaki, K., Kurihara, S., et al. (2014). Establishment of a new cell line susceptible to Cyprinid herpesvirus 3 (CyHV-3) and possible latency of CyHV-3 by temperature shift in the cells. *Journal of Fish Diseases*. http://dx.doi.org/10.1111/jfd.12252.

Ishioka, T., Yoshizumi, M., Izumi, S., Suzuki, K., Suzuki, H., Kozawa, K., et al. (2005). Detection and sequence analysis of DNA polymerase and major envelope protein genes in koi herpesviruses derived from *Cyprinus carpio* in Gunma prefecture, Japan. *Veterinary Microbiology*, *110*(1–2), 27–33. http://dx.doi.org/10.1016/j.vetmic.2005.07.002.

Ito, T., Kurita, J., & Yuasa, K. (2014). Differences in the susceptibility of Japanese indigenous and domesticated Eurasian common carp (*Cyprinus carpio*), identified by mitochondrial DNA typing, to cyprinid herpesvirus 3 (CyHV-3). *Veterinary Microbiology*, *171*(1–2), 31–40. http://dx.doi.org/10.1016/j.vetmic.2014.03.002.

Ito, T., & Maeno, Y. (2014). Effects of experimentally induced infections of goldfish *Carassius auratus* with cyprinid herpesvirus 2 (CyHV-2) at various water temperatures. *Diseases of Aquatic Organisms*, *110*(3), 193–200. http://dx.doi.org/10.3354/dao02759.

Ito, T., Sano, M., Kurita, J., Yuasa, K., & Iida, T. (2007). Carp larvae are not susceptible to Koi Herpesvirus. *Fish Pathology*, *42*(2), 107–109.

Kattlun, J., Menanteau-Ledouble, S., & El-Matbouli, M. (2014). Non-structural protein pORF 12 of cyprinid herpesvirus 3 is recognized by the immune system of the common carp *Cyprinus carpio*. *Diseases of Aquatic Organisms*, *111*(3), 269–273. http://dx.doi.org/10.3354/dao02793.

Kelley, G. O., Waltzek, T. B., McDowell, T. S., Yun, S. C., LaPatra, S. E., & Hedrick, R. P. (2005). Genetic relationships among Herpes-like viruses isolated from sturgeon. *Journal of Aquatic Animal Health*, *17*(4), 297–303. http://dx.doi.org/10.1577/h05-002.1.

Kempter, J., & Bergmann, S. M. (2007). Detection of koi herpesvirus (KHV) genome in wild and farmed fish from Northern Poland. *Aquaculture*, *272*(Suppl. 1), S275.

Kempter, J., Kielpinski, M., Panicz, R., Sadowski, J., Myslowski, B., & Bergmann, S. M. (2012). Horizontal transmission of koi herpes virus (KHV) from potential vector species to common carp. *Bulletin of the European Association of Fish Pathologists*, *32*(6), 212–219.

Kempter, J., Sadowski, J., Schütze, H., Fischer, U., Dauber, M., Fichtner, D., et al. (2009). Koi herpes virus: Do acipenserid restitution programs pose a threat to carp farms in the disease-free zones? *Acta Ichthyologica Et Piscatoria*, *39*(2), 119–126. http://dx.doi.org/10.3750/aip2009.39.2.06.

Kielpinski, M., Kempter, J., Panicz, R., Sadowski, J., Schütze, H., Ohlemeyer, S., et al. (2010). Detection of KHV in freshwater mussels and crustaceans from ponds with KHV history in common carp (*Cyprinus carpio*). *Israeli Journal of Aquaculture-Bamidgeh*, *62*(1), 28–37.

Kim, H. J., & Kwon, S. R. (2013). Evidence for two koi herpesvirus (KHV) genotypes in South Korea. *Diseases of Aquatic Organisms*, *104*(3), 197–202. http://dx.doi.org/10.3354/dao02590.

Kongchum, P., Sandel, E., Lutzky, S., Hallerman, E. M., Hulata, G., David, L., et al. (2011). Association between IL-10a single nucleotide polymorphisms and resistance to cyprinid herpesvirus-3 infection in common carp (*Cyprinus carpio*). *Aquaculture*, *315*(3–4), 417–421. http://dx.doi.org/10.1016/j.aquaculture.2011.02.035.

Kucuktas, H., & Brady, Y. J. (1999). Molecular biology of channel catfish virus. *Aquaculture*, *172*(1–2), 147–161. http://dx.doi.org/10.1016/s0044-8486(98)00442-6.

Kurita, J., Yuasa, K., Ito, T., Sano, M., Hedrick, R. P., Engelsma, M. Y., et al. (2009). Molecular epidemiology of koi herpesvirus. *Fish Pathology*, *44*(2), 59–66.

Kurobe, T., Kelley, G. O., Waltzek, T. B., & Hedrick, R. P. (2008). Revised phylogenetic relationships among herpesviruses isolated from sturgeons. *Journal of Aquatic Animal Health*, *20*(2), 96–102. http://dx.doi.org/10.1577/H07-028.1.

Lee, N. S., Jung, S. H., Park, J. W., & Do, J. W. (2012). *In situ* hybridization detection of koi herpesvirus in paraffin-embedded tissues of common carp *Cyprinus carpio* collected in 1998 in Korea. *Fish Pathology*, *47*(3), 100–103. http://dx.doi.org/10.3147/jsfp.47.100.

Li, W., Lee, X., Weng, S., He, J., & Dong, C. (2015). Whole-genome sequence of a novel Chinese cyprinid herpesvirus 3 isolate reveals the existence of a distinct European genotype in East Asia. *Veterinary Microbiology*, *175*(2–4), 185–194. http://dx.doi.org/10.1016/j.vetmic.2014.11.022.

Lin, S. L., Cheng, Y. H., Wen, C. M., & Chen, S. N. (2013). Characterization of a novel cell line from the caudal fin of koi carp *Cyprinus carpio*. *Journal of Fish Biology*, *82*(6), 1888–1903. http://dx.doi.org/10.1111/jfb.12116.

Lio-Po, G. D. (2011). Recent developments in the study and surveillance of koi herpesvirus (KHV) in Asia. In M. G. Bondad-Reantaso, J. B. Jones, F. Corsin, & T. Aoki (Eds.), *Proceedings of the seventh symposium on diseases in Asian Aquaculture* (385 p.). Selangor, Malaysia: Fish Health Section, Asian Fisheries Society.

Liu, Z., Ke, H., Ma, Y., Hao, L., Feng, G., Ma, J., et al. (2014). Oral passive immunization of carp *Cyprinus carpio* with anti-CyHV-3 chicken egg yolk immunoglobulin (IgY). *Fish Pathology, 49*(3), 113–120.

Loret, S., Guay, G., & Lippe, R. (2008). Comprehensive characterization of extracellular herpes simplex virus type 1 virions. *Journal of Virology, 82*(17), 8605–8618. http://dx.doi.org/10.1128/JVI.00904-08.

Mabuchi, K., Senou, H., Suzuki, T., & Nishida, M. (2005). Discovery of an ancient lineage of *Cyprinus carpio* from Lake Biwa, central Japan, based on mtDNA sequence data, with reference to possible multiple origins of koi. *Journal of Fish Biology, 66*(6), 1516–1528. http://dx.doi.org/10.1111/j.0022-1112.2005.00676.x.

Marcos-Lopez, M., Waltzek, T. B., Hedrick, R. P., Baxa, D. V., Garber, A. F., Liston, R., et al. (2012). Characterization of a novel alloherpesvirus from Atlantic cod (*Gadus morhua*). *Journal of Veterinary Diagnostic Investigation, 24*(1), 65–73. http://dx.doi.org/10.1177/1040638711416629.

McCleary, S., Ruane, N. M., Cheslett, D., Hickey, C., Rodger, H. D., Geoghegan, F., et al. (2011). Detection of koi herpesvirus (KHV) in koi carp (*Cyprinus carpio* L.) imported into Ireland. *Bulletin of the European Association of Fish Pathologists, 31*(3), 124–128.

McColl, K. A., Cooke, B. D., & Sunarto, A. (2014). Viral biocontrol of invasive vertebrates: Lessons from the past applied to cyprinid herpesvirus-3 and carp (*Cyprinus carpio*) control in Australia. *Biological Control, 72*, 109–117. http://dx.doi.org/10.1016/j.biocontrol.2014.02.014.

McDermott, C., & Palmeiro, B. (2013). Selected emerging infectious diseases of ornamental fish. *Veterinary Clinics of North America: Exotic Animal Practice, 16*(2), 261–282. http://dx.doi.org/10.1016/j.cvex.2013.01.006.

McGeoch, D. J., Dolan, A., & Ralph, A. C. (2000). Toward a comprehensive phylogeny for mammalian and avian herpesviruses. *Journal of Virology, 74*(22), 10401–10406.

McGeoch, D. J., & Gatherer, D. (2005). Integrating reptilian herpesviruses into the family *Herpesviridae. Journal of Virology, 79*(2), 725–731.

McGeoch, D. J., Rixon, F. J., & Davison, A. J. (2006). Topics in herpesvirus genomics and evolution. *Virus Research, 117*(1), 90–104. http://dx.doi.org/10.1016/j.virusres.2006.01.002.

McKinnell, R. G., & Tarin, D. (1984). Temperature-dependent metastasis of the Lucke renal carcinoma and its significance for studies on mechanisms of metastasis. *Cancer and Metastasis Reviews, 3*(4), 373–386.

Meeusen, E. N., Walker, J., Peters, A., Pastoret, P. P., & Jungersen, G. (2007). Current status of veterinary vaccines. *Clinical Microbiology Reviews, 20*(3), 489–510. http://dx.doi.org/10.1128/CMR.00005-07.

Mettenleiter, T. C. (2004). Budding events in herpesvirus morphogenesis. *Virus Research, 106*(2), 167–180. http://dx.doi.org/10.1016/j.virusres.2004.08.013.

Mettenleiter, T. C., Klupp, B. G., & Granzow, H. (2009). Herpesvirus assembly: An update. *Virus Research, 143*(2), 222–234. http://dx.doi.org/10.1016/j.virusres.2009.03.018.

Michel, B., Fournier, G., Lieffrig, F., Costes, B., & Vanderplasschen, A. (2010). Cyprinid herpesvirus 3. *Emerging Infectious Diseases, 16*(12), 1835–1843. http://dx.doi.org/10.3201/eid1612.100593.

Michel, B., Leroy, B., Stalin Raj, V., Lieffrig, F., Mast, J., Wattiez, R., et al. (2010). The genome of cyprinid herpesvirus 3 encodes 40 proteins incorporated in mature virions. *Journal of General Virology, 91*(Pt 2), 452–462. http://dx.doi.org/10.1099/vir.0.015198-0.

Midtlyng, P. J., Storset, A., Michel, C., Slierendrecht, W. J., & Okamoto, N. (2002). Breeding for disease resistance in fish. *Bulletin of the European Association of Fish Pathologists, 22*(2), 166–172.

Miest, J. J., Adamek, M., Pionnier, N., Harris, S., Matras, M., Rakus, K. L., et al. (2015). Differential effects of alloherpesvirus CyHV-3 and rhabdovirus SVCV on apoptosis in fish cells. *Veterinary Microbiology, 176*(1–2), 19–31. http://dx.doi.org/10.1016/j.vetmic.2014.12.012.

Minamoto, T., Honjo, M. N., & Kawabata, Z. (2009). Seasonal distribution of cyprinid herpesvirus 3 in Lake Biwa, Japan. *Applied and Environmental Microbiology*, 75(21), 6900–6904. http://dx.doi.org/10.1128/AEM.01411-09.

Minamoto, T., Honjo, M. N., Uchii, K., Yamanaka, H., Suzuki, A. A., Kohmatsu, Y., et al. (2009). Detection of cyprinid herpesvirus 3 DNA in river water during and after an outbreak. *Veterinary Microbiology*, 135(3–4), 261–266. http://dx.doi.org/10.1016/j.vetmic.2008.09.081.

Minamoto, T., Honjo, M. N., Yamanaka, H., Tanaka, N., Itayama, T., & Kawabata, Z. (2011). Detection of cyprinid herpesvirus-3 DNA in lake plankton. *Research in Veterinary Science*, 90(3), 530–532. http://dx.doi.org/10.1016/j.rvsc.2010.07.006.

Minamoto, T., Honjo, M. N., Yamanaka, H., Uchii, K., & Kawabata, Z. (2012). Nationwide Cyprinid herpesvirus 3 contamination in natural rivers of Japan. *Research in Veterinary Science*, 93(1), 508–514. http://dx.doi.org/10.1016/j.rvsc.2011.06.004.

Miwa, S., Ito, T., & Sano, M. (2007). Morphogenesis of koi herpesvirus observed by electron microscopy. *Journal of Fish Diseases*, 30(12), 715–722. http://dx.doi.org/10.1111/j.1365-2761.2007.00850.x.

Miwa, S., Kiryu, I., Yuasa, K., Ito, T., & Kaneko, T. (2014). Pathogenesis of acute and chronic diseases caused by cyprinid herpesvirus-3. *Journal of Fish Diseases*. http://dx.doi.org/10.1111/jfd.12282.

Miyazaki, T., Kuzuya, Y., Yasumoto, S., Yasuda, M., & Kobayashi, T. (2008). Histopathological and ultrastructural features of Koi herpesvirus (KHV)-infected carp *Cyprinus carpio*, and the morphology and morphogenesis of KHV. *Diseases of Aquatic Organisms*, 80(1), 1–11. http://dx.doi.org/10.3354/dao01929.

Miyazaki, T., Yasumoto, S., Kuzuya, Y., & Yoshimura, T. (2005). In *A primary study on oral vaccination with liposomes entrapping koi herpesvirus (KHV) antigens against KHV infection in carp. Paper presented at the diseases in Asian aquaculture, Colombo, Sri Lanka*.

Monaghan, S. J., Thompson, K. D., Adams, A., & Bergmann, S. M. (2015). Sensitivity of seven PCRs for early detection of koi herpesvirus in experimentally infected carp, *Cyprinus carpio* L., by lethal and non-lethal sampling methods. *Journal of Fish Diseases*, 38(3), 303–319. http://dx.doi.org/10.1111/jfd.12235.

Monaghan, S. J., Thompson, K. D., Adams, A., Kempter, J., & Bergmann, S. M. (2015). Examination of the early infection stages of koi herpesvirus (KHV) in experimentally infected carp Cyprinus carpio L. using in situ hybridization. *Journal of Fish Diseases*, 38(5), 477–489. http://dx.doi.org/10.1111/jfd.12260.

Musa, N., Leong, N. K., & Sunarto, A. (2005). *Koi herpesvirus (KHV)—An emerging pathogen in koi. Paper presented at the Colloquium on Viruses of Veterinary and Public Health Importance*. Malaysia: Bangi.

Neukirch, M., Böttcher, K., & Bunnajrakul, S. (1999). Isolation of a virus from koi with altered gills. *Bulletin of the European Association of Fish Pathologists*, 19, 221–224.

Neukirch, M., & Kunz, U. (2001). Isolation and preliminary characterization of several viruses from koi (*Cyprinus carpio*) suffering gill necrosis and mortality. *Bulletin of the European Association of Fish Pathologists*, 21(4), 125–135.

O'Connor, M. R., Farver, T. B., Malm, K. V., Yun, S. C., Marty, G. D., Salonius, K., et al. (2014). Protective immunity of a modified-live cyprinid herpesvirus 3 vaccine in koi (*Cyprinus carpio koi*) 13 months after vaccination. *American Journal of Veterinary Research*, 75(10), 905–911. http://dx.doi.org/10.2460/ajvr.75.10.905.

Ødegård, J., Olesena, I., Dixonb, P., Jeneyc, Z., Nielsena, H.-M., Wayb, K., et al. (2010). Genetic analysis of common carp (*Cyprinus carpio*) strains II: Resistance to koi herpesvirus and Aeromonas hydrophila and their relationship with pond survival. *Aquaculture*, 304(1–4), 7–13.

OIE. (2012). Chapter 2.3.6. Koi herpesvirus disease. In *Manual of Diagnostic Tests for Aquatic Animals* (pp. 328–344).

Ouyang, P., Rakus, K., Boutier, M., Reschner, A., Leroy, B., Ronsmans, M., et al. (2013). The IL-10 homologue encoded by cyprinid herpesvirus 3 is essential neither for viral replication *in vitro* nor for virulence *in vivo*. *Veterinary Research*, *44*(1), 53.

Ouyang, P., Rakus, K., van Beurden, S. J., Westphal, A. H., Davison, A. J., Gatherer, D., et al. (2014). IL-10 encoded by viruses: A remarkable example of independent acquisition of a cellular gene by viruses and its subsequent evolution in the viral genome. *Journal of General Virology*, *95*(Pt 2), 245–262. http://dx.doi.org/10.1099/vir.0.058966-0.

Pellett, P. E., Davison, A. J., Eberle, R., Ehlers, B., Hayward, G. S., Lacoste, V., et al. (2011a). Alloherpesviridae. In A. M. Q. King, E. Lefkowitz, M. J. Adams, & E. B. Carstens (Eds.), *Virus taxonomy: Ninth report of the International Committee on Taxonomy of Viruses* (pp. 108–110). London, United Kingdom: Elsevier Academic Press.

Pellett, P. E., Davison, A. J., Eberle, R., Ehlers, B., Hayward, G. S., Lacoste, V., et al. (2011b). Herpesvirales. In A. M. Q. King, E. Lefkowitz, M. J. Adams, & E. B. Carstens (Eds.), *Virus taxonomy: Ninth report of the International Committee on Taxonomy of Viruses* (pp. 99–107). London, United Kingdom: Elsevier Academic Press.

Pellett, P. E., Davison, A. J., Eberle, R., Ehlers, B., Hayward, G. S., Lacoste, V., et al. (2011c). Herpesviridae. In A. M. Q. King, E. Lefkowitz, M. J. Adams, & E. B. Carstens (Eds.), *Virus taxonomy: Ninth report of the International Committee on Taxonomy of Viruses* (pp. 111–122). London, United Kingdom: Elsevier Academic Press.

Pellett, P. E., Davison, A. J., Eberle, R., Ehlers, B., Hayward, G. S., Lacoste, V., et al. (2011d). Malacoherpesviridae. In A. M. Q. King, E. Lefkowitz, M. J. Adams, & E. B. Carstens (Eds.), *Virus taxonomy: Ninth report of the International Committee on Taxonomy of Viruses* (p. 123). London, United Kingdom: Elsevier Academic Press.

Perelberg, A., Ilouze, M., Kotler, M., & Steinitz, M. (2008). Antibody response and resistance of *Cyprinus carpio* immunized with cyprinid herpes virus 3 (CyHV-3). *Vaccine*, *26*(29–30), 3750–3756. http://dx.doi.org/10.1016/j.vaccine.2008.04.057.

Perelberg, A., Ronen, A., Hutoran, M., Smith, Y., & Kotler, M. (2005). Protection of cultured *Cyprinus carpio* against a lethal viral disease by an attenuated virus vaccine. *Vaccine*, *23*(26), 3396–3403. http://dx.doi.org/10.1016/j.vaccine.2005.01.096.

Perelberg, A., Smirnov, M., Hutoran, M., Diamant, A., Bejerano, Y., & Kotler, M. (2003). Epidemiological description of a new viral disease afflicting cultured *Cyprinus carpio* in Israel. *Israeli Journal of Aquaculture-Bamidgeh*, *55*(1), 5–12.

Piackova, V., Flajshans, M., Pokorova, D., Reschova, S., Gela, D., Cizek, A., et al. (2013). Sensitivity of common carp, *Cyprinus carpio* L., strains and crossbreeds reared in the Czech Republic to infection by cyprinid herpesvirus 3 (CyHV-3; KHV). *Journal of Fish Diseases*, *36*(1), 75–80. http://dx.doi.org/10.1111/jfd.12007.

Piazzon, M. C., Savelkoul, H. F., Pietretti, D., Wiegertjes, G. F., & Forlenza, M. (2015). Carp Il10 has anti-inflammatory activities on phagocytes, promotes proliferation of memory T cells, and regulates B cell differentiation and antibody secretion. *Journal of Immunology*, *194*(1), 187–199. http://dx.doi.org/10.4049/jimmunol.1402093.

Pikarsky, E., Ronen, A., Abramowitz, J., Levavi-Sivan, B., Hutoran, M., Shapira, Y., et al. (2004). Pathogenesis of acute viral disease induced in fish by carp interstitial nephritis and gill necrosis virus. *Journal of Virology*, *78*(17), 9544–9551. http://dx.doi.org/10.1128/JVI.78.17.9544-9551.2004.

Pikulkaew, S., Meeyam, T., & Banlunara, W. (2009). The outbreak of koi herpesvirus (KHV) in koi (*Cyprinus carpio koi*) from Chiang Mai province, Thailand. *Thai Journal of Veterinary Medicine*, *39*(1), 53–58.

Pionnier, N., Adamek, M., Miest, J. J., Harris, S. J., Matras, M., Rakus, K. L., et al. (2014). C-reactive protein and complement as acute phase reactants in common carp *Cyprinus carpio* during CyHV-3 infection. *Diseases of Aquatic Organisms*, *109*(3), 187–199. http://dx.doi.org/10.3354/dao02727.

Plant, K. P., & Lapatra, S. E. (2011). Advances in fish vaccine delivery. *Developmental and Comparative Immunology*, *35*(12), 1256–1262. http://dx.doi.org/10.1016/j.dci.2011.03.007.

Pokorova, D., Reschova, S., Hulova, J., Vicenova, M., Vesely, T., & Piackova, V. (2010). Detection of cyprinid herpesvirus-3 in field samples of common and koi carp by various single-round and nested PCR methods. *Journal of the World Aquaculture Society*, *41*(5), 773–779. http://dx.doi.org/10.1111/j.1749-7345.2010.00419.x.

Pokorova, D., Vesely, T., Piackova, V., Reschova, S., & Hulova, J. (2005). Current knowledge on koi herpesvirus (KHV): A review. *Veterinarni Medicina*, *5*(4), 139–147.

Pretto, T., Manfrin, A., Ceolin, C., Dalla Pozza, M., Zelco, S., Quartesan, R., et al. (2013). First isolation of koi herpes virus (KHV) in Italy from imported koi (*Cyprinus carpio koi*). *Bulletin of the European Association of Fish Pathologists*, *33*(4), 126–133.

Radosavljevic, V., Jeremic, S., Cirkovic, M., Lako, B., Milicevic, V., Potkonjak, A., et al. (2012). Common fish species in polyculture with carp as cyprinid herpes virus 3 carriers. *Acta Veterinaria*, *62*(5–6), 675–681. http://dx.doi.org/10.2298/avb1206675r.

Raj, V. S., Fournier, G., Rakus, K., Ronsmans, M., Ouyang, P., Michel, B., et al. (2011). Skin mucus of *Cyprinus carpio* inhibits cyprinid herpesvirus 3 binding to epidermal cells. *Veterinary Research*, *42*, 92. http://dx.doi.org/10.1186/1297-9716-42-92.

Rakus, K. L., Irnazarow, I., Adamek, M., Palmeira, L., Kawana, Y., Hirono, I., et al. (2012). Gene expression analysis of common carp (*Cyprinus carpio* L.) lines during Cyprinid herpesvirus 3 infection yields insights into differential immune responses. *Developmental and Comparative Immunology*, *37*(1), 65–76. http://dx.doi.org/10.1016/j.dci.2011.12.006.

Rakus, K. L., Ouyang, P., Boutier, M., Ronsmans, M., Reschner, A., Vancsok, C., et al. (2013). Cyprinid herpesvirus 3: An interesting virus for applied and fundamental research. *Veterinary Research*, *44*, 85. http://dx.doi.org/10.1186/1297-9716-44-85.

Rakus, K. L., Wiegertjes, G. F., Adamek, M., Siwicki, A. K., Lepa, A., & Irnazarow, I. (2009). Resistance of common carp (*Cyprinus carpio* L.) to cyprinid herpesvirus-3 is influenced by major histocompatibility (MH) class II *B* gene polymorphism. *Fish & Shellfish Immunology*, *26*(5), 737–743. http://dx.doi.org/10.1016/j.fsi.2009.03.001.

Rakus, K. L., Wiegertjes, G. F., Jurecka, P., Walker, P. D., Pilarczyk, A., & Irnazarow, I. (2009). Major histocompatibility (MH) class II *B* gene polymorphism influences disease resistance of common carp (*Cyprinus carpio* L.). *Aquaculture*, *288*(1–2), 44–50. http://dx.doi.org/10.1016/j.aquaculture.2008.11.016.

Reed, A. N., Izume, S., Dolan, B. P., LaPatra, S., Kent, M., Dong, J., et al. (2014). Identification of B cells as a major site for cyprinid herpesvirus 3 latency. *Journal of Virology*, *88*(16), 9297–9309. http://dx.doi.org/10.1128/JVI.00990-14.

Rixon, F. J., & Schmid, M. F. (2014). Structural similarities in DNA packaging and delivery apparatuses in Herpesvirus and dsDNA bacteriophages. *Current Opinion in Virology*, *5*, 105–110. http://dx.doi.org/10.1016/j.coviro.2014.02.003.

Rombout, J. H., Yang, G., & Kiron, V. (2014). Adaptive immune responses at mucosal surfaces of teleost fish. *Fish & Shellfish Immunology*, *40*(2), 634–643. http://dx.doi.org/10.1016/j.fsi.2014.08.020.

Ronen, A., Perelberg, A., Abramowitz, J., Hutoran, M., Tinman, S., Bejerano, I., et al. (2003). Efficient vaccine against the virus causing a lethal disease in cultured *Cyprinus carpio*. *Vaccine*, *21*(32), 4677–4684.

Ronsmans, M., Boutier, M., Rakus, K., Farnir, F., Desmecht, D., Ectors, F., et al. (2014). Sensitivity and permissivity of *Cyprinus carpio* to cyprinid herpesvirus 3 during the early stages of its development: Importance of the epidermal mucus as an innate immune barrier. *Veterinary Research*, *45*(1), 100. http://dx.doi.org/10.1186/s13567-014-0100-0.

Rosenkranz, D., Klupp, B. G., Teifke, J. P., Granzow, H., Fichtner, D., Mettenleiter, T. C., et al. (2008). Identification of envelope protein pORF81 of koi herpesvirus. *Journal of General Virology*, *89*(Pt 4), 896–900. http://dx.doi.org/10.1099/vir.0.83565-0.

Rothenburg, S., Deigendesch, N., Dittmar, K., Koch-Nolte, F., Haag, F., Lowenhaupt, K., et al. (2005). A PKR-like eukaryotic initiation factor 2alpha kinase from zebrafish contains Z-DNA binding domains instead of dsRNA binding domains. *Proceedings of the National Academy of Sciences of the United States of America*, *102*(5), 1602–1607. http://dx.doi.org/10.1073/pnas.0408714102.

Rueckert, C., & Guzman, C. A. (2012). Vaccines: From empirical development to rational design. *PLoS Pathogens*, *8*(11), e1003001. http://dx.doi.org/10.1371/journal.ppat.1003001.

Ryo, S., Wijdeven, R. H., Tyagi, A., Hermsen, T., Kono, T., Karunasagar, I., et al. (2010). Common carp have two subclasses of bonyfish specific antibody IgZ showing differential expression in response to infection. *Developmental and Comparative Immunology*, *34*(11), 1183–1190. http://dx.doi.org/10.1016/j.dci.2010.06.012.

Sadler, J., Marecaux, E., & Goodwin, A. E. (2008). Detection of koi herpes virus (CyHV-3) in goldfish, *Carassius auratus* (L.), exposed to infected koi. *Journal of Fish Diseases*, *31*(1), 71–72. http://dx.doi.org/10.1111/j.1365-2761.2007.00830.x.

Sano, T., Fukuda, H., & Furukawa, M. (1985). Herpesvirus cyprini: Biological and oncogenic properties. *Fish Pathology*, *20*(2/3), 381–388. http://dx.doi.org/10.3147/jsfp.20.381.

Sano, M., Fukuda, H., & Sano, T. (1990). Isolation and characterization of a new herpesvirus from eel. In P. O. Perkins & T. C. Cheng (Eds.), *Pathology in marine sciences* (pp. 15–31). San Diego, CA: Academic Press.

Sano, M., Ito, T., Kurita, J., Yanai, T., Watanabe, N., Satoshi, M., et al. (2004). First detection of koi herpes virus in cultured common carp *Cyprinus carpio* in Japan. *Fish Pathology*, *39*, 165–168.

Schubert, G. H. (1966). The infective agent in carp pox. *Bulletin de l'Office International des Epizooties*, *65*(7), 1011–1022.

Shapira, Y., Magen, Y., Zak, T., Kotler, M., Hulata, G., & Levavi-Sivan, B. (2005). Differential resistance to koi herpes virus (KHV)/carp interstitial nephritis and gill necrosis virus (CNGV) among common carp (*Cyprinus carpio* L.) strains and crossbreds. *Aquaculture*, *245*(1–4), 1–11.

Shephard, K. L. (1994). Functions for fish mucus. *Reviews in Fish Biology and Fisheries*, *4*(4), 401–429. http://dx.doi.org/10.1007/bf00042888.

Shimizu, T., Yoshida, N., Kasai, H., & Yoshimizu, M. (2006). Survival of koi herpesvirus (KHV) in environmental water. *Fish Pathology*, *41*(4), 153–157.

Sievers, F., & Higgins, D. G. (2014). Clustal omega. *Current Protocols in Bioinformatics*, *48*, 3.13.11–13.13.16. http://dx.doi.org/10.1002/0471250953.bi0313s48.

Siwicki, A., Kazuń, K., Kazuń, B., & Majewicz-Zbikowska, E. (2012). Impact of cyprinid herpesvirus-3, which causes interstitial nephritis and gill necrosis, on the activity of carp (*Cyprinus carpio* L.) macrophages and lymphocytes. *Archives of Polish Fisheries*, *20*(2), 123–128. http://dx.doi.org/10.2478/v10086-012-0014-2.

Soliman, H., & El-Matbouli, M. (2010). Loop mediated isothermal amplification combined with nucleic acid lateral flow strip for diagnosis of cyprinid herpes virus-3. *Molecular and Cellular Probes*, *24*(1), 38–43. http://dx.doi.org/10.1016/j.mcp.2009.09.002.

Sommerset, I., Krossoy, B., Biering, E., & Frost, P. (2005). Vaccines for fish in aquaculture. *Expert Review of Vaccines*, *4*(1), 89–101. http://dx.doi.org/10.1586/14760584.4.1.89.

St-Hilaire, S., Beevers, N., Joiner, C., Hedrick, R. P., & Way, K. (2009). Antibody response of two populations of common carp, *Cyprinus carpio* L., exposed to koi herpesvirus. *Journal of Fish Diseases*, *32*(4), 311–320. http://dx.doi.org/10.1111/j.1365-2761.2008.00993.x.

St-Hilaire, S., Beevers, N., Way, K., Le Deuff, R. M., Martin, P., & Joiner, C. (2005). Reactivation of koi herpesvirus infections in common carp *Cyprinus carpio*. *Diseases of Aquatic Organisms*, *67*(1–2), 15–23. http://dx.doi.org/10.3354/dao067015.

Stingley, R. L., & Gray, W. L. (2000). Transcriptional regulation of the channel catfish virus genome direct repeat region. *Journal of General Virology*, *81*(Pt 8), 2005–2010.

Sunarto, A., Liongue, C., McColl, K. A., Adams, M. M., Bulach, D., Crane, M. S., et al. (2012). Koi herpesvirus encodes and expresses a functional interleukin-10. *Journal of Virology*, *86*(21), 11512–11520. http://dx.doi.org/10.1128/JVI.00957-12.

Sunarto, A., McColl, K. A., Crane, M. S., Schat, K. A., Slobedman, B., Barnes, A. C., et al. (2014). Characteristics of cyprinid herpesvirus 3 in different phases of infection: Implications for disease transmission and control. *Virus Research*, *188*, 45–53. http://dx.doi.org/10.1016/j.virusres.2014.03.024.

Sunarto, A., McColl, K. A., Crane, M. S., Sumiati, T., Hyatt, A. D., Barnes, A. C., et al. (2011). Isolation and characterization of koi herpesvirus (KHV) from Indonesia: Identification of a new genetic lineage. *Journal of Fish Diseases*, *34*(2), 87–101. http://dx.doi.org/10.1111/j.1365-2761.2010.01216.x.

Sunarto, A., Rukyani, A., & Itami, T. (2005). Indonesian experience on the outbreak of koi herpesvirus in koi and carp (*Cyprinus carpio*). *Bulletin of Fisheries Research Agency*, (Suppl. 2), 15–21.

Syakuri, H., Adamek, M., Brogden, G., Rakus, K. L., Matras, M., Irnazarow, I., et al. (2013). Intestinal barrier of carp (*Cyprinus carpio* L.) during a cyprinid herpesvirus 3-infection: Molecular identification and regulation of the mRNA expression of claudin encoding genes. *Fish & Shellfish Immunology*, *34*(1), 305–314. http://dx.doi.org/10.1016/j.fsi.2012.11.010.

Takahara, T., Honjo, M. N., Uchii, K., Minamoto, T., Doi, H., Ito, T., et al. (2014). Effects of daily temperature fluctuation on the survival of carp infected with Cyprinid herpesvirus 3. *Aquaculture*, *433*, 208–213. http://dx.doi.org/10.1016/j.aquaculture.2014.06.001.

Tamura, K., Stecher, G., Peterson, D., Filipski, A., & Kumar, S. (2013). MEGA6: Molecular evolutionary genetics analysis version 6.0. *Molecular Biology and Evolution*, *30*(12), 2725–2729. http://dx.doi.org/10.1093/molbev/mst197.

Taylor, N. G., Dixon, P. F., Jeffery, K. R., Peeler, E. J., Denham, K. L., & Way, K. (2010). Koi herpesvirus: Distribution and prospects for control in England and Wales. *Journal of Fish Diseases*, *33*(3), 221–230. http://dx.doi.org/10.1111/j.1365-2761.2009.01111.x.

Taylor, N. G., Norman, R. A., Way, K., & Peeler, E. J. (2011). Modelling the koi herpesvirus (KHV) epidemic highlights the importance of active surveillance within a national control policy. *Journal of Applied Ecology*, *48*(2), 348–355.

Taylor, N. G., Way, K., Jeffery, K. R., & Peeler, E. J. (2010). The role of live fish movements in spreading koi herpesvirus throughout England and Wales. *Journal of Fish Diseases*, *33*(12), 1005–1007.

Terhune, J. S., Grizzle, J. M., Hayden, K., McClenahan, S. D., Lamprecht, S. D., & White, M. G. (2004). First report of koi herpesvirus in wild common carp in the Western Hemisphere. *Fish Health Newsletter*, *32*(3), 8–9.

Tischer, B. K., & Kaufer, B. B. (2012). Viral bacterial artificial chromosomes: Generation, mutagenesis, and removal of mini-F sequences. *Journal of Biomedicine and Biotechnology*, *2012*, 472537. http://dx.doi.org/10.1155/2012/472537.

Tome, A. R., Kus, K., Correia, S., Paulo, L. M., Zacarias, S., de Rosa, M., et al. (2013). Crystal structure of a poxvirus-like zalpha domain from cyprinid herpesvirus 3. *Journal of Virology*, *87*(7), 3998–4004. http://dx.doi.org/10.1128/JVI.03116-12.

Tu, C., Lu, Y. P., Hsieh, C. Y., Huang, S. M., Chang, S. K., & Chen, M. M. (2014). Production of monoclonal antibody against ORF72 of koi herpesvirus isolated in Taiwan. *Folia Microbiologica*, *59*(2), 159–165. http://dx.doi.org/10.1007/s12223-013-0261-7.

Tu, C., Weng, M. C., Shiau, J. R., & Lin, S. Y. (2004). Detection of koi herpesvirus in koi *Cyprinus carpio* in Taiwan. *Fish Pathology*, *39*(2), 109–110.

Uchii, K., Matsui, K., Iida, T., & Kawabata, Z. (2009). Distribution of the introduced cyprinid herpesvirus 3 in a wild population of common carp, *Cyprinus carpio* L. *Journal of Fish Diseases*, *32*(10), 857–864. http://dx.doi.org/10.1111/j.1365-2761.2009.01064.x.

Uchii, K., Minamoto, T., Honjo, M. N., & Kawabata, Z. (2014). Seasonal reactivation enables Cyprinid herpesvirus 3 to persist in a wild host population. *FEMS Microbiology Ecology*, *87*(2), 536–542. http://dx.doi.org/10.1111/1574-6941.12242.

Uchii, K., Okuda, N., Minamoto, T., & Kawabata, Z. (2013). An emerging infectious pathogen endangers an ancient lineage of common carp by acting synergistically with conspecific exotic strains. *Animal Conservation*, *16*(3), 324–330. http://dx.doi.org/10.1111/j.1469-1795.2012.00604.x.

Uchii, K., Telschow, A., Minamoto, T., Yamanaka, H., Honjo, M. N., Matsui, K., et al. (2011). Transmission dynamics of an emerging infectious disease in wildlife through host reproductive cycles. *ISME Journal*, *5*(2), 244–251. http://dx.doi.org/10.1038/ismej.2010.123.

van Beurden, S. J., Bossers, A., Voorbergen-Laarman, M. H., Haenen, O. L., Peters, S., Abma-Henkens, M. H., et al. (2010). Complete genome sequence and taxonomic position of anguillid herpesvirus 1. *Journal of General Virology*, *91*(Pt 4), 880–887. http://dx.doi.org/10.1099/vir.0.016261-0.

van Beurden, S. J., & Engelsma, M. Y. (2012). Herpesviruses of fish, amphibians and invertebrates. In G. D. Magel & S. Tyring (Eds.), *Herpesviridae—A look into this unique family of viruses* (pp. 217–242). Rijeka: InTech.

van Beurden, S. J., Engelsma, M. Y., Roozenburg, I., Voorbergen-Laarman, M. A., van Tulden, P. W., Kerkhoff, S., et al. (2012). Viral diseases of wild and farmed European eel Anguilla anguilla with particular reference to the Netherlands. *Diseases of Aquatic Organisms*, *101*(1), 69–86. http://dx.doi.org/10.3354/dao02501.

van Beurden, S. J., Forlenza, M., Westphal, A. H., Wiegertjes, G. F., Haenen, O. L., & Engelsma, M. Y. (2011). The alloherpesviral counterparts of interleukin 10 in European eel and common carp. *Fish & Shellfish Immunology*, *31*(6), 1211–1217. http://dx.doi.org/10.1016/j.fsi.2011.08.004.

van Beurden, S. J., Gatherer, D., Kerr, K., Galbraith, J., Herzyk, P., Peeters, B. P., et al. (2012). Anguillid herpesvirus 1 transcriptome. *Journal of Virology*, *86*(18), 10150–10161. http://dx.doi.org/10.1128/JVI.01271-12.

van Beurden, S. J., Leroy, B., Wattiez, R., Haenen, O. L., Boeren, S., Vervoort, J. J., et al. (2011). Identification and localization of the structural proteins of anguillid herpesvirus 1. *Veterinary Research*, *42*(1), 105. http://dx.doi.org/10.1186/1297-9716-42-105.

van Beurden, S. J., Peeters, B. P., Rottier, P. J., Davison, A. J., & Engelsma, M. Y. (2013). Genome-wide gene expression analysis of anguillid herpesvirus 1. *BMC Genomics*, *14*, 83. http://dx.doi.org/10.1186/1471-2164-14-83.

van der Marel, M., Adamek, M., Gonzalez, S. F., Frost, P., Rombout, J. H., Wiegertjes, G. F., et al. (2012). Molecular cloning and expression of two beta-defensin and two mucin genes in common carp (*Cyprinus carpio* L.) and their up-regulation after beta-glucan feeding. *Fish & Shellfish Immunology*, *32*(3), 494–501. http://dx.doi.org/10.1016/j.fsi.2011.12.008.

Vrancken, R., Boutier, M., Ronsmans, M., Reschner, A., Leclipteux, T., Lieffrig, F., et al. (2013). Laboratory validation of a lateral flow device for the detection of CyHV-3 antigens in gill swabs. *Journal of Virological Methods*, *193*(2), 679–682. http://dx.doi.org/10.1016/j.jviromet.2013.07.034.

Walster, C. I. (1999). Clinical observations of severe mortalities in koi carp, *Cyprinus carpio*, with gill disease. *Fish Veterinary Journal*, *3*, 54–58.

Waltzek, T. B., Kelley, G. O., Alfaro, M. E., Kurobe, T., Davison, A. J., & Hedrick, R. P. (2009). Phylogenetic relationships in the family *Alloherpesviridae*. *Diseases of Aquatic Organisms*, *84*(3), 179–194. http://dx.doi.org/10.3354/dao02023.

Waltzek, T. B., Kelley, G. O., Stone, D. M., Way, K., Hanson, L., Fukuda, H., et al. (2005). Koi herpesvirus represents a third cyprinid herpesvirus (CyHV-3) in the family *Herpesviridae. Journal of General Virology, 86*(Pt 6), 1659–1667. http://dx.doi.org/10.1099/vir.0.80982-0.

Wang, Y., Zeng, W., Li, Y., Liang, H., Liu, C., Pan, H., et al. (2015). Development and characterization of a cell line from the snout of koi (*Cyprinus carpio* L.) for detection of koi herpesvirus. *Aquaculture, 435,* 310–317. http://dx.doi.org/10.1016/j.aquaculture.2014.10.006.

Weber, E. P., 3rd, Malm, K. V., Yun, S. C., Campbell, L. A., Kass, P. H., Marty, G. D., et al. (2014). Efficacy and safety of a modified-live cyprinid herpesvirus 3 vaccine in koi (*Cyprinus carpio koi*) for prevention of koi herpesvirus disease. *American Journal of Veterinary Research, 75*(10), 899–904. http://dx.doi.org/10.2460/ajvr.75.10.899.

Whittington, R. J., Crockford, M., Jordan, D., & Jones, B. (2008). Herpesvirus that caused epizootic mortality in 1995 and 1998 in pilchard, *Sardinops sagax neopilchardus* (Steindachner), in Australia is now endemic. *Journal of Fish Diseases, 31*(2), 97–105. http://dx.doi.org/10.1111/j.1365-2761.2007.00869.x.

Winton, J., Batts, W., deKinkelin, P., LeBerre, M., Bremont, M., & Fijan, N. (2010). Current lineages of the *epithelioma papulosum cyprini* (EPC) cell line are contaminated with fathead minnow, *Pimephales promelas*, cells. *Journal of Fish Diseases, 33*(8), 701–704. http://dx.doi.org/10.1111/j.1365-2761.2010.01165.x.

Wolf, K. (1988). Channel catfish virus disease. In K. Wolf (Ed.), *Fish viruses and fish viral diseases* (pp. 21–42). Ithaca, New York: Cornell University Press.

Wolf, K., & Darlington, R. W. (1971). Channel catfish virus: A new herpesvirus of ictalurid fish. *Journal of Virology, 8*(4), 525–533.

Xu, J. R., Bently, J., Beck, L., Reed, A., Miller-Morgan, T., Heidel, J. R., et al. (2013). Analysis of koi herpesvirus latency in wild common carp and ornamental koi in Oregon, USA. *Journal of Virological Methods, 187*(2), 372–379. http://dx.doi.org/10.1016/j.jviromet.2012.11.015.

Xu, Z., Parra, D., Gomez, D., Salinas, I., Zhang, Y. A., von Gersdorff Jorgensen, L., et al. (2013). Teleost skin, an ancient mucosal surface that elicits gut-like immune responses. *Proceedings of the National Academy of Sciences of the United States of America, 110*(32), 13097–13102. http://dx.doi.org/10.1073/pnas.1304319110.

Yasumoto, S., Kuzuya, Y., Yasuda, M., Yoshimura, T., & Miyazaki, T. (2006). Oral immunization of common carp with a liposome vaccine fusing koi herpesvirus antigen. *Fish Pathology, 41*(4), 141–145.

Yi, Y., Zhang, H., Lee, X., Weng, S., He, J., & Dong, C. (2014). Extracellular virion proteins of two Chinese CyHV-3/KHV isolates, and identification of two novel envelope proteins. *Virus Research, 191,* 108–116. http://dx.doi.org/10.1016/j.virusres.2014.07.034.

Yoshida, N., Sasaki, R. K., Kasai, H., & Yoshimizu, M. (2013). Inactivation of koi-herpesvirus in water using bacteria isolated from carp intestines and carp habitats. *Journal of Fish Diseases, 36*(12), 997–1005. http://dx.doi.org/10.1111/j.1365-2761.2012.01449.x.

Yoshino, M., Watari, H., Kojima, T., & Ikedo, M. (2006). Sensitive and rapid detection of koi herpesvirus by LAMP method. *Fish Pathology, 41*(1), 19–27. http://dx.doi.org/10.3147/jsfp.41.19.

Yoshino, M., Watari, H., Kojima, T., Ikedo, M., & Kurita, J. (2009). Rapid, sensitive and simple detection method for koi herpesvirus using loop-mediated isothermal amplification. *Microbiology and Immunology, 53*(7), 375–383. http://dx.doi.org/10.1111/j.1348-0421.2009.00145.x.

Yuasa, K., Ito, T., & Sano, M. (2008). Effect of water temperature on mortality and virus shedding in carp experimentally infected with koi herpesvirus. *Fish Pathology, 43*(2), 83–85. http://dx.doi.org/10.3147/jsfp.43.83.

Yuasa, K., Kurita, J., Kawana, M., Kiryu, I., Oseko, N., & Sano, M. (2012). Development of mRNA-specific RT-PCR for the detection of koi herpesvirus (KHV) replication stage. *Diseases of Aquatic Organisms, 100*(1), 11–18. http://dx.doi.org/10.3354/dao02499.

Yuasa, K., & Sano, M. (2009). Koi herpesvirus: Status of outbreaks, diagnosis, surveillance, and research. *Israeli Journal of Aquaculture-Bamidgeh, 61*(3), 169–179.

Yuasa, K., Sano, M., Kurita, J., Ito, T., & Iida, T. (2005). Improvement of a PCR method with the Sph I-5 primer set for the detection of koi herpesvirus (KHV). *Fish Pathology, 40*(1), 37–39.

Yuasa, K., Sano, M., & Oseko, N. (2012). Effective procedures for culture isolation of koi herpesvirus (KHV). *Fish Pathology, 47*(3), 97–99. http://dx.doi.org/10.3147/jsfp.47.97.

Yuasa, K., Sano, M., & Oseko, N. (2013). Goldfish is not a susceptible host of koi herpesvirus (KHV) disease. *Fish Pathology, 48*(2), 52–55. http://dx.doi.org/10.3147/jsfp.48.52.

Zak, T., Perelberg, A., Magen, I., Milstein, A., & Joseph, D. (2007). Heterosis in the growth rate of hungarian-israeli common carp crossbreeds and evaluation of their sensitivity to koi herpes virus (KHV) disease. *Israeli Journal of Aquaculture-Bamidgeh, 59*(2), 63–72.

Zhang, Y. A., Salinas, I., Li, J., Parra, D., Bjork, S., Xu, Z., et al. (2010). IgT, a primitive immunoglobulin class specialized in mucosal immunity. *Nature Immunology, 11*(9), 827–835. http://dx.doi.org/10.1038/ni.1913.

Zhou, J., Wang, H., Li, X. W., Zhu, X., Lu, W., & Zhang, D. (2014). Construction of KHV-CJ ORF25 DNA vaccine and immune challenge test. *Journal of Fish Diseases, 37*(4), 319–325. http://dx.doi.org/10.1111/jfd.12105.

Zhou, J., Wang, H., Zhu, X., Li, X., Lv, W., & Zhang, D. (2013). The primary culture of mirror carp snout and caudal fin tissues and the isolation of koi herpesvirus. *In Vitro Cellular and Developmental Biology: Animal, 49*(9), 734–742. http://dx.doi.org/10.1007/s11626-013-9661-x.

Zhou, J., Xue, J., Wang, Q., Zhu, X., Li, X., Lv, W., et al. (2014). Vaccination of plasmid DNA encoding ORF81 gene of CJ strains of KHV provides protection to immunized carp. *Vitro Cellular and Developmental Biology: Animal, 50*(6), 489–495. http://dx.doi.org/10.1007/s11626-014-9737-2.

CHAPTER FOUR

Human Immunodeficiency Virus Type 1 Cellular Entry and Exit in the T Lymphocytic and Monocytic Compartments: Mechanisms and Target Opportunities During Viral Disease

Benjamas Aiamkitsumrit*,†,1, Neil T. Sullivan*,†,1,
Michael R. Nonnemacher*,†, Vanessa Pirrone*,†, Brian Wigdahl*,†,2
*Department of Microbiology and Immunology, Drexel University College of Medicine, Philadelphia, Pennsylvania, USA
†Center for Molecular Virology and Translational Neuroscience, Institute for Molecular Medicine and Infectious Disease, Drexel University College of Medicine, Philadelphia, Pennsylvania, USA
1Both authors contributed equally to this manuscript.
2Corresponding author: e-mail address: bwigdahl@drexelmed.edu

Contents

Abstract

During the course of human immunodeficiency virus type 1 infection, a number of cell types throughout the body are infected, with the majority of cells representing CD4+ T cells and cells of the monocyte–macrophage lineage. Both types of cells express, to varying levels, the primary receptor molecule, CD4, as well as one or both of the coreceptors, CXCR4 and CCR5. Viral tropism is determined by both the coreceptor utilized for entry and the cell type infected. Although a single virus may have the capacity to infect both a CD4+ T cell and a cell of the monocyte–macrophage lineage, the mechanisms involved in both the entry of the virus into the cell and the viral egress from the cell during budding and viral release differ depending on the cell type. These host–virus

Advances in Virus Research, Volume 93
ISSN 0065-3527
http://dx.doi.org/10.1016/bs.aivir.2015.04.001

interactions and processes can result in the differential targeting of different cell types by selected viral quasispecies and the overall amount of infectious virus released into the extracellular environment or by direct cell-to-cell spread of viral infectivity. This review covers the major steps of virus entry and egress with emphasis on the parts of the replication process that lead to differences in how the virus enters, replicates, and buds from different cellular compartments, such as CD4⁺ T cells and cells of the monocyte–macrophage lineage.

1. INTRODUCTION

Human immunodeficiency virus type 1 (HIV-1) was first identified as the causative agent of the acquired immunodeficiency syndrome (AIDS) in 1984 by identification and characterization of a retrovirus isolated from the peripheral blood of patients suffering from a progressive and often fatal disease involving the quantitative depletion of the CD4⁺ T-cell compartment, which leads to an array of defects in many arms of the immune system as well as the nervous system as a result of the neuroinvasive properties of this pathogenic retrovirus (Gallo & Montagnier, 2003). Since then, many legions of investigators have been drawn to study this virus in order to understand the immuno- and neuropathogenic mechanisms associated with HIV-1 replication in T cells, cells of the monocyte–macrophage lineage, as well as a number of other cellular targets within the brain and other end organs. Many research discoveries have now led to the development of new diagnostic tools, numerous combination therapeutic strategies, as well as a significant amount of information that will pave the way for new preventative therapeutics and vaccines, and possibly a cure for this infectious chromosomal invader.

2. OVERVIEW OF THE HIV-1 LIFE CYCLE WITH EMPHASIS ON VIRAL ENTRY AND EXIT

HIV-1 is a retrovirus and belongs to the lentivirus family. HIV-1 infection exhibits many of the features typically associated with the lentivirus family, including a long period of clinical latency, persistent viral replication, and multiorgan infection, including the brain (Anderson, Zink, Xiong, & Gendelman, 2002; Clements & Zink, 1996). Target cells infected by HIV-1 include cells within the immune system, primarily T cells and cells of the monocyte–macrophage lineage (including dendritic cells) and cells within the brain including perivascular macrophages, astrocytes, and microglial cells as well as the endothelial cells forming the blood–brain barrier (BBB) (Rubbert et al., 1998; Strizki et al., 1996; van't Wout et al., 1996). Viral entry into most cells is initiated by attachment of the virus-encoded gp120 to the host cell receptor, CD4 molecule, and subsequently by viral binding to host cell chemokine coreceptors (Fig. 1). While many coreceptors have been identified to play a role in HIV-1 entry, CXCR4 and CCR5 are the major coreceptors that have been characterized to support viral entry *in vivo* (Clapham & McKnight, 2001). Upon binding, HIV-1 gp120 undergoes structural rearrangements involving conformational changes, which lead to the change of HIV-1 gp41 from a nonfusogenic to a fusogenic state. This structural alteration brings the cellular and viral membranes into closer proximity and thus allows membrane fusion to occur (Blackard et al., 2000). Subsequently, the viral core enters into the host cell

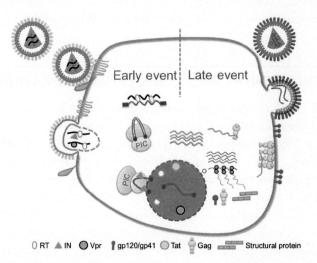

$\bigcirc$ RT $\blacktriangle$ IN $\bigcirc$ Vpr   ⚑ gp120/gp41 $\bigcirc$ Tat   Gag   Structural protein

Figure 1 Viral replication scheme. HIV-1 life cycle can be classified into an early and late phase. The early phase begins with virus attachment, followed by fusion of the viral envelope and host membrane. Once fused viral entry takes place by releasing the viral nucleocapsid into the host cell cytoplasm. The reverse transcription process occurs quickly by synthesizing the double-stranded proviral DNA from single-stranded RNA utilizing viral reverse transcriptase. The preintegration complex is then formed. Subsequently, HIV-1 double-stranded DNA is integrated into host chromosome via the viral integrase enzyme. In the late phase of the HIV-1 life cycle, the viral genome is transcribed and all viral proteins are synthesized. The viral proteins Tat and Vpr have been shown to be involved within the gene regulation process within the nucleus. Once the structural proteins are expressed, viral core particles subsequently assemble and are released from the cell by a budding process to the exterior environment.

cytoplasm, and during this entire process, the viral enzyme reverse transcriptase (RT) begins converting viral genomic RNA into a double-stranded DNA proviral genome. The proviral genome is subsequently integrated into the host cell genome by another virus-encoded particle-associated enzymatic activity, integrase (IN) (Simon-Loriere, Rossolillo, & Negroni, 2011). Once integrated, the viral promoter, also known as the long terminal repeat, is responsible for guiding transcription of the viral genome in conjunction with a large number of cellular proteins and several viral proteins. With the continued accumulation of the viral protein Tat, the transcription process is enhanced with a greater production of full-length transcripts that ultimately leads to greater concentrations of Tat and a number of structural proteins that drive the replication process to completion. The major steps of virus entry and viral egress with emphasis on the parts of the replication process that lead to differences in how the virus enters, replicates, and buds from different cellular compartments, such as $CD4^+$ T cells and cells of the monocyte–macrophage lineage, will form the central theme of this review.

3. HIV-1 ENTRY AND CORECEPTOR UTILIZATION

HIV-1 infects susceptible target cells via direct interactions between viral gp120 and the host cell CD4 receptor molecule as well as one of the number of coreceptor molecules (Dalgleish et al., 1984; Maddon et al., 1986; Moore, 1997; Stein et al., 1987). Shortly after the discovery of HIV-1, the CD4 molecule was shown to serve as the primary cellular receptor for HIV-1 entry based on the observation that a series of monoclonal antibodies directed against the CD4 molecule were shown to block syncytia formation and inhibit vesicular stomatitis virus (VSV) pseudotype production involving the use of the HIV-1 envelope entry protein to facilitate viral entry into selected susceptible cell types (Dalgleish et al., 1984; Klatzmann, Barre-Sinoussi, et al., 1984; Klatzmann, Champagne, et al., 1984; Sattentau & Weiss, 1988). Furthermore, preincubation of $CD4^+$ T cells with three different antibodies directed against different epitopes of the CD4 molecule was shown to block HIV-1 infection *in vitro* (Klatzmann, Champagne, et al., 1984). Interaction of the viral glycoprotein gp120 and the cell surface receptor CD4 has been shown to induce a subsequent conformational change within the viral gp120 molecule to increase the exposure of the V3 loop and the region of gp41 needed for fusion of the virus to the cell surface. This interaction was demonstrated in experiments showing an increase of antibody binding to gp120/V3 loops and gp41 when HIV-1-infected cells were shown to form complexes with soluble CD4 (Sattentau & Moore, 1991).

The interaction between the viral V3 loop and the coreceptor was first demonstrated by the physical association of fusin and the gp120–CD4 complex on the cell membrane which suggested that this association possibly contributed to the exposure of the gp41 hydrophobic NH_2-terminus (Lapham et al., 1996). Fusin was identified as the coreceptor that was required for entry of a subfraction of HIV-1 isolates and subsequently renamed CXCR4 (Bleul, Farzan, et al., 1996) since it was shown to be the C-X-C chemokine coreceptor CXCR4. HIV-1 isolates preferentially utilizing this coreceptor have been designated CXCR4-utilizing isolates or X4 viruses (Feng, Broder, Kennedy, & Berger, 1996). The chemokine receptor CCR5 was subsequently discovered as a coreceptor, and HIV-1 viral isolates using this coreceptor were designated R5 viruses or CCR5-utilizing viruses (Alkhatib et al., 1996; Choe et al., 1996; Deng et al., 1996; Doranz et al., 1996; Dragic et al., 1996). Some viral isolates have been shown to utilize both CXCR4 and CCR5 (X4/R5) and as a result have been referred to as dual-tropic viruses because of this capability. Other chemokine receptors in addition to CCR5 and CXCR4 have been show to facilitate HIV-1 entry; however, the level of utilization of these other receptors is far less than the two dominant chemokine receptors (Choe et al., 1996, 1998; Clapham & McKnight, 2001; Liao et al., 1997). The seven-transmembrane coreceptors, Apj and CCR9, were discovered in 1998 and have also been shown to efficiently support the entry of several primary X4-utilizing and dual X4/R5-utilizing HIV-1 isolates (Choe et al., 1998). In addition, the beta-chemokine G protein-coupled receptor CCR3 has been reported to facilitate HIV-1 infection by primary isolates (Choe et al., 1996), while the beta-chemokine receptor CCR-2b has been shown to serve as a coreceptor for some macrophage-tropic (M-tropic) virus strains and the dual-tropic 89.6 strain (Doranz et al., 1996).

4. SELECTIVE PRESSURES DRIVE HIV-1 EVOLUTION, ADAPTATION, AND TROPISM

HIV-1 infects a variety of cells and tissues; however, the majority of identified susceptible target cells are T lymphocytes and cells of the monocyte–macrophage lineage (Killian & Levy, 2011). Activated cells support viral gene expression and production of viral progeny, while quiescent cells promote integration and the establishment of viral latency and/or persistence (Garcia-Blanco & Cullen, 1991). The quiescent cells such as nonactivated T memory cells and cells of the monocyte–macrophage lineage are considered important viral reservoirs; however, these cells can be potentially activated to different metabolic and immunologic states by a number of stimuli and produce viral progeny virus at different levels as a result of differential activation of the productive replication cycle (Schrier, McCutchan, Venable, Nelson, & Wiley, 1990). Circulating infected monocytes are able to travel and differentiate into macrophages in different tissues and compartments seeding infectious HIV-1 to additional organ systems and cellular targets (Gonzalez-Scarano & Martin-Garcia, 2005).

Viral tropism can be defined by the ability of different viral strains or isolates to infect different cell types or tissues and to induce syncytia formation and/or acute or chronic infectious virus production as a result of infection. T-tropic viruses are viruses that can infect T-cell lines and induce syncytia formation in the MT2 cell line (a human T-cell leukemia virus type 1-infected cell line isolated from cord blood lymphocytes from patients with adult T-cell leukemia; Berger, Murphy, & Farber, 1999). Monocytotropic (M-tropic) viruses have been shown to be able to infect monocytic cells, are unable to infect most T-cell lines, and do not form syncytia (Berger et al., 1999). Most, if not all, HIV-1 primary isolates are able to infect activated primary CD4$^+$ T cells with varying degrees of efficiency; however, the ability to infect T-cell lines and monocytic cells depends on the type of viral strain (Berger et al., 1999).

The ability to use one or more host cell surface molecules as coreceptors for entry along with the primary CD4 receptor is intrinsically linked to the phenotypic properties of the virus. The coreceptor families that have been identified include the CC chemokine receptor, the CXC chemokine receptor, and chemokine receptor-like orphan molecules, as described above. HIV-1 preferentially uses either CXCR4 or CCR5 or both as a coreceptor(s) (Alkhatib et al., 1996; Bjorndal et al., 1997; Choe et al., 1996; Deng et al., 1996; Rucker et al., 1997; Zhang et al., 1998). In general, the phenotypic nature of viral tropism has been categorized by the use of a specific coreceptor(s) to facilitate viral entry; the CCR5-utilizing (R5) virus designation has been used for a virus that can use the CCR5 coreceptor for viral entry, whereas the CXCR4-utilizing (X4) virus designation has been used mainly for viruses that use the CXCR4 coreceptor for viral entry. The dual-tropic (X4R5) virus designation is used for viruses that can utilize coreceptor molecule during the viral entry process (Moyle et al., 2005; Thielen et al., 2010).

The tropism for cells of T-cell origin and cells of the monocyte–macrophage lineage has great bearing on the overall nature of HIV-1 pathogenesis and the course of disease. Although monocytes themselves do not support productive HIV-1 infection *in vitro*, it was demonstrated that HIV-1 could be isolated from mature monocytes, especially from cells expressing CD14 and CD16 (Fischer-Smith et al., 2001; Wang, Xu, et al., 2013). Subsets of cell populations derived from monocytes, such as macrophages and dendritic cells, have been shown to support productive HIV-1 infection (Gendelman et al., 1988; Gordon & Taylor, 2005) and possess a number of biological properties that may impact HIV-1 infection, pathogenesis, and disease including the observations that (i) macrophages have a long life span (Gordon & Taylor, 2005), (ii) macrophages have the ability to suppress the cytopathic effect resulting from viral infection and thus prevent cell death (Swingler, Mann, Zhou, Swingler, & Stevenson, 2007; Wang, Xu, et al., 2013), (iii) HIV-1 particles can reside in intracellular vesicles or virus-containing compartments within macrophages and dendritic cells, which benefits the virus with respect to evasion of the immune response (Benaroch, Billard, Gaudin, Schindler, & Jouve, 2010; Koppensteiner, Banning, Schneider, Hohenberg, & Schindler, 2012), (iv) cells of the monocyte–macrophage lineage can act as a vehicle for HIV-1 to enter and seed infectious virus within the central nervous system (CNS) and thereby infect other cells within the brain (Fischer-Smith et al., 2001; Gonzalez-Scarano & Martin-Garcia, 2005), (v) cells of the monocyte–macrophage lineage may serve as a cellular reservoir for HIV-1 in the bone marrow, peripheral blood, and tissues during long-term therapy, and (vi) cells of the monocytic lineage, especially dendritic cells, may also be key players in facilitating mucosal HIV-1 transmission. T-cell populations also represent a major target for HIV-1 infection and actively produce HIV-1 particles because of the high density of CD4 molecules and the expression of both coreceptors, CXCR4 and CCR5, to different levels depending on the T-cell subset (Davenport, Zaunders, Hazenberg, Schuitemaker, & van Rij, 2002; Joseph et al., 2014; Stevenson, 2003; Wightman et al., 2010). Activated proliferating CD4$^+$ T cells are highly susceptible to infection and actively produce infectious virus, whereas resting CD4$^+$ T cells restrict HIV-1 replication due to a blockage in the reverse transcription process; however, the presence of several chemokine coreceptor molecules including CCR5 has been shown to facilitate the infection and lead to the establishment of a latent viral reservoir (Cameron et al., 2010; Chege et al., 2011; Pan, Baldauf, Keppler, & Fackler, 2013; Stevenson, 2003). In addition to direct infection of resting CD4$^+$ T cells, the resting T–cell reservoir can also be formed by the reversion of activated cells to a resting state (Eisele & Siliciano, 2012). HIV-1 infection causes the death of activated CD4$^+$ T cells via a caspase-3-mediated apoptosis pathway; in addition, a caspase-1-mediated pyroptosis pathway has been recently proposed to correspond to quiescent CD4$^+$ T-cell death (Doitsh et al., 2014; Gougeon et al., 1996; Grivel, Malkevitch, & Margolis, 2000). The theory of cell turnover during HIV-1 disease progression has also been proposed, according to which, as the loss of CD4$^+$ T cells occurs, the numbers of naïve and memory T cells are relatively increased (Davenport et al., 2002). Especially in HIV-1-infected patients who have undergone successful ART, a normal CD4$^+$ T-cell count has been observed because of the increase in the number of naïve and memory CD4$^+$ T cells, but not of HIV-1-specific CD4$^+$ T cells (Wightman et al., 2010). Both naïve and memory CD4$^+$ T cells exhibit a dominant role during the late stages of disease (Blaak et al., 2000). Naïve CD4$^+$ T cells divide slowly, while memory CD4$^+$ T cells maintain the latent reservoir population by undergoing homeostatic cell division; therefore, these cells seem to be responsible for producing infectious viral particles (Davenport et al., 2002).

During HIV-1 transmission, it has been demonstrated that T lymphocytes may be a major target for HIV-1 infection because transmitted viruses have been shown to preferentially use CCR5 as a coreceptor, require high levels of CD4, and are able to replicate efficiently in primary CD4$^+$ T cells, suggesting an R5-utilizing T-tropic virus phenotype (Dejucq, Simmons, & Clapham, 1999; Ochsenbauer et al., 2012; Parrish et al., 2013; Peters et al., 2004, 2006; Ping et al., 2013). Subsequently, these transmitted/founder viruses have been shown to adapt to be able to infect cells of the monocyte–macrophage lineage that results in the development of a wider target cell population and one that consists of target cells expressing higher levels of CCR5 than T cells (Ochsenbauer et al., 2012). R5 viruses have gained more importance in HIV-1 pathogenesis because they are able to infect at least a subset of T cells and cells of monocyte–macrophage lineage. Cells of the monocyte–macrophage lineage serve as reservoirs for chronic infection and have the ability to travel across the blood–brain barrier, contributing to HIV-1-associated neuropathogenesis (Gartner, Markovits, Markovitz, Betts, & Popovic, 1986; Gonzalez-Scarano & Martin-Garcia, 2005; Peters et al., 2004). It has been reported that the R5 viruses isolated from blood, semen, and different tissue samples such as brain, lung, and lymph nodes of patients throughout the course of disease exhibited different levels of M-tropic capacity due to the different levels of expression of both CD4 and CCR5 molecules on specific target cell populations (Peters, Duenas-Decamp, Sullivan, & Clapham, 2007; Peters et al., 2006). Recent studies utilizing Affinofile technology with cells which are engineered to express inducible and titratable levels of CD4 and CCR5 but constitutively express CXCR4 have shown that the M-tropic virus can infect cells expressing low levels of CD4 and high levels of CCR5, whereas the T-tropic virus exhibited less ability to infect these cells (Joseph et al., 2014). Furthermore, the R5 M-tropic and paired R5 T-tropic viruses exhibited no difference in ability to infect cells expressing low levels of CCR5 and high levels of CD4, suggesting the special ability of R5 virus in utilizing low levels of CCR5 for their target cell infections (Joseph et al., 2014).

The relationship between coreceptor usage and replication capacity in trafficking leukocytes, perivascular macrophages, and microglial cells, with respect to primary viruses isolated from brain and cerebrospinal fluid (CSF), has also been investigated to distinguish between CCR5 utilization and M-tropism from the perspective of neurotropism. Similarly, these studies have demonstrated that HIV-1 entry is restricted by a mechanism independent of coreceptor utilization (Gorry et al., 2001). The highly M-tropic HIV-1 variants isolated from brain tissue have been associated with neurovirulence; however, this is in contrast to virus derived from blood, semen, and lymph node tissue, which require high levels of CD4 for infection (Peters et al., 2007). Perivascular macrophages and microglial cells expressing both CD4 and CCR5 are infected and represent a major reservoir of HIV-1 within the brain (Gonzalez-Scarano & Martin-Garcia, 2005). The R5 isolates from brain tissue exhibited compartmentalization, were shown to be highly M-tropic, and had the ability to induce syncytia formation in primary macrophage cultures, and these characteristics are different from those of R5 viruses isolated from semen, blood, lymph nodes, and spleen (Gonzalez-Perez et al., 2012; Peters et al., 2007, 2006). Other cell types within the brain, such as astrocytes and oligodendrocytes, also exhibit HIV-1 susceptibility but do not support high levels of virus production (Gonzalez-Scarano & Martin-Garcia, 2005). In order to further study neuropathogenesis and the interrelatedness of coreceptor utilization and cellular tropism, the HIV Brain Sequence Database has compiled and annotated HIV envelope sequences sampled from the brain, or from patients from whom the database already contains a brain sequence (Holman, Mefford, O'Connor, & Gabuzda, 2010).

During late-stage infection, which involves loss of large numbers of T cells, the emergence of X4 virus occurs possibly due to viral adaptation to naïve CD4$^+$ T cells, which express high levels of CXCR4 (Arrildt, Joseph, & Swanstrom, 2012; Shankarappa et al., 1999; Zamarchi et al., 2002). In addition, R5 T-tropic virus was found in late-stage disease in lymph nodes and was able to infect memory CD4$^+$ T cells and actively produced virus (Davenport et al., 2002; Peters et al., 2004, 2006), suggesting that during the rapid loss of T cells and progression to AIDS, both the X4 and R5 viruses play a role, but in different populations of T cells.

In conclusion, it would appear that HIV-1 adapts to different cellular compartments during the course of disease with the process of viral adaptation involving changes in the ability to utilize different levels of CD4 as well as changes in the differential utilization of coreceptors present on the plasma membrane of different cell populations. In other words, the differential infection profiles have been shown to be based on the expression levels of cellular CD4, CXCR4, and CCR5 on specific target cell populations. T-cell populations play a major role during viral transmission and disease progression with respect to R5 and X4 viruses, respectively. During clinical latency, the R5 virus predominantly plays a role by virtue of its ability to reside within the monocyte–macrophage lineage and to infect resting memory CD4$^+$ T cells, which contribute to the periodic production of virus in low levels even during ART. Although effective ART has greatly increased life expectancy, HIV-1 reservoirs within the monocyte–macrophage lineage and the resting memory CD4$^+$ T-cell compartment contribute to long-term survival of the virus due to their low turnover (Perelson, Neumann, Markowitz, Leonard, & Ho, 1996), resistance to apoptosis (Le Douce, Herbein, Rohr, & Schwartz, 2010), and continuous low-level replication as indicated by continued viral evolution during the course of therapy (Lambotte et al., 2000; Zhu, 2000, 2002). With respect to patients on ART, the CD16$^+$ monocyte is preferentially infected by CCR5-utilizing strains in addition to being more permissive to replication (Ellery et al., 2007). Throughout the acute and chronic stages of infection, macrophages have facilitated the evasion of the immune system through multiple mechanisms, thereby contributing to the failure of the host to clear HIV-1 infection (Koppensteiner, Banning, et al., 2012; Koppensteiner, Brack-Werner, & Schindler, 2012). Cells of the monocyte–macrophage lineage, particularly perivascular macrophages and latently infected monocytes, are thought to be the most important cells with respect to dissemination of the virus throughout the body, including the brain, due to their migratory nature (Crowe, Zhu, & Muller, 2003; Strazza, Pirrone, Wigdahl, & Nonnemacher, 2011).

5. DEFINING THE GENOTYPIC AND PHENOTYPIC DIFFERENCES IN THE HIV-1 ENVELOPE

The HIV-1 gp120 subunit is composed of a 60,000-Da core polypeptide that contains highly N-linked glycosylation modifications that have been shown to increase the molecular weight up to 120,000 Da (Lasky et al., 1986). The gp120 molecule contains five variable regions (V1–5) that alternate between five constant regions (C1–5), each region separated by a disulfide bond (Leonard et al., 1990). The amino acid sequence of the variable regions can be highly modified by insertion, deletion, and substitution. Even though it is highly variable, the overall structure and functional elements are still relatively well conserved.

The first and second variable loops (V1 and V2) represent the largest part of gp120 and play an important role in envelope assembly and function, viral entry, antibody-induced neutralization, CD4 binding, coreceptor usage, and viral spread (Collins-Fairclough et al., 2011; Da, Quan, & Wu, 2009; Doria-Rose et al., 2012; McLellan et al., 2011; Saunders et al., 2005; Wan et al., 2009). Despite being hypervariable, V1 and V2 contain a cluster of relatively well-conserved residues, among which are nine charged amino acids, a greater number than are present in other regions of gp120 (Myers, Korber, Wain-Hobson, Smith, & Pavlakis, 1993). In addition, within these nine charged amino acids, the aspartic acid at position 180 has been considered highly conserved among HIV-1 isolates (Myers et al., 1993) and is crucial in the early steps of viral entry, as confirmed by alanine substitution studies (Wang, Essex, & Lee, 1995). Mutagenesis of the aspartic acid at position 180 to alanine results in an alteration in viral infectivity and slower viral growth kinetics. However, the incorporation of gp120 and gp41 into the viral particle and their ability to subsequently bind CD4 were not affected by this sequence alteration (Wang et al., 1995). Other studies have shown that deletion of either V1 or V2 alone had a minimal effect on envelope function and viral replication in a cell type-specific manner (Stamatatos, Wiskerchen, & Cheng-Mayer, 1998). However, deletion of both the V1 and V2 loops impaired envelope function, including impaired viral entry (Wyatt et al., 1993, 1995), and replication in primary CD4[+] T cells (Bontjer et al., 2009). This was explained by the absence of an α4β7 recognition motif, which has been shown to be located within the V2 loop region. The lack of this motif resulted in the downregulation of expression of the lymphocyte function-associated antigen 1 (LFA-1) adhesion molecule in infected cells, resulting in decreased viral spread. However, an alternative explanation of these results has suggested that the deletion of the V1/V2 region altered the interaction of the viral particle with CD4 and/or CXCR4 (Bontjer et al., 2009).

In conjunction with previous observations, it has been reported that alteration of the V1/V2 loops can modulate the viral phenotype with regard to coreceptor usage, CD4 binding, antibody neutralization, and its association with changes in HIV-1 disease severity (Cao et al., 1997; Nabatov et al., 2004). A mutated primary R5 HIV-1 isolate, SF162ΔV1ΔV2, which has been shown to lack the V1 and V2 loops, exhibited an overall normal CD4 binding site conformation on viral surfaces because the binding pattern of antibody directed against the CD4 molecule was similar to that of cells infected with the parental virus (Stamatatos et al., 1998). Nevertheless, the antibody recognition profiles were different between V1–negative virus, V2–negative virus, and the parental virus, SF162, suggesting a different structure and exposure of epitopes located on the viral gp120 molecule (Stamatatos et al., 1998). These V1- and V2-mutant viruses exhibited a comparable level of replication within peripheral blood mononuclear cells as compared with the wild-type parental virus (Stamatatos et al., 1998). The decrease in viral replication in macrophages has been observed with V1- or V2-deleted viruses, and it is therefore correlated with a postentry step because the CD4 binding and coreceptor usage profile were not altered in these viruses, as compared with the parental virus, further indicating that the role of V1 and V2 in viral entry is cell type-specific (Stamatatos et al., 1998). The neutralizing antibody susceptibility profiles of V1-deleted SF162 (SF162ΔV1), V2-deleted SF162 (SF162ΔV2), V1/V2-deleted SF162 (SF162ΔV1ΔV2), and parental SF162 were shown to be different. Removal of the V2 loop resulted in increased neutralization susceptibility to most of the antibodies examined, which might be because the overall V2 region was involved in a number of interactions with different gp120 epitopes (Saunders et al., 2005). The susceptibility of SF162ΔV2 and SF162ΔV1ΔV2 to antibody against the CD4 binding site was increased potentially due to the CD4 binding site of these viruses being more exposed based on the lack of the V2 loop region (Saunders et al., 2005). In contrast, SF162ΔV1 exhibited resistance to neutralization by antibody against CD4i epitopes as well as the CD4 binding site and V3 loop, potentially due to readjustment of the gp120 glycoprotein resulting in decreased epitope exposure, however, with no disruption in the ability of the virus to enter and replicate (Saunders et al., 2005). Another study has shown that HIV-1 lacking the V1 and V2 regions exhibited approximately 30% less entry capability with respect to Jurkat T cells, was syncytium-forming incompetent, exhibited delayed replication kinetics, and had increased sensitivity to neutralization antibodies as compared with the wild-type parental virus (Cao et al., 1997). A recent vaccine efficacy study focused on preventing HIV-1 infection also showed that amino acid positions 169 and 181 within the V1/V2 region were involved in elevating the immune response to HIV-1, suggesting that specific genetic signatures within the V1/V2 region could contribute to the exposure of epitopes targeted by neutralizing antibodies (Rolland et al., 2012). Additionally, a molecular structure study has proposed a model in which V1/V2 functions as a shield for the V3 region because deletion of V1/V2 regions leads to elevation of neutralizing antibodies directed against V3 epitopes (Liu, Cimbro, Lusso, & Berger, 2011).

In summary, V1 and V2 serve as concealers of the CD4 binding site; thus, deletion of both of these regions results in increasing the sensitivity to neutralization antibodies. Deletion of either V1 or V2 alone appears to display opposite effects in defining HIV-1 neutralization susceptibility; deletion of the V2 region enhances the exposure of epitopes on gp120, whereas deletion of only the V1 region has no effect on the level of neutralizing antibody sensitivity. Overall, viral gp120 lacking operational V1 and/or V2 loops affects viral infectivity and replication potential in a cell type-specific manner (Cao et al., 1997; Stamatatos & Cheng-Mayer, 1998).

The variable loop region 3 (V3) has been intensively studied because it was identified as the principal neutralizing domain (PND) on the viral envelope glycoprotein (Goudsmit et al., 1988; Javaherian et al., 1989) and was considered to be hypervariable with regard to amino acid sequence (LaRosa et al., 1990). The V3 loop consists of approximately 34–36 amino acid residues, at positions 296–331 of the HIV-1 HXB2 strain (Sirois, Sing, & Chou, 2005). Based on nuclear magnetic resonance (NMR) analysis, V3-derived peptides exhibit two distinctive sequences and conformations, which are similar to two different groups of natural ligands of CCR5 and CXCR4. Based on this structural information, it has been suggested that the V3 loop was responsible for the selective interaction with the two different coreceptors (CXCR4 and CCR5) (Sharon et al., 2003).

Several studies of V3 have focused on its role in viral entry and coreceptor usage. A number of investigators have examined V3 loop sequence variation with respect to viral phenotype. In this regard, two major approaches have been taken to analysis of the sequence variation that has been observed. First, studies were performed to determine the covariant sequence alterations within the V3 loops themselves, and second, the sequences were compared to a specific strain or consensus sequence (Sirois et al., 2005). Analysis of the covariant sequence alterations within the V3 loop has focused on determining the mutual information between amino acid residues. The most commonly encountered covariant residues detected when 308 distinctive V3 loop sequences were aligned consisted of the residue pairs of 11/13, 11/25, 13/19, 13/24, 13/25, 20/25, and 24/25 (Korber, Farber, Wolpert, & Lapedes, 1993). However, the amino acid positions 11 and 25 were shown to be consistent with the charge rule, and this pair of residues has been widely used for predicting viral phenotypes (De Jong, De Ronde, Keulen, Tersmette, & Goudsmit, 1992). When the V3 loop region of HXB2 was replaced with isolates that exhibited nonsyncytium-inducing (NSI) and syncytium-inducing (SI) phenotypes from the same patient, the NSI virus displayed a serine (S) at position 11 and aspartic acid (D) at position 25; however, arginine (R) at position 11 and glutamine (Q) at position 25 were identified in the SI viruses (De Jong et al., 1992). As previously described, the charge rule has been applied to the total net charge of V3 sequences, which varies globally between +2 and +10 (Cormier & Dragic, 2002). NSI viruses exhibited less V3 positive charge than SI viruses (De Jong et al., 1992), and similar results were demonstrated when comparing M-tropic versus non-M-tropic viral isolates (De Jong et al., 1992).

The viral phenotype transition has been shown to involve coreceptor usage. In this regard, a T-cell tropic (T-tropic) isolate or SI virus was discovered to use primarily CXCR4 as a coreceptor for viral entry. In contrast, the primary NSI isolate-exhibiting phenotype was shown to preferentially use CCR5 as the entry coreceptor (Deng et al., 1996; Feng et al., 1996). The 11/25 and net charge rules have previously been used to identify the viral phenotype in the context of coreceptor selection, SI capability, and tropism. These studies have demonstrated that if a positively charged amino acid was located at position 11 and/or 25 within the V3 loop and the total net V3 charge resided between 5 and 8, the virus would exhibit CXR4 usage and designated an X4 virus with an SI phenotype (Briggs, Tuttle, Sleasman, & Goodenow, 2000; De Jong et al., 1992). Conversely, if the amino acid at position 11 and/or 25 was neutral or negatively charged and was accompanied by a total V3 net charge that fell between 2 and 4, the virus would exhibit CCR5 usage and be designated an R5 virus with an NSI phenotype (Briggs et al., 2000; Fouchier et al., 1992; Xiao et al., 1998). Dual-tropic virus (X4/R5) has been defined by its ability to use either CXCR4 or CCR5 as a coreceptor, and these viruses generally score like an

X4 virus based on the application of both the 11/25 and net charge rules (Briggs et al., 2000).

The strategy of amino acid charge and coreceptor usage in phenotyping viral isolates has been supported by protein crystal structures and NMR studies that have indicated that amino acid positions 11 and 25 were located at the ends of the V3 crown (Cormier & Dragic, 2002) and are found opposite of each other with their charges capable of leading to an electrostatic interaction that can bind chemokine ligand-like structures (Dobrowsky, Zhou, Sun, Siliciano, & Wirtz, 2008; Huang et al., 2005). Amino acid residues 1–10 and 26–35 were designated as the N- and C-termini of the V3 stem, respectively. Residues 11–25 were designated as the V3 crown (Cormier & Dragic, 2002). The V3 stem and crown have distinct functional domains (Sirois et al., 2005). The V3 stem alone, for example, has been shown to mediate soluble gp120 binding to CCR5 amino-terminal domain-based sulfopeptides; however, both the stem and crown are necessary for binding of soluble gp120 to cell-associated CCR5 (Cormier & Dragic, 2002). Nevertheless, within the context of the virion, the V3 crown alone was sufficient for determination of coreceptor usage (Cormier & Dragic, 2002). The charges of amino acid residues 11/25, which are part of the crown, are likely involved in the selection of the coreceptor via electrostatic interaction. The extracellular loops of CXCR4 have been shown to contain higher negative charge than that of CCR5 (Dimitrov, Xiao, Chabot, & Broder, 1998). It has been suggested that the higher total positive charge of an X4 V3 would interact with CXCR4 more efficiently than with CCR5. Even though the V3 loops are highly variable, the glycine, proline, glycine (GPG) motif located in the middle of the PND, within the crown, has been shown to be well conserved among HIV-1 isolates (LaRosa et al., 1990). The changes that occur around this GPG motif alter the structure of the V3 crown and/or surface accessibility and thereby may guide coreceptor selection (Sirois et al., 2005).

Variable loop 4 (V4) of HIV-1 gp120 is approximately 19–26 amino acid residues long and is overall negatively charged (Douglas, Munro, & Daniels, 1997). According to the sequence alignment, the V4 loop has been shown to be as variable as V1 in all HIV-1 subtypes, except subtype E (Douglas et al., 1997). In studies using the simian immunodeficiency virus (SIV)-infected macaque model, neutralizing epitopes recognizing the V4 region were detected, and changes in the amino acid profile within V4, especially at positions 412–418, were shown to alter the N-linked glycosylation pattern, resulting in alteration of antibody recognition and the overall neutralizing profile (Kinsey et al., 1996). In addition, one of the amino acid mutations found within V4 resulted in the production of a new site for glycan modification, and this alteration was shown to play a role in disease progression in simian AIDS (Overbaugh & Rudensey, 1992). However, in HIV-1 disease, neutralizing antibody against V4 has not yet been described (Douglas et al., 1997). This may be due to the frequency of glycosylation modification detected within V4. The potential N-linked glycosylation found in V4 in SIV is less than that of the HIV-1 V4; consequently, less carbohydrate masking has been observed during the course of SIV infection, thereby resulting in a greater response to neutralization antibody as compared with HIV-1 (Douglas et al., 1997; Overbaugh & Rudensey, 1992; Torres et al., 1993). Given this observation, neutralizing antibody against V4 has not been found in HIV-1 infection in humans; the core motif of the superantigen (SAg)-binding site has been identified on a discontinuous epitope spanning V4 (Karray & Zouali, 1997). The SAg can engage B cells and activate VH3 B cells *in vitro* (Berberian, Shukla, Jefferis, & Braun, 1994; Zouali, 1995). Mutational analysis of V4 demonstrated significant binding of defective SAg, suggesting a role for V4 in gp120 SAg binding to normal human immunoglobulins (Karray & Zouali, 1997). Furthermore, the binding of SAg and VH3 B cells to the V4 region may directly explain the dysfunction and depletion of B cells during HIV disease progression. The potential sites for glycosylation on HIV-1 gp120 have been identified (Leonard et al., 1990), and based on the high degree of glycosylation, the virus has been shown to exhibit altered receptor binding sites and neutralization recognition (Kwong et al., 1998). Within the V4 loop, amino acid position 385 has been shown to be involved in the formation of a disulfide bond with amino acid position 418 and the glycan at position 386, which has been located at the base of the V4 loop, demonstrating their importance in neutralization sensitivity with their removal resulting in resistance to the 2G12 antibody (Sanders et al., 2008). N386 does not have an impact on envelope folding, but N-linked glycosylation has been shown to facilitate immune evasion by protecting the CD4 binding site against antibodies (Sanders et al., 2008). In addition, removal of this glycan by mutating amino acid N386K has been shown to enhance viral infection, and this process is CD4 independent (Edwards et al., 2001). Therefore, removal of the glycan at N386 may enhance therapeutic strategies by increasing neutralizing antibody against the CD4 binding site and the SAg motif on V4 and may serve as a potential candidate vaccine target to elicit antibody production.

Variable loop 5 (V5) of HIV-1 gp120 has been shown to contain approximately 7–11 amino acid residues, occasionally found to be 11–15 amino acids in length (Chamberland et al., 2011), with an average of two N-linked glycosylation sites (Douglas et al., 1997). Very little functional information has been obtained concerning V5. Sequence analysis of the HIV-1 gp120 V5 region has shown this loop structure to be hydrophilic with a net negative charge (Douglas et al., 1997). Sequence variation or mutations at the glyco-sylation sites within V5 enhance viral escape from neutralizing antibodies, which result from the alteration of gp120 envelope structure and epitope exposure (Cheng-Mayer, Brown, Harouse, Luciw, & Mayer, 1999; Sagar, Wu, Lee, & Overbaugh, 2006; Wei et al., 2003). One study has shown that removal of a V5 N-linked glycosylation site by mutagenesis at N462 led to resistance to anti-gp41 antibodies (Wang, Nie, et al., 2013). The level of genetic diversity studied within V5 demonstrated a small increase during the first year of infection in patients undergoing antiretro-viral therapy (ART), and a higher level of genetic diversity was observed in patients naïve to ART (Chamberland et al., 2011). With respect to the production of HIV-1 gp120 V5 region neutralizing antibody during human infection, antibodies directed at the V5 region have not been reported within an HIV-1-infected hemophiliac cohort (Douglas et al., 1997). How-ever, in a recent vaccine study, neutralizing activity against V5 was identified in sera from rabbits that were immunized with the HIV-1 gp120 subunit derived from the JR-CSF envelope (Narayan et al., 2013).

6. HIV-1 ASSEMBLY AND RELEASE

Once integrated into the host cell chromosome, the proviral DNA genome becomes the template for the production of virus-specific RNAs and proteins needed to drive the viral assembly process utilizing much of the host cell transcriptional machinery. HIV-1 assembly and egress at the plasma membrane is a multistep process involving multiple host–viral protein interactions to coordinate the synthesis, assembly, and release of a nascent virion without the disintegration of the host cell, followed by processing and maturation to form a fully infectious HIV-1 virion particle (Morita & Sundquist, 2004). Briefly, the process of HIV-1 assembly and release can be described in seven steps: (i) synthesis of viral polyproteins, (ii) targeting of the three polyproteins within the infected cell, (iii) intracellular membrane binding of viral components, (iv) viral envelope protein incorporation into microdomains of the plasma membrane, (v) multimerization of Gag at the plasma membrane, (vi) invagination of plasma membrane/release, and (vii) maturation/processing (Bello et al., 2012).

Prior to the release of a fully infectious HIV-1 virion, translation of viral envelope (Env) and core structural (Gag) proteins must occur. Translation of viral mRNA by endoplasmic reticulum (ER)-bound ribosomes is initiated to form the Env (gp160) precursor polyprotein, which is modified as it proceeds from the ER to the Golgi apparatus and subsequently to the plasma membrane. The Env precursor polyprotein is processed within the Golgi by a number of host proteases (PRs), such as furin, into an external CD4 binding domain (gp120) and an internal transmembrane (gp41) spanning region (Falcigno et al., 2004; Haim, Salas, & Sodroski, 2013). On the other hand, the Gag (Pr55) precursor polypeptide, a critical structural viral protein involved in guiding assembly and release, is synthesized in the cytosol and is cleaved by the virus-encoded PR following budding, to form a mature, infectious virus particle. Cleavage of this polyprotein has been shown to produce three major proteins: (i) matrix protein (MA, p17), (ii) capsid protein (CA, p24), and (iii) nucleocapsid protein (NC, p7) (Fig. 2A; Bello et al., 2012; Freed, 1998). In addition, altering the reading frame by 1 base pair results in a longer Gag–Pol transcript and precursor polyprotein (p160), which codes for the key HIV-1-specific enzymes PR, RT, and IN

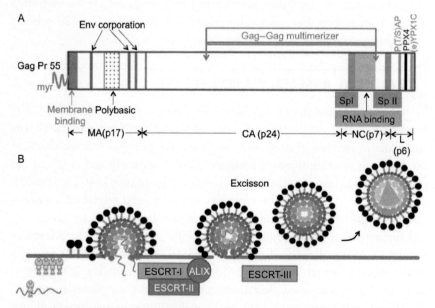

Figure 2 Gag protein function in viral assembly and release. (A) The Gag polyprotein precursor with matrix (MA), capsid (CA), nucleocapsid (NC), and p6 proteins depicted and important protein domains highlighted. (B) Major proteins involved in the budding process of the ESCRT (endocytic sorting complex required for transport) pathway.

(Nguyen & Hildreth, 2000). Owing to its unique ability to coordinate assembly and release, the Pr55 precursor polypeptide is also divided into six functional domains that are involved with (i) membrane binding, (ii) Env incorporation, (iii) plasma membrane targeting, (iv) Gag–Gag interaction multimerization, (v) RNA binding, and (vi) particle release (Ono, 2010). It is apparent that many of the stages involved in assembly and release are dictated by the Gag polyprotein itself. After cleavage of the Gag polyprotein, MA remains associated with viral lipid envelope, which is derived from cellular plasma membrane, while CA condenses and rearranges to form a cone-shaped structure around the viral RNA/NC complex to become a mature viral capsid. NC has high affinity for the vRNA (genomic HIV-1 RNA) and facilitates the condensation of the vRNA, dimerization, and interaction with the packaging signal ψ (psi), located within the 5′-UTR of the vRNA, for the encapsidation of viral RNA (Freed, 1998; Gelderblom, 1991; Weiss & Gottlinger, 2011).

In addition to these three major proteins, the cleavage of the Gag precursor also produces a small peptide from its C-terminus, p6, which is also found within the virion core (Bello et al., 2012; Freed, 1998). The p6 protein, or the late (L) domain, is reported to be involved and required for the late stages of viral maturation, budding, and separation from infected cells and is the predominate phosphoprotein encoded within the virion (Bello et al., 2012). As the HIV-1 virion buds off from host plasma membrane, it acquires the host lipid bilayer to cover the capsid, a process that does not occur spontaneously (Demirov & Freed, 2004; Demirov, Orenstein, & Freed, 2002; Weiss & Gottlinger, 2011). P6, and the proteins it interacts with, is essential for the separation and a number of subsequent events, which seems to be conserved across other retroviruses (Nguyen & Hildreth, 2000).

It has been shown that deletion and/or mutations within p6 resulted in inhibition of virus production in a cell type-dependent manner during the late stages of virus infection (Demirov & Freed, 2004; Demirov et al., 2002; Weiss & Gottlinger, 2011). Infection of HeLa and COS-7 cells with pseudotyped HIV-1 virus lacking p6 demonstrated that virus budding was initiated, but virions remained attached to the cell surface (Gottlinger, Dorfman, Sodroski, & Haseltine, 1991; Weiss & Gottlinger, 2011). Similarly, infection of primary monocyte-derived macrophages (MDMs) with the pseudotyped virus carrying mutated p6 caused a production of tethered virions on the membrane, whereas T lymphocytes and the Jurkat T-cell line were circumstantially affected by the mutant p6 but, in general, do not require p6 for release. In addition to the cell type-specific differences in p6 mutations, particular mutations within p6 also inhibited replication, while not necessarily inhibiting release, whereas other mutations inhibited both processes. These results indicated the importance of the p6 motifs, while alterations outside these motifs did not impair release, and a possible differential effect and/or requirement of p6 within specific immune cell subsets was also observed (Demirov et al., 2002). Interestingly, the accessory protein viral protein U (Vpu) has been documented to stimulate the release of nascent virions and yet seems to be independent of p6-mediated mechanisms. In this regard, further investigation will be required to define in detail the roles that p6 and Vpu play in cell types important in the pathogenesis of HIV-1 disease including $CD4^+$ T lymphocytes or cells of the monocyte–macrophage lineage (Schwartz, Geraghty, & Panganiban, 1996). However, to the best of our knowledge, it has not been determined whether targeting p6 or Vpu has provided any clinical therapeutic potential or exerted any additive or synergistic activity in conjunction with other antiviral agents. This possibility is not limited to HIV-1 since many other retroviruses that express p6 can be exchanged between different viruses (Schwartz et al., 1996; Weiss & Gottlinger, 2011).

Several studies have molecularly characterized the L domain. It is composed of three conserved amino acid regions that are essential for viral release: (i) P(T/S)AP, (ii) PPXY, and (iii) LYPX$_n$L (Bello et al., 2012; Martin-Serrano, Eastman, Chung, & Bieniasz, 2005; Parent et al., 1995; Puffer, Parent, Wills, & Montelaro, 1997; Strack, Calistri, Craig, Popova, & Gottlinger, 2003; Weiss & Gottlinger, 2011). These three regions are important for interacting primarily with cellular proteins of the ESCRT (endocytic sorting complex required for transport) pathway (Fig. 2B) (Bello et al., 2012; Martin-Serrano, Yarovoy, Perez-Caballero, & Bieniasz, 2003; Strack et al., 2003; von Schwedler et al., 2003). The ESCRT pathway is found in all eukaryotes and is a catalytic process coordinating a number of cellular processes associated with vesicular trafficking and sorting, including the trafficking of multivesicular bodies (MVBs) to the lysosome system, trafficking of cytosolic constituents, such as ubiquitinated proteins, and alteration of membrane to initiate the processes associated with abscission (Hurley, 2008). The ESCRT pathway is composed of five different complexes, which are ESCRT-0, ESCRT-I, ESCRT-II, ESCRT-III, and VPS4, all of which are required for viral assembly by promoting the invagination of the endosomal membrane, thereby facilitating the viral budding process (Hurley, 2008; Weiss & Gottlinger, 2011). The ESCRT pathway plays an important role in HIV-1 budding by mediating membrane remodeling, cytokinesis, and intraluminal endosomal vesicle formation (Bello et al., 2012; Kieffer et al., 2008; Weiss & Gottlinger, 2011). These processes are bridged by the small p6 peptide that contains a number of phosphorylated residues, which are phosphorylated by atypical protein kinase C, providing interaction points for the viral protein Vpr, as well as with cellular proteins (Solbak et al., 2013). In addition, inhibiting the phosphorylation of p6 also disrupts HIV-1 infectivity by interfering with its protein interactions and more specifically with Vpr packaging (Kudoh et al., 2014). Specifically, the HIV-1 PTAP region of the L domain binds directly to Tsg101, a cellular component of ESCRT-I, which is involved in the recruitment of other downstream factors to the site (Hurley, 2008; Weiss & Gottlinger, 2011). The recruitment of the ESCRT-III complex, which is formed by charged multivesicular body proteins (CHMPs), has been shown to represent a key for membrane scission because CHMP isoforms promote the release of nascent virions from the cell by recruiting the VPS4 ATPase enzyme to the site and directly binding to CHMP subunits (Babst, Katzmann, Estepa-Sabal, Meerloo, & Emr, 2002; Bello et al., 2012; Kieffer et al., 2008; Morita et al., 2011). HIV-1 budding can be strongly inhibited by blocking or depleting any individual CHMP subunit, VPS4 protein, Tsg101 interaction, and numerous other components that associate with the ESCRT pathway (Kieffer et al., 2008; Morita et al., 2011). As previously described, $LYPX_nL$ regions expressed on the L domain serve as docking sites for the cellular factor ALIX, which has been shown to be another factor that plays a role in the early stages of the ESCRT pathway by directly binding to CHMPs and compensate for disrupted PTAP–Tsg101 interactions (Bello et al., 2012; Weiss & Gottlinger, 2011). The budding of HIV-1, once believed to rely heavily on the PTAP region, can also use the $LYPX_nL$-L domain as an alternative because the releasing of mutated PTAP-L domain virus can be rescued by an overexpression of ALIX, suggesting that HIV-1 may be able to replicate in cells that express a low level of cellular Tsg101 protein or compensate by interacting with other binding partners (Dussupt et al., 2009, 2011; Popov, Popova, Inoue, & Gottlinger, 2008; Strack et al., 2003; Usami, Popov, & Gottlinger, 2007; Wang et al., 2014). In addition, these results suggest that even though virus may carry divergent NC sequences, it chooses to utilize a common mechanism and/or cellular factors for viral release, but the overall mechanism remains unclear (Bello et al., 2012). The process of hijacking the ESCRT pathway, among others, for viral budding appears to be a common theme for a number of retroviruses

(Haim et al., 2013), therefore supporting the importance in HIV-1 release (Chamanian et al., 2013; Kudoh et al., 2014).

The role of p6 in HIV-1 assembly is only one of the many proteins needed to coordinate assembly and release. In this regard, the viral protein primarily responsible for directing the assembly of immature virions is Gag and, more specifically, the N-terminal MA portion of Gag. MA is located at the N-terminal region of the Gag precursor polypeptide and the majority of the MA molecules are not cleaved/processed until the budding process has initiated (Kaplan, Manchester, & Swanstrom, 1994). The MA protein associates with the lipid bilayer, and this has been shown to be important for the formation of nascent virions (Freed, 1998; Gelderblom, 1991; Weiss & Gottlinger, 2011). However, the reason why the MA protein remains associated with plasma membrane after cleavage is unknown but is currently under investigation by a number of research groups. Previous studies have suggested that this is because of the myristoylation at the N-terminus of the Gag precursor polypeptide. Mutational analysis of these myristoylation sites has indicated that impairment of Gag protein transport to the plasma membrane subsequently results in inhibition of virus assembly, although the myristoylation process does not appear to be solely necessary for PR cleavage and membrane association (Bryant & Ratner, 1990; Freed, Orenstein, Buckler-White, & Martin, 1994). In fact, it seems to be important for membrane stabilization during virion formation (Bryant & Ratner, 1990; Zhou, Parent, Wills, & Resh, 1994). The myristoylation modification is not the only requirement for MA to travel to the plasma membrane because not all myristoylated proteins travel to the plasma membrane; moreover, HIV-1 carrying myristoylation-deficient Gag–Pol polyproteins can still be packaged and incorporated into virus-like particles (Park & Morrow, 1992). Subsequently, it has been demonstrated that the myristoylated protein carries hydrophobic domains that can interact with the plasma membrane and the electrostatic interaction between basic residues of MA with the acidic phospholipids is reported to enhance the association of MA with the cellular plasma membrane (Peitzsch & McLaughlin, 1993; Zhou et al., 1994). In addition, the polybasic region is located at the N-terminus of MA and mutations within this region have been shown to result in blockage of Gag transport to the plasma membrane thereby inhibiting virus production, suggesting that this region might also be involved in signaling events associated with virion production (Yuan, Yu, Lee, & Essex, 1993; Zhou et al., 1994). These mutational studies have demonstrated residues 4 and 6 in the polybasic region of MA when mutated caused defects in viral assembly;

however, second-site compensatory changes found in the middle of MA or near the C-terminus can reverse this phenotypic assembly defect. Interaction of MA with the plasma membrane via myristoylation and the electrostatic interaction of the basic regions seem to be important, therefore demonstrating the involvement of more than just one domain of MA in membrane binding (Freed, 1998; Ono, Huang, & Freed, 1997).

Most retroviruses have been shown to assemble and release nascent virions at the plasma membrane. The dynamic nature of the plasma membrane itself can possibly influence this process (Fig. 3). The plasma membrane is a mosaic of multiple functional microdomains that promote several physiological processes associated with the cell surface (Laude & Prior, 2004; Maxfield, 2002; Ono, 2010). Microdomains exist in different sizes, forms, shapes, compositions, dynamics, and organizations and vary in their biological functions (Maxfield, 2002). Among all plasma membrane microdomains, lipid raft and tetraspanin-enriched microdomains (TEMs) have been shown to be involved in the HIV-1 entry and exit processes (Martin-Serrano & Neil, 2011; Ono, 2010).

Lipid rafts are dense, freely moving microdomains within the plasma membrane, composed primarily of cholesterol and sphingolipids (Hemler, 2005; Ono, 2010). They are also characterized as containing high concentrations of glycosylphosphatidylinositol-linked protein, caveolin, various cellular receptors, and Src-family kinases that play a pivotal role in signaling, endocytosis, and, in particular here, entry of HIV. However, these components are not commonly found in TEMs (Hemler, 2005). Furthermore, lipid rafts can also be differentiated from TEMs because they can be disrupted at 37 °C by cholesterol depletion reagents and are typically insoluble in non-ionic detergent such as Triton X-100 (Fantini, Garmy, Mahfoud, & Yahi, 2002; Hemler, 2005). It is widely accepted that lipid rafts play a vital role in cellular signaling and intracellular protein trafficking because of the high expression of several signaling proteins found within the structures (Fantini et al., 2002; Ono, 2010). Numerous viruses, as well as the bacterial cholera toxin (Campbell, Crowe, & Mak, 2001), utilize lipid rafts as the gateway for entering and/or exiting. In the context of HIV-1, lipid rafts are employed for both entry and egress. Additionally, a number of other pathogens have been reported to be able to control host cell functions and deploy their pathogenic effects by utilizing lipid rafts, cholesterol, and caveolae (Campbell et al., 2001).

Many investigators have reported the importance of lipid rafts in budding and assembly of HIV-1 virion particles (Aloia, Tian, & Jensen, 1993;

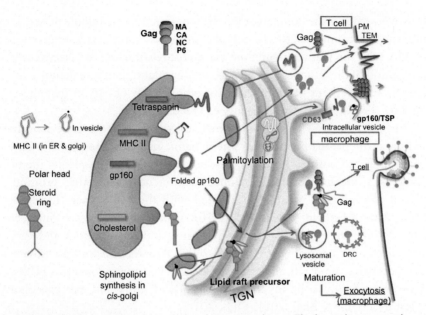

Figure 3 Budding of HIV-1 through the plasma membrane. The host plasma membrane is composed of multiple functional microdomains. Lipid raft and tetraspanin-enriched microdomains (TEMs) have been shown to be involved in HIV-1 budding. HIV-1 exhibits budding in a cell type-specific manner. Budding occurs via the plasma membrane of T lymphocytes and other cell types, whereas budding into multivesicular bodies (MVBs) followed by the extracellular releasing by exocytosis occurs primarily in macrophages. The type of microdomain from which HIV-1 buds also results in a different composition of the viral envelope. Lipid raft precursors are composed of cholesterol, which has been shown to be synthesized in the ER, and sphingolipid, which has been shown to be synthesized in the cis-Golgi. During trafficking within the Golgi, the lipid raft precursor, which has been shown to be synthesized in the trans-Golgi network (TGN), and HIV-1 gp160, which has been shown to be synthesized in the ER, become associated together and travel to the cell surface of macrophages by DRC (detergent-resistant complex), and subsequently released by exocytosis in primary macrophages. Within T cells, the lipid raft, gp120/gp41, and the Gag protein are found to be associated as an intermediate complex before reaching the plasma membrane for assembly and egress. Tetraspanin, synthesized in the ER, has been shown to be associated with the major histocompatibility complex (MHC) II molecule, gp160, and CD63 within an intracellular vesicle before exocytosis in macrophages. However, Gag directly transports to the plasma membrane where gp160 and tetraspanin are located and budding occurs.

Campbell et al., 2001; Fantini et al., 2002; Nguyen & Hildreth, 2000; Ono, 2010). Their importance has been rectified by studies that molecularly characterized the composition of the HIV-1 envelope. The HIV-1 envelope is similar to other enveloped viruses, as well as the erythrocyte plasma membrane; however, the molar ratio of cholesterol to phospholipid is higher than that of the host cell plasma membranes, thereby indicating a strong correlation with lipid rafts (Aloia et al., 1993). Treating HIV-1 virions with a cholesterol-binding agent or cholesterol-chelating agent causes a loss in viral structural stability in a dose-dependent manner, suggesting that cholesterol is incorporated into the virion and helps to maintain its overall structure (Campbell et al., 2001). Treating cells with either nystatin, a polyene antifungal, which has been shown to disrupt lipid raft content in the plasma membrane, or β-methyl cyclodextrin, which extracts cholesterol from the plasma membrane, causes a reduction in virus production (Pickl, Pimentel-Muinos, & Seed, 2001).

Cells of the monocyte–macrophage lineage have shown that HIV-1 seems to assemble within and bud from MVBs, subsequently releasing the virus extracellularly by exocytosis (Jolly & Sattentau, 2007; Nydegger, Khurana, Krementsov, Foti, & Thali, 2006; Pelchen-Matthews, Kramer, & Marsh, 2003). Immunolabeling technologies in conjugation with cryosections from pseudotyped HIV-1 BaL-infected primary human MDMs demonstrated that these cryosections were positive for both p17 and p24 and more specifically EM images displayed virions scattered within small vesicles that labeled positively for CD63 and were located in close proximity to positively labeled LAMP-1 on the cell surface, a late endosomal (LE) membrane marker (Pelchen-Matthews et al., 2003). These intracellular vesicles also exhibited positive immunoreactivity to antibodies directed against late endosome proteins LAMP-1 and LAMP-2, as well as the macrophage-specific lysosomal membrane glycoprotein, CD68 (Pelchen-Matthews et al., 2003). CD63 has been identified as a tetraspanin protein that is usually expressed at high levels intracellularly with low levels on the cell surface and was shown to be present on primary MDM-derived HIV-1 particles in association with gp120 (Bryant & Ratner, 1990). On the other hand, very small amounts of assembling virions were present at the plasma membrane. Additionally, physical protein–protein interactions were shown to support the immunocytochemistry results. Most of the above results have been supported by studies utilizing pseudotyped virions; yet few fully infectious viruses were used to support the results obtained with the pseudotyped particles. Consequently, these results have suggested that HIV-1 may be produced in an intracellular compartment which is characteristic of late endosomes and subsequently secreted into extracellular medium; however, other proteins involved in cellular ESCRT machineries were not examined in this study (Pelchen-Matthews et al., 2003). Additional studies are needed to examine in greater detail the molecular basis of the assembly of HIV in endosomal vesicles in macrophages. In this regard, it has been demonstrated by using live cell imaging, immunocytochemistry, electron microscopy, and coimmunoprecipitation analyses that HIV-1 Gag proteins colocalize at the membrane of late endocytic compartments in many cell types, not just macrophages (Nydegger, Foti, Derdowski, Spearman, & Thali, 2003).

In both HEK293T cells and HeLa cells, Gag directly interacts with δ subunit of the AP-3 complex, which is involved in trafficking pathways from the Golgi to lysosome (Dong et al., 2005; Odorizzi, Cowles, & Emr, 1998). Therefore, it has been hypothesized that HIV-1 Gag protein was associated with an endosomal compartment, which has been shown to contain all protein complexes essential for budding. Gag itself may contain important signal(s) that could promote or redirect this endosomal compartment to different destinations in different cell types (Benaroch et al., 2010; Dong et al., 2005; Finzi, Brunet, Xiao, Thibodeau, & Cohen, 2006; Resh, 2005). In CD4$^+$ T cells and other cell types, Gag may rapidly transport the complex directly to the plasma membrane, and budding steps continue as normal (Resh, 2005). In macrophages, the endocytic compartment complex may stay in the endolysosomal pathway, while nascent virion assembly is completed and subsequently released from the cell via exocytosis; yet the mechanism regulating this is still not understood (Resh, 2005). A study utilizing HEK293T cells expressing MHC II demonstrated that the Gag protein relocates to intracellular compartment, which has been shown to express LE/MVB markers. These results support the hypothesis that HIV-1 assembly and budding occur in the LE body localizing intracellularly in macrophages which express the MHC II molecule (Finzi et al., 2006). Furthermore, while T cells have been shown to bud at the plasma membrane, analyzing EM profiles of HIV-1-infected macrophages has revealed no evidence of budding at the plasma membrane. However, this is controversial because other research utilizing both EM and immunofluorescence has shown minimal binding at the plasma membrane (Benaroch et al., 2010). This is an interesting phenotype, and additional research is needed to determine if this is the primary pathway for assembly in macrophages, as the research seems to present, or a secondary pathway, and/or one that exists only under discrete environmental condition such as intense HAART therapy that would not be replicated in vitro. It has also been hypothesized that nascent virion-containing intracellular compartments connect to the plasma membrane through thin microchannels or an extensive network of tubules, and these are not detected in HIV-1-infected T cells

(Benaroch et al., 2010; Bennett et al., 2009; Deneka, Pelchen-Matthews, Byland, Ruiz-Mateos, & Marsh, 2007). The physiological significance and mechanism of these cell type-specific differences have yet to be elucidated.

It is believed that the cellular antigens that are incorporated into the viral particle can be represented as markers for virus cellular origin and can be used as a tool to capture virus (Abbate et al., 2005; Capobianchi et al., 1994; Lawn, Roberts, Griffin, Folks, & Butera, 2000; Rozera et al., 2009; Saarloos, Sullivan, Czerniewski, Parameswar, & Spear, 1997). The expression of HLA-DR (MHC II) was the first molecule to be detected from *in vivo* plasma-derived isolates and *in vitro* virus isolated from infected primary MDM (Saarloos et al., 1997). Subsequently, several reports have demonstrated the expression of a number of host-derived cell membrane proteins (CMPs) on the viral particle, including CD36, CD14, CD3, CD26, CD45RO, CD58, ICAM-1, CD11a, and CD44 (Abbate et al., 2005; Capobianchi et al., 1994; Lawn et al., 2000; Rozera et al., 2009; Tremblay, Fortin, & Cantin, 1998). By utilizing specific antibodies directed against these proteins, it has been shown that CD36 is predominantly expressed on macrophage-derived virus, while CD25, CD26, and CD3 molecules have been shown to be highly expressed on T lymphocyte-derived virus (Lawn et al., 2000; Rozera et al., 2009). This can be of advantage clinically and therapeutically since the circulating virus can be tracked with respect to cellular compartmentalization. The mechanism of incorporating host CMP into the virus particle has remained unclear; however, incorporating host protein into the viral particle provides several advantages to the virus, for instance, host immune evasion, enhanced viral entry, and replication (Tremblay et al., 1998). Furthermore, it would be of interest to determine if the tropism of the virus or the activation state of macrophages impacts this assembly process or if it is totally dependent on the physiological differences between macrophages and T cells.

On the other hand, HIV-1-infected $CD4^+$ T cells have demonstrated that HIV-1 Gag and Gag–Pol precursor proteins were associated with two intermediate complexes, detergent-resistant complex (DRC) and detergent-sensitive complex (DSC), prior to arriving at the plasma membrane and before the release of a viral particle. As their names imply, these intermediate complexes were identified on the basis of detergent treatment. Using density centrifugation coupled with Western blots and TEM, Lee and Yu (1998) and Lee, Liu, and Yu (1999) demonstrated that the DRC was composed of Gag, Gag–Pol precursor proteins, two precursors, and a mature

RT protein product. This viral amalgamate was shown to be held together noncovalently within the cytosol prior to traveling to the plasma membrane. However, before completing the assembly process and egress at the plasma membrane, the DRC appears to undergo some level of HIV-1 PR-independent processing, from the immature precursor polyproteins to mature, Gag-derived CA, MA, and RT as well as becoming covalently attached via myristoylation to the plasma membrane; this complex is now referred to as DSC. The DSC begins to resemble a mature virion as opposed to DRC (Fig. 3) (Lee et al., 1999; Lee & Yu, 1998). At this point, the two HIV-1 genomic RNAs associate with the other viral components followed by budding and maturation via activities that involve HIV-1 PR-dependent processing. Interestingly, Lee and Yu (1998) and Lee et al. (1999) did not show the presence of gp160 in either intermediate complex, indicating the possibility that multiple complexes are targeted to the plasma membrane with specific viral components to possibly coordinate the assembly process. Disruption of the association of these intermediate complexes or post-translation modifications of the viral protein can significantly interfere with assembly and egress. Alteration of the fatty acid composition, by replacing the myristic saturated acids with unsaturated fatty acids, at the N-terminus of Gag leads to a reduction in Gag interactions with the plasma membrane lipid rafts and inhibition of Gag-driven virion assembly, but not the membrane in general (Lindwasser & Resh, 2002), thus demonstrating that lipidation of Env and Gag polypeptides is necessary for proper assembly and release of an infectious HIV-1 virion particle from the plasma membrane. On the same note, while $CD4^+$ T cells have been shown to assemble with other complexes at the cytosolic face of the plasma membrane, it is still unclear why HIV-1-infected macrophages assemble in vesicles.

As previously indicated, lipid rafts are not the only area of the plasma membrane where Gag is targeted and HIV-1 can be released. Similarly, there has been colocalization of the Gag protein with TEMs, which seems to be conserved in selected cell types, such as in T cells, macrophages, and other cells (Grigorov et al., 2009; Jolly & Sattentau, 2007; Nydegger et al., 2006). Tetraspanin proteins encompass a large family of membrane proteins, which are characterized by a 200–350-amino acid polypeptide, four transmembrane domains, two extracellular domains, and an intracellular N- and C-termini (Hemler, 2005; Nydegger et al., 2006). Tetraspanins are widely expressed in eukaryotes and are involved in many biological functions such as cell adhesion, migration, and proliferation as well as signaling and antigen presentation within the immune system (Hemler, 2005; Levy & Shoham,

2005; Nydegger et al., 2006). Tetraspanins interact with neighboring proteins with high specificity to form scaffolds called TEMs and have several features that are involved in functional cellular platforms (Hemler, 2005; Levy & Shoham, 2005). Experimentally, TEMs can be recognized by utilizing antibodies directed against CD9, CD63, CD81, CD82, CD53, and CD151 (Horvath et al., 1998; Pelchen-Matthews et al., 2003). It has also been shown that TEMs are expressed as visible patches either on the cell surface or intracellularly (Horvath et al., 1998; Ono, 2010; Pelchen-Matthews et al., 2003). Utilizing immunohistochemistry, Nydegger et al. (2006) have shown Gag and Env polyproteins to be associated with the tetraspanin domain with subsequent release leading to incorporation of some of the TEM within the virion (Bryant & Ratner, 1990).

As indicated previously, tetraspanin proteins accumulate at budding sites where they are subsequently incorporated into viral particles, thereby playing important roles in cellular fusion processes (Jolly & Sattentau, 2007; Nguyen & Hildreth, 2000; Pelchen-Matthews et al., 2003; Sato et al., 2008; Tachibana & Hemler, 1999; Weng, Krementsov, Khurana, Roy, & Thali, 2009). The level of tetraspanin expression in virions has also been reported to reflect the ability of envelope-induced cell–cell fusion and syncytium formation (Weng et al., 2009). Overexpression of tetraspanin in virus-producing cells caused a reduction in either syncytium formation or infectivity of cell-free virus. On the other hand, knockdown of tetraspanin protein production increased the fusion activity but had no effect on cell-free infectivity (Krementsov, Weng, Lambele, Roy, & Thali, 2009; Weng et al., 2009). The syncytium formation ability depends on the presence of Gag, the recruitment of Env protein to TEMs, and the viral strain itself (Murakami & Freed, 2000; Weng et al., 2009). Based on these observations, it would appear that HIV-1 assembly occurring at TEMs prevents infected cell contact with uninfected cells, thereby also preventing the initiation of premature fusion (Ono, 2010; Weng et al., 2009). However, because the reduction in tetraspanin has been reported in HIV-1-infected T cells and this down-regulation of tetraspanin in virus-producing cells exhibited increasing fusion activity, it was possible that HIV-1 may modulate the expression of tetraspanin in order to increase the efficiency of viral transmission within the host (Krementsov et al., 2009). Research aimed at elucidating the impact of tetraspanin on fusion or syncytium formation has focused on the types of cells involved, viral strain, and the overall kinetics of this process (Ono, 2010).

Utilizing immunofluorescence and cryo-immunoelectron microscopy techniques have broadened the understanding of HIV-1 egress in certain cell types and the specific role of TEMs in this process. Transient transfection of both HeLa and Jurkat cell lines with pseudotyped HIV-1 genomic constructs resulted in the colocalization of HIV-1 Env and Gag proteins along with CD63 at the plasma membrane. This discovery supported the hypothesis that HIV-1 Gag proteins accumulate at the cellular surface where TEMs have been shown to localize (Nydegger et al., 2006). This prompted the question as to whether the cellular ESCRT machineries are involved in virion accumulation at TEMs. In this regard, it was shown that cellular ESCRT proteins, Tsg101 and VPS28, were found colocalized with CD63 and very close to the Gag localizing patches at the plasma membrane (Nydegger et al., 2006). Furthermore, both Gag and Env proteins have been colocalized to TEM, specifically CD63, CD81, and CD9, in both Jurkat cells and primary T cells. Despite colocalization with TEMs, there was no association with the lysosomal marker LAMP2 (Nydegger et al., 2006). To support the immunofluorescence images, coimmunoprecipitations against the selected membrane markers were performed. It was determined that viral particles form at the plasma membrane and they are coimmunoprecipitated with p24 protein, an observation consistent with HIV-1 assembly and budding from T cells via TEMs, but not lysosomal compartments (Jolly & Sattentau, 2007). Thus, supporting the fact that nascent virions assemble at the plasma membrane as well as bud off through the TEMs, in addition, or secondary to lipid rafts (Nydegger et al., 2006). All in all, morphological differences in assembly between macrophages and CD4$^+$ T cells are apparent. However, there are also differences in lipid rafts and TEMs with regard to T-cell assembly. The molecular mechanism governing this and the functional impact on HIV pathogenesis is still poorly understood and a fertile area in HIV-1 research.

7. SUMMARY OF VIRAL ENTRY, CORECEPTOR UTILIZATION, VIRAL EXIT, AND TRANSLATIONAL MEDICINE

The viral surface glycoprotein gp120 plays an important role in the initial step of HIV-1 infection as it mediates the attachment and fusion to the host cellular membranes (Bour, Geleziunas, & Wainberg, 1995). This step starts with the attachment of viral gp120 and host cellular surface receptor CD4 molecule, which triggers the conformational change of gp120 that subsequently allows the viral V3 loop to interact with one of the seven-transmembrane G proteins of the chemokine receptor family expressed on the target cells (Brelot et al., 1999; Broder & Dimitrov, 1996; Cocchi et al., 1996; Lifson & Engleman, 1989; Nara, Hwang, Rausch, Lifson, & Eiden, 1989; Stein et al., 1987). It has been demonstrated that HIV-1 is able to utilize several chemokine receptors such as CCR3, CCR7, and CCR8, but several lines of evidence suggest that CXCR4 and CCR5 are the most relevant coreceptors used *in vivo* (Alkhatib et al., 1996; Bleul, Farzan, et al., 1996; Bleul, Fuhlbrigge, Casasnovas, Aiuti, & Springer, 1996; Doranz et al., 1996; Edinger et al., 1998; He et al., 1997; Samson et al., 1998; Zhang et al., 1997). The discovery of coreceptor proteins, including their mechanisms with respect to the viral entry process, has led to a better understanding of HIV-1 pathogenesis by the classification of virus tropism according to their ability to utilize one of these coreceptors during the entry step, with the three most studied types of viruses being the CCR5-utilizing (R5), CXCR4-utilizing (X4), and dual-tropic (X4/R5) viruses (Clapham & McKnight, 2001; Sacktor et al., 2001). These three viral types exhibit their own preference with respect to infecting target cells due to the relative level of receptor and coreceptor proteins expressed on the cell surface. However, this coreceptor tropism classification frequently leads to confusion with the HIV-1 cellular tropism classification that classifies viral types based on primary infected target cells, which are monocyte–macrophage (M)-tropic and T-cell line (T)-tropic viruses as previously reviewed (Aiamkitsumrit et al., 2014; Gorry et al., 2004). The R5 virus preferentially utilizes CCR5 as a coreceptor and this molecule is often found on cells of monocyte–macrophage lineage as well as selected T-cell subsets, while the X4 virus favors the use of the CXCR4 molecule, which is primarily located on T cells (Feng et al., 1996). Therefore, R5 virus is sometimes mistakenly referred to as an M-tropic virus, while the X4 virus is often referred to as a T-tropic virus. In addition, some viruses exhibit ability to infect both T cells and monocytes–macrophages because it can use either CXCR4 or CCR5 as its coreceptor, and this virus is referred to as a dual-tropic or X4/R5 virus (Collman et al., 1992; Glushakova et al., 1999). Many aspects relevant to the cellular tropism of HIV have had a significant impact on the overall nature of HIV-1 pathogenesis and the development of HIV disease in general. The interaction of HIV-1 with specific target cell populations will also have a great impact on our ability to define HIV-1-infected cellular reservoirs that remain after long-term combination therapy and our overall efforts to achieve a cure. Selected T-cell subpopulations have been shown to be the major targets for HIV-1 infection and subsequently the source of highly productive replication because of the high density of CD4 and the presence of one or more of the coreceptor molecules (CXCR4 and CCR5 depending on the T-cell subset) expressed on the cell surface, with the resting memory CD4$^+$ T cells identified as an important reservoir for virus during the course of combination therapy. Along with T cells present in mucosal tissues, cells of monocyte–macrophage lineage are also associated with HIV-1 transmission and probable a secondary latent reservoir. The development of this viral reservoir seems to take place shortly after initial infection, whereby infected cells traverse the BBB to reside in the CNS, have tropism for low CD4$^+$-expressing cells, impair neurological function, and in some cases augment the development of the more severe forms of HIV-associated neurocognitive disorders (Cameron et al., 2010; Davenport et al., 2002; Fischer-Smith et al., 2001; Gonzalez-Scarano & Martin-Garcia, 2005; Gordon & Taylor, 2005; Joseph et al., 2014; Stevenson, 2003; Wightman et al., 2010). The relationship between cellular tropism and coreceptor usage has been considered to be more delicate because there is not a clear-cut distinction between the universal presence of CXCR4 and CCR5 on T cell and cells of monocyte origin. In addition, some HIV-1 quasispecies have been shown to infect CD4$^+$ cells in the absence of coreceptor engagement and some cells in the absence of the receptor CD4.

Research and development in HIV-1/AIDS therapy has been focusing on interrupting many steps in HIV-1 life cycle, and the use of current combination antiretroviral therapy (cART) has been relatively successful with respect to controlling the plasma viral load level to below the limits of current assay detection capabilities (Doherty, Ford, Vitoria, Weiler, & Hirnschall, 2013; Gulick et al., 1997; Hammerle, Himmelspach, Dorner, & Falkner, 1997; Hirnschall, Harries, Easterbrook, Doherty, & Ball, 2013; Marconi et al., 2010; Williams, Lima, & Gouws, 2011). One of the cART regiments has included an entry inhibitor, with the inclusion of one of the two current U.S. Food and Drug (FDA)-approved entry inhibitors used in humans either Enfuvirtide or Maraviroc (LaBonte, Lebbos, & Kirkpatrick, 2003; Lieberman-Blum, Fung, & Bandres, 2008; MacArthur & Novak, 2008). Enfuvirtide (also known as T-20) is a peptide homolog to HIV-1 gp41 heptad repeat (HR) two regions and was developed by Roche and Trimeris (Greenberg & Cammack, 2004; LaBonte et al., 2003). Enfuvirtide has been shown to be able to bind to the gp41 HR 1 region, therefore inhibiting the formation of six-helix bundle essentially for fusion process (Greenberg & Cammack, 2004; Joly, Jidar, Tatay, & Yeni, 2010). Enfuvirtide has been recommended for treatment of HIV-1 infection because it has demonstrated low cytotoxicity and less problems with resistance (Jang et al., 2014; Mink et al., 2005; Reis, de Alcantara, Cardoso, & Stefani, 2014; Xu et al., 2005). Maraviroc is a small molecule that has been shown to selectively bind to a human chemokine CCR5 coreceptor molecule, preventing the interaction and binding of HIV-1 gp120 with the cellular CCR5 coreceptor molecule (Lieberman-Blum et al., 2008; Pfizer, 2007). Maraviroc is the first FDA-approved CCR5 antagonist and has been used in combination therapeutic strategies of patients infected with only R5 virus (Pfizer, 2007). However, it shows no activity against the X4 or X4/R5 dual-tropic viruses and, therefore, the general tropism pattern of the virus in a given infected patient must be validated prior to utilization in any treatment protocol (Lieberman-Blum et al., 2008; Pfizer, 2007). In addition, an HIV-1-infected individual who has been demonstrated to be homozygous for CCR5 delta 32, and exhibiting resistance to infection by an R5-tropic virus, would need to be carefully evaluated with respect to the utility of employing a CCR5 inhibitor as a component of any combination therapeutic strategy for effective management of HIV disease in this type of patient (Doranz et al., 1996; Zimmerman et al., 1997).

The tropism of HIV-1 exhibits a major impact with regard to viral pathogenesis and the course of HIV disease. Determination of virus coreceptor tropism has become more necessary for HIV/AIDS therapy and treatment monitoring because of the limitation in the number of effective entry inhibitors (Enfuvirtide and Maraviroc). Resistance to entry inhibitors could arise due to coreceptor usage switching and the outgrowth of preexisting X4 virus (Panos & Watson, 2014; Westby et al., 2006). Although the mechanisms associated with coreceptor usage switching have not been clearly established, a number of studies have associated changes in coreceptor utilization with major changes in the clinical course of HIV disease. In addition, the emergence of HIV-1 CD4-independent infection pathways has been shown and these alterations have also led to changes in the nature of HIV disease (Borsetti et al., 2000; Zerhouni, Nelson, & Saha, 2004).

Entry inhibitors are usually administered in triple combination with other classes of antivirals. These antivirals target other critical steps in the HIV life cycle, such as reverse transcription, integration, and processing of the viral polyproteins, thereby reducing the replicative pool of infectious virus. In addition to entry inhibitors, antiretroviral compounds will fall into one of the four main groups, the nucleoside reverse transcriptase inhibitors and nonnucleoside reverse transcriptase inhibitors that interfere with reverse transcription, IN strand transfer inhibitors that block the integration of proviral DNA into the host genome, and protease inhibitors (PI) that prevent the processing and maturation of the Gag and Gag–Pol proteins to form a fully infectious virions (Solbak et al., 2013). Despite the potency of these antiviral agents, they do not combat noninducible and/or latent provirus but do allow an individual at times, to maintain viral loads, at undetectable levels. A detailed and up-to-date list of all the anti-HIV therapeutic agents in clinical use as well as those in the experimental phases is cited with a URL provided in Engelman et al. (2012).

With the advent of high-throughput screening of small-molecule inhibitors, numerous antivirals have been discovered that specifically target these key aspects of the viral life cycle. Yet, increases in resistance have led to the necessity of combination therapy with antivirals targeting multiple stages of the viral life cycle with the continued screening of novel compounds. Currently, there are 5 entry inhibitors in use or in development and 11 PIs targeting exit (Solbak et al., 2013). Through the efforts of discovering new antiviral compounds with anti-HIV activity, an overall increase in HIV pathogenesis has been gained and vice versa. By studying and targeting host–viral protein interactions as possible therapeutics, it could minimize the development of resistance in the ever-evolving viral quasispecies and elucidate the cell type-specific differences in HIV entry, assembly, and egress.

Of interest to both the entry and exit stages of HIV replication cycle are a number of host-dependent HIV restriction factors and host-dependent factors that either combat or are crucial for HIV infectivity. These interactions are specific host–viral protein interactions that provide another avenue for antiviral drug development and have been contemplated by others as well (Solbak et al., 2013). Following entry, uncoating, and reverse transcription, the virus is challenged by the host restriction factors TRIM5α, SAMDH1, and APOBEC3G (Pham, Bouchard, Grutter, & Berthoux, 2010; Solbak et al., 2013). APOBEC3G is a cytidine deaminase that is packaged into the virion during assembly and functions as a host restriction factor by catalytically deaminating cytidine to uracil which therefore induced mutations throughout the viral genome. This has therefore pressured viral evolution to favor the viral protein Vif as an antagonist to the antiviral effects of APOBEC3G. Results reported by Porcellini et al. (2009) have demonstrated that mutant Vif can be used to override consensus Vif and inhibit integration of HIV in both CD4$^+$ T cells and macrophages, implying the possibility of designing Vif agonists to prevent HIV infection in a non-cell type-specific manner.

Similarly, for virions that successfully integrate their genomes into the host cell chromosome, two general pathways are available, viral expression or transcriptional latency. During assembly, the host restriction factor tetherin, a transmembrane receptor, causes attachment of viral particles to the plasma membrane of the infected cell and is antagonized by the viral protein Vpu (Neil, Zang, & Bieniasz, 2008; Solbak et al., 2013). The specific tetherin–Vpu interaction has haphazardly shed light on a possible mechanism relating to the difference between egress of HIV in macrophages versus CD4$^+$ T cells. In this regard, Schindler et al. (2010) demonstrated that residue 52 in Vpu is important for antagonizing tetherin and therefore replication in macrophages but not in T cells. In addition, Giese and Marsh (2014) purposed tetherin as a potent restriction factor relative to cell–cell transmission from primary macrophages to CD4$^+$ T cells. However, there is some disagreement as to whether tetherin is necessary for HIV-intracellular vesicle formation in macrophages (Chu et al., 2012; Giese & Marsh, 2014). Furthermore, it is plausible that M-tropic viruses have specific Vpu sequences that could interact with ESCRT proteins in order to shuttle them to endocytic vesicles, possible in combination with p6, to enhance assembly and exocytosis in macrophages rather than assembly at the plasma membrane in CD4$^+$ T cells. Understanding the basics of cell type-specific difference between entry and egress in macrophages and T cells in the context of HIV infection is imperative with respect to discovering and druggable targets that could be cell type-specific and possibly more efficacious with a reduced rate with respect to developing viral resistance.

In conclusion, elucidation of the host–viral protein interactions could aid in the development of new therapies to augment current HAART therapy and increase the number of druggable targets, additionally augmenting the antiretroviral state in chronically infected HIV individuals by enhancing or boosting the activity and/or concentration of host restriction factors that could synergistically ablate replication and infection of naïve immune cells. All in all, it is apparent that numerous factors influence the pathogenesis of HIV, and further understanding of host regulatory mechanisms in combination with HAART, by targeting the interface between host–viral protein interactions and further elucidating the mechanism of entry and exit, could drastically reduce the rate of resistance and the ever-growing need for novel antiretroviral therapeutic agents.

REFERENCES

Abbate, I., Cappiello, G., Longo, R., Ursitti, A., Spano, A., Calcaterra, S., et al. (2005). Cell membrane proteins and quasispecies compartmentalization of CSF and plasma HIV-1 from aids patients with neurological disorders. *Infection, Genetics and Evolution, 5*(3),

247–253. http://dx.doi.org/10.1016/j.meegid.2004.08.006. PubMed PMID: 15737916, Epub 2005/03/02; S1567-1348(04)00120-0 [pii].

Aiamkitsumrit, B., Dampier, W., Antell, G., Rivera, N., Martin-Garcia, J., Pirrone, V., et al. (2014). Bioinformatic analysis of HIV-1 entry and pathogenesis. *Current HIV Research*, *12*(2), 132–161. PubMed PMID: 24862329.

Alkhatib, G., Combadiere, C., Broder, C. C., Feng, Y., Kennedy, P. E., Murphy, P. M., et al. (1996). CC CKR5: A RANTES, MIP-1alpha, MIP-1beta receptor as a fusion cofactor for macrophage-tropic HIV-1. *Science*, *272*(5270), 1955–1958. PubMed PMID: 8658171.

Aloia, R. C., Tian, H., & Jensen, F. C. (1993). Lipid composition and fluidity of the human immunodeficiency virus envelope and host cell plasma membranes. *Proceedings of the National Academy of Sciences of the United States of America*, *90*(11), 5181–5185. PubMed PMID: 8389472.

Anderson, E., Zink, W., Xiong, H., & Gendelman, H. E. (2002). HIV-1-associated dementia: A metabolic encephalopathy perpetrated by virus-infected and immune-competent mononuclear phagocytes. *Journal of Acquired Immune Deficiency Syndromes*, *31*(Suppl. 2), S43–S54. PubMed PMID: 12394782.

Arrildt, K. T., Joseph, S. B., & Swanstrom, R. (2012). The HIV-1 env protein: A coat of many colors. *Current HIV/AIDS Reports*, *9*(1), 52–63. http://dx.doi.org/10.1007/s11904-011-0107-3. PubMed PMID: 22237899. PubMed Central PMCID: PMC3658113.

Babst, M., Katzmann, D. J., Estepa-Sabal, E. J., Meerloo, T., & Emr, S. D. (2002). Escrt-III: An endosome-associated heterooligomeric protein complex required for mvb sorting. *Developmental Cell*, *3*(2), 271–282. PubMed PMID: 12194857, Epub 2002/08/27; S1534580702002204 [pii].

Bello, N. F., Dussupt, V., Sette, P., Rudd, V., Nagashima, K., Bibollet-Ruche, F., et al. (2012). Budding of retroviruses utilizing divergent L domains requires nucleocapsid. *Journal of Virology*, *86*(8), 4182–4193. http://dx.doi.org/10.1128/JVI.07105-11. PubMed PMID: 22345468, Epub 2012/02/22; JVI.07105-11 [pii].

Benaroch, P., Billard, E., Gaudin, R., Schindler, M., & Jouve, M. (2010). HIV-1 assembly in macrophages. *Retrovirology*, *7*, 29. http://dx.doi.org/10.1186/1742-4690-7-29. PubMed PMID: 20374631. PubMed Central PMCID: PMC2861634, Epub 2010/04/09; 1742-4690-7-29 [pii].

Bennett, A. E., Narayan, K., Shi, D., Hartnell, L. M., Gousset, K., He, H., et al. (2009). Ion-abrasion scanning electron microscopy reveals surface-connected tubular conduits in HIV-infected macrophages. *PLoS Pathogens*, *5*(9), e1000591. http://dx.doi.org/10.1371/journal.ppat.1000591. PubMed PMID: 19779568. PubMed Central PMCID: PMC2743285, Epub 2009/09/26.

Berberian, L., Shukla, J., Jefferis, R., & Braun, J. (1994). Effects of HIV infection on VH3 (D12 idiotope) B cells in vivo. *Journal of Acquired Immune Deficiency Syndromes*, *7*(7), 641–646. PubMed PMID: 8207642, Epub 1994/07/01.

Berger, E. A., Murphy, P. M., & Farber, J. M. (1999). Chemokine receptors as HIV-1 coreceptors: Roles in viral entry, tropism, and disease. *Annual Review of Immunology*, *17*, 657–700. http://dx.doi.org/10.1146/annurev.immunol.17.1.657. PubMed PMID: 10358771, Epub 1999/06/08.

Bjorndal, A., Deng, H., Jansson, M., Fiore, J. R., Colognesi, C., Karlsson, A., et al. (1997). Coreceptor usage of primary human immunodeficiency virus type 1 isolates varies according to biological phenotype. *Journal of Virology*, *71*(10), 7478–7487. PubMed PMID: 9311827. PubMed Central PMCID: PMC192094.

Blaak, H., van't Wout, A. B., Brouwer, M., Hooibrink, B., Hovenkamp, E., & Schuitemaker, H. (2000). In vivo HIV-1 infection of CD45RA(+)CD4(+) T cells is established primarily by syncytium-inducing variants and correlates with the rate of CD4(+) T cell decline. *Proceedings of the National Academy of Sciences of the United States of America, 97*(3), 1269–1274. PubMed PMID: 10655520. PubMed Central PMCID: PMC15592.

Blackard, J. T., Renjifo, B., Chaplin, B., Msamanga, G., Fawzi, W., & Essex, M. (2000). Diversity of the HIV-1 long terminal repeat following mother-to-child transmission. *Virology, 27*(2), 402–411. http://dx.doi.org/10.1006/viro.2000.0466. PubMed PMID: 10964782, S0042-6822(00)90466-5 [pii]; Epub 2000/08/31.

Bleul, C. C., Farzan, M., Choe, H., Parolin, C., Clark-Lewis, I., Sodroski, J., et al. (1996). The lymphocyte chemoattractant SDF-1 is a ligand for LESTR/fusin and blocks HIV-1 entry. *Nature, 382*(6594), 829–833. http://dx.doi.org/10.1038/382829a0. PubMed PMID: 8752280, Epub 1996/08/29.

Bleul, C. C., Fuhlbrigge, R. C., Casasnovas, J. M., Aiuti, A., & Springer, T. A. (1996). A highly efficacious lymphocyte chemoattractant, stromal cell-derived factor 1 (SDF-1). *The Journal of Experimental Medicine, 184*(3), 1101–1109. PubMed PMID: 9064327. PubMed Central PMCID: PMC2192798, Epub 1996/09/01.

Bontjer, I., Land, A., Eggink, D., Verkade, E., Tuin, K., Baldwin, C., et al. (2009). Optimization of human immunodeficiency virus type 1 envelope glycoproteins with V1/V2 deleted, using virus evolution. *Journal of Virology, 83*(1), 368–383. http://dx.doi.org/10.1128/JVI.01404-08. PubMed PMID: 18922866. PubMed Central PMCID: PMC2612307, Epub 2008/10/17; JVI.01404-08 [pii].

Borsetti, A., Parolin, C., Ridolfi, B., Sernicola, L., Geraci, A., Ensoli, B., et al. (2000). CD4-independent infection of two CD4(-)/CCR5(-)/CXCR4(+) pre-T-cell lines by human and simian immunodeficiency viruses. *Journal of Virology, 74*(14), 6689–6694. PubMed PMID: 10864687. PubMed Central PMCID: PMC112183.

Bour, S., Geleziunas, R., & Wainberg, M. A. (1995). The human immunodeficiency virus type 1 (HIV-1) CD4 receptor and its central role in promotion of HIV-1 infection. *Microbiological Reviews, 59*(1), 63–93. PubMed PMID: 7708013. PubMed Central PMCID: PMC239355.

Brelot, A., Heveker, N., Adema, K., Hosie, M. J., Willett, B., & Alizon, M. (1999). Effect of mutations in the second extracellular loop of CXCR4 on its utilization by human and feline immunodeficiency viruses. *Journal of Virology, 73*(4), 2576–2586. PubMed PMID: 10074102. PubMed Central PMCID: PMC104012, Epub 1999/03/12.

Briggs, D. R., Tuttle, D. L., Sleasman, J. W., & Goodenow, M. M. (2000). Envelope V3 amino acid sequence predicts HIV-1 phenotype (co-receptor usage and tropism for macrophages). *AIDS, 14*(18), 2937–2939. PubMed PMID: 11153675, Epub 2001/01/12.

Broder, C. C., & Dimitrov, D. S. (1996). HIV and the 7-transmembrane domain receptors. *Pathobiology: Journal of Immunopathology, Molecular and Cellular Biology, 64*(4), 171–179. PubMed PMID: 9031325.

Bryant, M., & Ratner, L. (1990). Myristoylation-dependent replication and assembly of human immunodeficiency virus 1. *Proceedings of the National Academy of Sciences of the United States of America, 87*(2), 523–527. PubMed PMID: 2405382. PubMed Central PMCID: PMC53297, Epub 1990/01/01.

Cameron, P. U., Saleh, S., Sallmann, G., Solomon, A., Wightman, F., Evans, V. A., et al. (2010). Establishment of HIV-1 latency in resting CD4 + T cells depends on chemokine-induced changes in the actin cytoskeleton. *Proceedings of the National Academy of Sciences of the United States of America, 107*(39), 16934–16939. http://dx.doi.org/10.1073/pnas.1002894107. PubMed PMID: 20837531. PubMed Central PMCID: PMC2947912.

Campbell, S. M., Crowe, S. M., & Mak, J. (2001). Lipid rafts and HIV-1: From viral entry to assembly of progeny virions. *Journal of Clinical Virology, 22*(3), 217–227. PubMed PMID: 11564586, Epub 2001/09/21; S1386653201001937 [pii].

Cao, J., Sullivan, N., Desjardin, E., Parolin, C., Robinson, J., Wyatt, R., et al. (1997). Replication and neutralization of human immunodeficiency virus type 1 lacking the V1 and

V2 variable loops of the gp120 envelope glycoprotein. *Journal of Virology*, *71*(12), 9808–9812. PubMed PMID: 9371651. PubMed Central PMCID: PMC230295.

Capobianchi, M. R., Fais, S., Castilletti, C., Gentile, M., Ameglio, F., & Dianzani, F. (1994). A simple and reliable method to detect cell membrane proteins on infectious human immunodeficiency virus type 1 particles. *The Journal of Infectious Diseases*, *169*(4), 886–889. PubMed PMID: 7907644, Epub 1994/04/01.

Chamanian, M., Purzycka, K. J., Wille, P. T., Ha, J. S., McDonald, D., Gao, Y., et al. (2013). A cis-acting element in retroviral genomic RNA links Gag-Pol ribosomal frameshifting to selective viral RNA encapsidation. *Cell Host & Microbe*, *13*(2), 181–192. http://dx.doi.org/10.1016/j.chom.2013.01.007. PubMed PMID: 23414758. PubMed Central PMCID: PMC3587049.

Chamberland, A., Sylla, M., Boulassel, M. R., Baril, J. G., Cote, P., Thomas, R., et al. (2011). Effect of antiretroviral therapy on HIV-1 genetic evolution during acute infection. *International Journal of STD & AIDS*, *22*(3), 146–150. http://dx.doi.org/10.1258/ijsa.2010.010292. PubMed PMID: 21464451.

Chege, D., Sheth, P. M., Kain, T., Kim, C. J., Kovacs, C., Loutfy, M., et al. (2011). Sigmoid Th17 populations, the HIV latent reservoir, and microbial translocation in men on long-term antiretroviral therapy. *AIDS*, *25*(6), 741–749. http://dx.doi.org/10.1097/QAD.0b013e328344cefb. PubMed PMID: 21378536.

Cheng-Mayer, C., Brown, A., Harouse, J., Luciw, P. A., & Mayer, A. J. (1999). Selection for neutralization resistance of the simian/human immunodeficiency virus SHIVSF33A variant in vivo by virtue of sequence changes in the extracellular envelope glycoprotein that modify N-linked glycosylation. *Journal of Virology*, *73*(7), 5294–5300. PubMed PMID: 10364275. PubMed Central PMCID: PMC112584.

Choe, H., Farzan, M., Konkel, M., Martin, K., Sun, Y., Marcon, L., et al. (1998). The orphan seven-transmembrane receptor apj supports the entry of primary T-cell-line-tropic and dualtropic human immunodeficiency virus type 1. *Journal of Virology*, *72*(7), 6113–6118. PubMed PMID: 9621075, Epub 1998/06/17.

Choe, H., Farzan, M., Sun, Y., Sullivan, N., Rollins, B., Ponath, P. D., et al. (1996). The beta-chemokine receptors CCR3 and CCR5 facilitate infection by primary HIV-1 isolates. *Cell*, *85*(7), 1135–1148. PubMed PMID: 8674119, Epub 1996/06/28; S0092-8674(00)81313-6 [pii].

Chu, H., Wang, J. J., Qi, M., Yoon, J. J., Chen, X., Wen, X., et al. (2012). Tetherin/BST-2 is essential for the formation of the intracellular virus-containing compartment in HIV-infected macrophages. *Cell Host & Microbe*, *12*(3), 360–372. http://dx.doi.org/10.1016/j.chom.2012.07.011. PubMed PMID: 22980332. PubMed Central PMCID: PMC3444820.

Clapham, P. R., & McKnight, A. (2001). HIV-1 receptors and cell tropism. *British Medical Bulletin*, *58*, 43–59. PubMed PMID: 11714623.

Clements, J. E., & Zink, M. C. (1996). Molecular biology and pathogenesis of animal lentivirus infections. *Clinical Microbiology Reviews*, *9*(1), 100–117. PubMed PMID: 8665473. PubMed Central PMCID: PMC172884.

Cocchi, F., DeVico, A. L., Garzino-Demo, A., Cara, A., Gallo, R. C., & Lusso, P. (1996). The V3 domain of the HIV-1 gp120 envelope glycoprotein is critical for chemokine-mediated blockade of infection. *Nature Medicine*, *2*(11), 1244–1247. PubMed PMID: 8898753.

Collins-Fairclough, A. M., Charurat, M., Nadai, Y., Pando, M., Avila, M. M., Blattner, W. A., et al. (2011). Significantly longer envelope V2 loops are characteristic of heterosexually transmitted subtype B HIV-1 in Trinidad. *PLoS One*, *6*(6), e19995. http://dx.doi.org/10.1371/journal.pone.0019995. PubMed PMID: 21698149. PubMed Central PMCID: PMC3117786.

Collman, R., Balliet, J. W., Gregory, S. A., Friedman, H., Kolson, D. L., Nathanson, N., et al. (1992). An infectious molecular clone of an unusual macrophage-tropic and highly cytopathic strain of human immunodeficiency virus type 1. *Journal of Virology*, *66*(12), 7517–7521. PubMed PMID: 1433527. PubMed Central PMCID: PMC240461.

Cormier, E. G., & Dragic, T. (2002). The crown and stem of the V3 loop play distinct roles in human immunodeficiency virus type 1 envelope glycoprotein interactions with the CCR5 coreceptor. *Journal of Virology*, *76*(17), 8953–8957. PubMed PMID: 12163614, Epub 2002/08/07.

Crowe, S., Zhu, T., & Muller, W. A. (2003). The contribution of monocyte infection and trafficking to viral persistence, and maintenance of the viral reservoir in HIV infection. *Journal of Leukocyte Biology*, *74*(5), 635–641. http://dx.doi.org/10.1189/jlb.0503204. PubMed PMID: 12960232.

Da, L. T., Quan, J. M., & Wu, Y. D. (2009). Understanding of the bridging sheet formation of HIV-1 glycoprotein gp120. *The Journal of Physical Chemistry B*, *113*(43), 14536–14543. http://dx.doi.org/10.1021/jp9081239. PubMed PMID: 19813706.

Dalgleish, A. G., Beverley, P. C., Clapham, P. R., Crawford, D. H., Greaves, M. F., & Weiss, R. A. (1984). The CD4 (T4) antigen is an essential component of the receptor for the AIDS retrovirus. *Nature*, *312*(5996), 763–767. PubMed PMID: 6096719, Epub 1984/12/20.

Davenport, M. P., Zaunders, J. J., Hazenberg, M. D., Schuitemaker, H., & van Rij, R. P. (2002). Cell turnover and cell tropism in HIV-1 infection. *Trends in Microbiology*, *10*(6), 275–278. PubMed PMID: 12088663.

De Jong, J. J., De Ronde, A., Keulen, W., Tersmette, M., & Goudsmit, J. (1992). Minimal requirements for the human immunodeficiency virus type 1 V3 domain to support the syncytium-inducing phenotype: Analysis by single amino acid substitution. *Journal of Virology*, *66*(11), 6777–6780. PubMed PMID: 1404617. PubMed Central PMCID: PMC240176, Epub 1992/11/01.

Dejucq, N., Simmons, G., & Clapham, P. R. (1999). Expanded tropism of primary human immunodeficiency virus type 1 R5 strains to CD4(+) T-cell lines determined by the capacity to exploit low concentrations of CCR5. *Journal of Virology*, *73*(9), 7842–7847. PubMed PMID: 10438877. PubMed Central PMCID: PMC104314.

Demirov, D. G., & Freed, E. O. (2004). Retrovirus budding. *Virus Research*, *106*(2), 87–102. http://dx.doi.org/10.1016/j.virusres.2004.08.007. PubMed PMID: 15567490, Epub 2004/11/30; S0168-1702(04)00318-1 [pii].

Demirov, D. G., Orenstein, J. M., & Freed, E. O. (2002). The late domain of human immunodeficiency virus type 1 p6 promotes virus release in a cell type-dependent manner. *Journal of Virology*, *76*(1), 105–117. PubMed PMID: 11739676. PubMed Central PMCID: PMC135729, Epub 2001/12/12.

Deneka, M., Pelchen-Matthews, A., Byland, R., Ruiz-Mateos, E., & Marsh, M. (2007). In macrophages, HIV-1 assembles into an intracellular plasma membrane domain containing the tetraspanins CD81, CD9, and CD53. *The Journal of Cell Biology*, *177*(2), 329–341. http://dx.doi.org/10.1083/jcb.200609050. PubMed PMID: 17438075. PubMed Central PMCID: PMC2064140, Epub 2007/04/18; jcb.200609050 [pii].

Deng, H., Liu, R., Ellmeier, W., Choe, S., Unutmaz, D., Burkhart, M., et al. (1996). Identification of a major co-receptor for primary isolates of HIV-1. *Nature*, *381*(6584), 661–666. http://dx.doi.org/10.1038/381661a0. PubMed PMID: 8649511, Epub 1996/06/20.

Dimitrov, D. S., Xiao, X., Chabot, D. J., & Broder, C. C. (1998). HIV coreceptors. *The Journal of Membrane Biology*, *166*(2), 75–90. PubMed PMID: 9841733, Epub 1998/12/05.

Dobrowsky, T. M., Zhou, Y., Sun, S. X., Siliciano, R. F., & Wirtz, D. (2008). Monitoring early fusion dynamics of human immunodeficiency virus type 1 at single-molecule resolution. *Journal of Virology*, *82*(14), 7022–7033. http://dx.doi.org/10.1128/JVI.00053-08.

PubMed PMID: 18480458. PubMed Central PMCID: PMC2446967, Epub 2008/05/16; JVI.00053-08 [pii].

Doherty, M., Ford, N., Vitoria, M., Weiler, G., & Hirnschall, G. (2013). The 2013 WHO guidelines for antiretroviral therapy: Evidence-based recommendations to face new epidemic realities. *Current Opinion in HIV and AIDS*, *8*(6), 528–534. http://dx.doi.org/10.1097/COH.0000000000000008. PubMed PMID: 24100873.

Doitsh, G., Galloway, N. L., Geng, X., Yang, Z., Monroe, K. M., Zepeda, O., et al. (2014). Cell death by pyroptosis drives CD4 T-cell depletion in HIV-1 infection. *Nature*, *505*(7484), 509–514. http://dx.doi.org/10.1038/nature12940. PubMed PMID: 24356306.

Dong, X., Li, H., Derdowski, A., Ding, L., Burnett, A., Chen, X., et al. (2005). AP-3 directs the intracellular trafficking of HIV-1 Gag and plays a key role in particle assembly. *Cell*, *120*(5), 663–674. http://dx.doi.org/10.1016/j.cell.2004.12.023. PubMed PMID: 15766529, Epub 2005/03/16; S0092-8674(04)01248-6 [pii].

Doranz, B. J., Rucker, J., Yi, Y., Smyth, R. J., Samson, M., Peiper, S. C., et al. (1996). A dual-tropic primary HIV-1 isolate that uses fusin and the beta-chemokine receptors CKR-5, CKR-3, and CKR-2b as fusion cofactors. *Cell*, *85*(7), 1149–1158. PubMed PMID: 8674120, Epub 1996/06/28; S0092-8674(00)81314-8 [pii].

Doria-Rose, N. A., Georgiev, I., O'Dell, S., Chuang, G. Y., Staupe, R. P., McLellan, J. S., et al. (2012). A short segment of the HIV-1 gp120 V1/V2 region is a major determinant of resistance to V1/V2 neutralizing antibodies. *Journal of Virology*, *86*(15), 8319–8323. http://dx.doi.org/10.1128/JVI.00696-12. PubMed PMID: 22623764. PubMed Central PMCID: PMC3421697.

Douglas, N. W., Munro, G. H., & Daniels, R. S. (1997). HIV/SIV glycoproteins: Structure-function relationships. *Journal of Molecular Biology*, *273*(1), 122–149. http://dx.doi.org/10.1006/jmbi.1997.1277. PubMed PMID: 9367752, Epub 1997/11/21; S0022-2836(97)91277-8 [pii].

Dragic, T., Litwin, V., Allaway, G. P., Martin, S. R., Huang, Y., Nagashima, K. A., et al. (1996). HIV-1 entry into CD4+ cells is mediated by the chemokine receptor CC-CKR-5. *Nature*, *381*(6584), 667–673. PubMed PMID: 8649512.

Dussupt, V., Javid, M. P., Abou-Jaoude, G., Jadwin, J. A., de La Cruz, J., Nagashima, K., et al. (2009). The nucleocapsid region of HIV-1 Gag cooperates with the PTAP and LYPXnL late domains to recruit the cellular machinery necessary for viral budding. *PLoS Pathogens*, *5*(3), e1000339. http://dx.doi.org/10.1371/journal.ppat.1000339. PubMed PMID: 19282983. PubMed Central PMCID: PMC2651531.

Dussupt, V., Sette, P., Bello, N. F., Javid, M. P., Nagashima, K., & Bouamr, F. (2011). Basic residues in the nucleocapsid domain of Gag are critical for late events of HIV-1 budding. *Journal of Virology*, *85*(5), 2304–2315. http://dx.doi.org/10.1128/JVI.01562-10. PubMed PMID: 21159863. PubMed Central PMCID: PMC3067763.

Edinger, A. L., Hoffman, T. L., Sharron, M., Lee, B., Yi, Y., Choe, W., et al. (1998). An orphan seven-transmembrane domain receptor expressed widely in the brain functions as a coreceptor for human immunodeficiency virus type 1 and simian immunodeficiency virus. *Journal of Virology*, *72*(10), 7934–7940. PubMed PMID: 9733831. PubMed Central PMCID: PMC110125.

Edwards, T. G., Hoffman, T. L., Baribaud, F., Wyss, S., LaBranche, C. C., Romano, J., et al. (2001). Relationships between CD4 independence, neutralization sensitivity, and exposure of a CD4-induced epitope in a human immunodeficiency virus type 1 envelope protein. *Journal of Virology*, *75*(11), 5230–5239. http://dx.doi.org/10.1128/JVI.75.11.5230-5239.2001. PubMed PMID: 11333905. PubMed Central PMCID: PMC114929, Epub 2001/05/03.

Eisele, E., & Siliciano, R. F. (2012). Redefining the viral reservoirs that prevent HIV-1 eradication. *Immunity*, *37*(3), 377–388. http://dx.doi.org/10.1016/j.immuni.2012.08.010. PubMed PMID: 22999944. PubMed Central PMCID: PMC3963158.

Ellery, P. J., Tippett, E., Chiu, Y. L., Paukovics, G., Cameron, P. U., Solomon, A., et al. (2007). The CD16+ monocyte subset is more permissive to infection and preferentially harbors HIV-1 in vivo. *Journal of Immunology*, *178*(10), 6581–6589. PubMed PMID: 17475889.

Engelman, A., & Cherepanov, P. (2012). The structural biology of HIV-1: mechanistic and therapeutic insights. *Nature Reviews Microbiology*, *16;10*(4), 279–290. PubMed PMID: 22421880. PubMed Central PMCID: PMC 3588166.

Falcigno, L., Oliva, R., D'Auria, G., Maletta, M., Dettin, M., Pasquato, A., et al. (2004). Structural investigation of the HIV-1 envelope glycoprotein gp160 cleavage site 3: Role of site-specific mutations. *Chembiochem: A European Journal of Chemical Biology*, *5*(12), 1653–1661. http://dx.doi.org/10.1002/cbic.200400181. PubMed PMID: 15526330.

Fantini, J., Garmy, N., Mahfoud, R., & Yahi, N. (2002). Lipid rafts: Structure, function and role in HIV, Alzheimer's and prion diseases. *Expert Reviews in Molecular Medicine*, *4*(27), 1–22. http://dx.doi.org/10.1017/S1462399402005392. PubMed PMID: 14987385, Epub 2004/02/28; S1462399402005392 [pii].

Feng, Y., Broder, C. C., Kennedy, P. E., & Berger, E. A. (1996). HIV-1 entry cofactor: Functional cDNA cloning of a seven-transmembrane, G protein-coupled receptor. *Science*, *272*(5263), 872–877. PubMed PMID: 8629022, Epub 1996/05/10.

Finzi, A., Brunet, A., Xiao, Y., Thibodeau, J., & Cohen, E. A. (2006). Major histocompatibility complex class II molecules promote human immunodeficiency virus type 1 assembly and budding to late endosomal/multivesicular body compartments. *Journal of Virology*, *80*(19), 9789–9797. http://dx.doi.org/10.1128/JVI.01055-06. PubMed PMID: 16973583. PubMed Central PMCID: PMC1617259, Epub 2006/09/16; 80/19/9789 [pii].

Fischer-Smith, T., Croul, S., Sverstiuk, A. E., Capini, C., L'Heureux, D., Regulier, E. G., et al. (2001). CNS invasion by CD14+/CD16+ peripheral blood-derived monocytes in HIV dementia: Perivascular accumulation and reservoir of HIV infection. *Journal of Neurovirology*, *7*(6), 528–541. http://dx.doi.org/10.1080/135502801753248114. PubMed PMID: 11704885.

Fouchier, R. A., Groenink, M., Kootstra, N. A., Tersmette, M., Huisman, H. G., Miedema, F., et al. (1992). Phenotype-associated sequence variation in the third variable domain of the human immunodeficiency virus type 1 gp120 molecule. *Journal of Virology*, *66*(5), 3183–3187. PubMed PMID: 1560543. PubMed Central PMCID: PMC241084, Epub 1992/05/01.

Freed, E. O. (1998). HIV-1 gag proteins: Diverse functions in the virus life cycle. *Virology*, *251*(1), 1–15. http://dx.doi.org/10.1006/viro.1998.9398. PubMed PMID: 9813197, Epub 1998/11/14; S0042-6822(98)99398-9 [pii].

Freed, E. O., Orenstein, J. M., Buckler-White, A. J., & Martin, M. A. (1994). Single amino acid changes in the human immunodeficiency virus type 1 matrix protein block virus particle production. *Journal of Virology*, *68*(8), 5311–5320. PubMed PMID: 8035531. PubMed Central PMCID: PMC236481, Epub 1994/08/01.

Gallo, R. C., & Montagnier, L. (2003). The discovery of HIV as the cause of AIDS. *The New England Journal of Medicine*, *349*(24), 2283–2285. http://dx.doi.org/10.1056/NEJMp038194. PubMed PMID: 14668451.

Garcia-Blanco, M. A., & Cullen, B. R. (1991). Molecular basis of latency in pathogenic human viruses. *Science*, *254*(5033), 815–820. PubMed PMID: 1658933, Epub 1991/11/08.

Gartner, S., Markovits, P., Markovitz, D. M., Betts, R. F., & Popovic, M. (1986). Virus isolation from and identification of HTLV-III/LAV-producing cells in brain tissue from a patient with AIDS. *JAMA*, *256*(17), 2365–2371. PubMed PMID: 3490587.

Gelderblom, H. R. (1991). Assembly and morphology of HIV: Potential effect of structure on viral function. *AIDS*, *5*(6), 617–637. PubMed PMID: 1652977, Epub 1991/06/01.

Gendelman, H. E., Orenstein, J. M., Martin, M. A., Ferrua, C., Mitra, R., Phipps, T., et al. (1988). Efficient isolation and propagation of human immunodeficiency virus on recombinant colony-stimulating factor 1-treated monocytes. *The Journal of Experimental Medicine, 167*(4), 1428–1441. PubMed PMID: 3258626. PubMed Central PMCID: PMC2188914.

Giese, S., & Marsh, M. (2014). Tetherin can restrict cell-free and cell-cell transmission of HIV from primary macrophages to T cells. *PLoS Pathogens, 10*(7), e1004189. http://dx.doi.org/10.1371/journal.ppat.1004189. PubMed PMID: 24991932. PubMed Central PMCID: PMC4081785.

Glushakova, S., Yi, Y., Grivel, J. C., Singh, A., Schols, D., De Clercq, E., et al. (1999). Preferential coreceptor utilization and cytopathicity by dual-tropic HIV-1 in human lymphoid tissue ex vivo. *The Journal of Clinical Investigation, 104*(5), R7–R11. http://dx.doi.org/10.1172/JCI7403. PubMed PMID: 10487781. PubMed Central PMCID: PMC408546.

Gonzalez-Perez, M. P., O'Connell, O., Lin, R., Sullivan, W. M., Bell, J., Simmonds, P., et al. (2012). Independent evolution of macrophage-tropism and increased charge between HIV-1 R5 envelopes present in brain and immune tissue. *Retrovirology, 9*, 20. http://dx.doi.org/10.1186/1742-4690-9-20. PubMed PMID: 22420378. PubMed Central PMCID: PMC3362761.

Gonzalez-Scarano, F., & Martin-Garcia, J. (2005). The neuropathogenesis of AIDS. *Nature Reviews. Immunology, 5*(1), 69–81. http://dx.doi.org/10.1038/nri1527. PubMed PMID: 15630430, Epub 2005/01/05; nri1527 [pii].

Gordon, S., & Taylor, P. R. (2005). Monocyte and macrophage heterogeneity. *Nature Reviews. Immunology, 5*(12), 953–964. http://dx.doi.org/10.1038/nri1733. PubMed PMID: 16322748.

Gorry, P. R., Bristol, G., Zack, J. A., Ritola, K., Swanstrom, R., Birch, C. J., et al. (2001). Macrophage tropism of human immunodeficiency virus type 1 isolates from brain and lymphoid tissues predicts neurotropism independent of coreceptor specificity. *Journal of Virology, 75*(21), 10073–10089. http://dx.doi.org/10.1128/JVI.75.21.10073-10089.2001. PubMed PMID: 11581376. PubMed Central PMCID: PMC114582.

Gorry, P. R., Sterjovski, J., Churchill, M., Witlox, K., Gray, L., Cunningham, A., et al. (2004). The role of viral coreceptors and enhanced macrophage tropism in human immunodeficiency virus type 1 disease progression. *Sexual Health, 1*(1), 23–34. PubMed PMID: 16335478.

Gottlinger, H. G., Dorfman, T., Sodroski, J. G., & Haseltine, W. A. (1991). Effect of mutations affecting the p6 gag protein on human immunodeficiency virus particle release. *Proceedings of the National Academy of Sciences of the United States of America, 88*(8), 3195–3199. PubMed PMID: 2014240. PubMed Central PMCID: PMC51412.

Goudsmit, J., Debouck, C., Meloen, R. H., Smit, L., Bakker, M., Asher, D. M., et al. (1988). Human immunodeficiency virus type 1 neutralization epitope with conserved architecture elicits early type-specific antibodies in experimentally infected chimpanzees. *Proceedings of the National Academy of Sciences of the United States of America, 85*(12), 4478–4482. PubMed PMID: 2454471. PubMed Central PMCID: PMC280453, Epub 1988/06/01.

Gougeon, M. L., Lecoeur, H., Dulioust, A., Enouf, M. G., Crouvoiser, M., Goujard, C., et al. (1996). Programmed cell death in peripheral lymphocytes from HIV-infected persons: Increased susceptibility to apoptosis of CD4 and CD8 T cells correlates with lymphocyte activation and with disease progression. *Journal of Immunology, 156*(9), 3509–3520. PubMed PMID: 8617980.

Greenberg, M. L., & Cammack, N. (2004). Resistance to enfuvirtide, the first HIV fusion inhibitor. *The Journal of Antimicrobial Chemotherapy, 54*(2), 333–340. http://dx.doi.org/10.1093/jac/dkh330. PubMed PMID: 15231762.

Grigorov, B., Attuil-Audenis, V., Perugi, F., Nedelec, M., Watson, S., Pique, C., et al. (2009). A role for CD81 on the late steps of HIV-1 replication in a chronically infected T cell line. *Retrovirology*, *6*, 28. http://dx.doi.org/10.1186/1742-4690-6-28. PubMed PMID: 19284574. PubMed Central PMCID: PMC2657109, Epub 2009/03/17; 1742-4690-6-28 [pii].

Grivel, J. C., Malkevitch, N., & Margolis, L. (2000). Human immunodeficiency virus type 1 induces apoptosis in CD4(+) but not in CD8(+) T cells in ex vivo-infected human lymphoid tissue. *Journal of Virology*, *74*(17), 8077–8084. PubMed PMID: 10933717. PubMed Central PMCID: PMC112340.

Gulick, R. M., Mellors, J. W., Havlir, D., Eron, J. J., Gonzalez, C., McMahon, D., et al. (1997). Treatment with indinavir, zidovudine, and lamivudine in adults with human immunodeficiency virus infection and prior antiretroviral therapy. *The New England Journal of Medicine*, *337*(11), 734–739. http://dx.doi.org/10.1056/NEJM199709113371102. PubMed PMID: 9287228.

Haim, H., Salas, I., & Sodroski, J. (2013). Proteolytic processing of the human immunodeficiency virus envelope glycoprotein precursor decreases conformational flexibility. *Journal of Virology*, *87*(3), 1884–1889. http://dx.doi.org/10.1128/JVI.02765-12. PubMed PMID: 23175369. PubMed Central PMCID: PMC3554131.

Hammerle, T., Himmelspach, M., Dorner, F., & Falkner, F. G. (1997). A sensitive PCR assay system for the quantitation of viral genome equivalents: Human immunodeficiency virus type 1 (HIV-1) and hepatitis B virus (HBV). *Archives of Virology*, *142*(7), 1297–1306. PubMed PMID: 9267444.

He, J., Chen, Y., Farzan, M., Choe, H., Ohagen, A., Gartner, S., et al. (1997). CCR3 and CCR5 are co-receptors for HIV-1 infection of microglia. *Nature*, *385*(6617), 645–649. http://dx.doi.org/10.1038/385645a0. PubMed PMID: 9024664, Epub 1997/02/13.

Hemler, M. E. (2005). Tetraspanin functions and associated microdomains. *Nature Reviews. Molecular Cell Biology*, *6*(10), 801–811. http://dx.doi.org/10.1038/nrm1736. PubMed PMID: 16314869, Epub 2005/11/30; nrm1736 [pii].

Hirnschall, G., Harries, A. D., Easterbrook, P. J., Doherty, M. C., & Ball, A. (2013). The next generation of the World Health Organization's global antiretroviral guidance. *Journal of the International AIDS Society*, *16*, 18757. http://dx.doi.org/10.7448/IAS.16.1.18757. PubMed PMID: 23819908. PubMed Central PMCID: PMC3699697.

Holman, A. G., Mefford, M. E., O'Connor, N., & Gabuzda, D. (2010). HIVBrainSeqDB: A database of annotated HIV envelope sequences from brain and other anatomical sites. *AIDS Research and Therapy*, *7*, 43. http://dx.doi.org/10.1186/1742-6405-7-43. PubMed PMID: 21156070. PubMed Central PMCID: PMC3018377.

Horvath, G., Serru, V., Clay, D., Billard, M., Boucheix, C., & Rubinstein, E. (1998). CD19 is linked to the integrin-associated tetraspans CD9, CD81, and CD82. *The Journal of Biological Chemistry*, *273*(46), 30537–30543. PubMed PMID: 9804823, Epub 1998/11/07.

Huang, C. C., Tang, M., Zhang, M. Y., Majeed, S., Montabana, E., Stanfield, R. L., et al. (2005). Structure of a V3-containing HIV-1 gp120 core. *Science*, *310*(5750), 1025–1028. http://dx.doi.org/10.1126/science.1118398. PubMed PMID: 16284180. PubMed Central PMCID: PMC2408531, Epub 2005/11/15; 310/5750/1025 [pii].

Hurley, J. H. (2008). ESCRT complexes and the biogenesis of multivesicular bodies. *Current Opinion in Cell Biology*, *20*(1), 4–11. http://dx.doi.org/10.1016/j.ceb.2007.12.002. PubMed PMID: 18222686. PubMed Central PMCID: PMC2282067, Epub 2008/01/29; S0955-0674(07)00191-3 [pii].

Jang, D. H., Yoon, C. H., Choi, B. S., Chung, Y. S., Kim, H. Y., Chi, S. G., et al. (2014). Characterization of Gp41 polymorphisms in the fusion peptide domain and T-20 (Enfuvirtide) resistance-associated regions in Korean HIV-1 isolates. *Journal of Korean Medical Science*, *29*(3), 456–459. http://dx.doi.org/10.3346/jkms.2014.29.3.456. PubMed PMID: 24616600. PubMed Central PMCID: PMC3945146.

Javaherian, K., Langlois, A. J., McDanal, C., Ross, K. L., Eckler, L. I., Jellis, C. L., et al. (1989). Principal neutralizing domain of the human immunodeficiency virus type 1 envelope protein. *Proceedings of the National Academy of Sciences of the United States of America, 86*(17), 6768–6772. PubMed PMID: 2771954. PubMed Central PMCID: PMC297927, Epub 1989/09/01.

Jolly, C., & Sattentau, Q. J. (2007). Human immunodeficiency virus type 1 assembly, budding, and cell-cell spread in T cells take place in tetraspanin-enriched plasma membrane domains. *Journal of Virology, 81*(15), 7873–7884. http://dx.doi.org/10.1128/JVI.01845-06. PubMed PMID: 17522207. PubMed Central PMCID: PMC1951303, Epub 2007/05/25; JVI.01845-06 [pii].

Joly, V., Jidar, K., Tatay, M., & Yeni, P. (2010). Enfuvirtide: From basic investigations to current clinical use. *Expert Opinion on Pharmacotherapy, 11*(16), 2701–2713. http://dx. doi.org/10.1517/14656566.2010.522178. PubMed PMID: 20977403.

Joseph, S. B., Arrildt, K. T., Swanstrom, A. E., Schnell, G., Lee, B., Hoxie, J. A., et al. (2014). Quantification of entry phenotypes of macrophage-tropic HIV-1 across a wide range of CD4 densities. *Journal of Virology, 88*(4), 1858–1869. http://dx.doi.org/10.1128/ JVI.02477-13. PubMed PMID: 24307580.

Kaplan, A. H., Manchester, M., & Swanstrom, R. (1994). The activity of the protease of human immunodeficiency virus type 1 is initiated at the membrane of infected cells before the release of viral proteins and is required for release to occur with maximum efficiency. *Journal of Virology, 68*(10), 6782–6786. PubMed PMID: 8084015. PubMed Central PMCID: PMC237104, Epub 1994/10/01.

Karray, S., & Zouali, M. (1997). Identification of the B cell superantigen-binding site of HIV-1 gp120. *Proceedings of the National Academy of Sciences of the United States of America, 94*(4), 1356–1360. PubMed PMID: 9037057. PubMed Central PMCID: PMC19795, Epub 1997/02/18.

Kieffer, C., Skalicky, J. J., Morita, E., De Domenico, I., Ward, D. M., Kaplan, J., et al. (2008). Two distinct modes of ESCRT-III recognition are required for VPS4 functions in lysosomal protein targeting and HIV-1 budding. *Developmental Cell, 15*(1), 62–73. http://dx.doi.org/10.1016/j.devcel.2008.05.014. PubMed PMID: 18606141. PubMed Central PMCID: PMC2586299, Epub 2008/07/09; S1534-5807(08)00239-6 [pii].

Killian, M. S., & Levy, J. A. (2011). HIV/AIDS: 30 years of progress and future challenges. *European Journal of Immunology, 41*(12), 3401–3411. http://dx.doi.org/10.1002/ eji.201142082. PubMed PMID: 22125008, Epub 2011/11/30.

Kinsey, N. E., Anderson, M. G., Unangst, T. J., Joag, S. V., Narayan, O., Zink, M. C., et al. (1996). Antigenic variation of SIV: Mutations in V4 alter the neutralization profile. *Virology, 221*(1), 14–21. http://dx.doi.org/10.1006/viro.1996.0348. PubMed PMID: 8661410, Epub 1996/07/01; S0042-6822(96)90348-7 [pii].

Klatzmann, D., Barre-Sinoussi, F., Nugeyre, M. T., Danquet, C., Vilmer, E., Griscelli, C., et al. (1984). Selective tropism of lymphadenopathy associated virus (LAV) for helper-inducer T lymphocytes. *Science, 225*(4657), 59–63. PubMed PMID: 6328660, Epub 1984/07/06.

Klatzmann, D., Champagne, E., Chamaret, S., Gruest, J., Guetard, D., Hercend, T., et al. (1984). T-lymphocyte T4 molecule behaves as the receptor for human retrovirus LAV. *Nature, 312*(5996), 767–768. PubMed PMID: 6083454, Epub 1984/12/20.

Koppensteiner, H., Banning, C., Schneider, C., Hohenberg, H., & Schindler, M. (2012). Macrophage internal HIV-1 is protected from neutralizing antibodies. *Journal of Virology, 86*(5), 2826–2836. http://dx.doi.org/10.1128/JVI.05915-11. PubMed PMID: 22205742. PubMed Central PMCID: PMC3302290.

Koppensteiner, H., Brack-Werner, R., & Schindler, M. (2012). Macrophages and their relevance in human immunodeficiency virus type I infection. *Retrovirology, 9*, 82. http://dx. doi.org/10.1186/1742-4690-9-82. PubMed PMID: 23035819. PubMed Central PMCID: PMC3484033.

Korber, B. T., Farber, R. M., Wolpert, D. H., & Lapedes, A. S. (1993). Covariation of mutations in the V3 loop of human immunodeficiency virus type 1 envelope protein: An
information theoretic analysis. *Proceedings of the National Academy of Sciences of the United
States of America*, *90*(15), 7176–7180. PubMed PMID: 8346232. PubMed Central
PMCID: PMC47099, Epub 1993/08/01.

Krementsov, D. N., Weng, J., Lambele, M., Roy, N. H., & Thali, M. (2009). Tetraspanins
regulate cell-to-cell transmission of HIV-1. *Retrovirology*, *6*, 64. http://dx.doi.org/
10.1186/1742-4690-6-64. PubMed PMID: 19602278. PubMed Central PMCID:
PMC2714829, Epub 2009/07/16; 1742-4690-6-64 [pii].

Kudoh, A., Takahama, S., Sawasaki, T., Ode, H., Yokoyama, M., Okayama, A., et al.
(2014). The phosphorylation of HIV-1 Gag by atypical protein kinase C facilitates viral
infectivity by promoting Vpr incorporation into virions. *Retrovirology*, *11*, 9. http://dx.
doi.org/10.1186/1742-4690-11-9. PubMed PMID: 24447338. PubMed Central
PMCID: PMC3905668.

Kwong, P. D., Wyatt, R., Robinson, J., Sweet, R. W., Sodroski, J., & Hendrickson, W. A.
(1998). Structure of an HIV gp120 envelope glycoprotein in complex with the CD4
receptor and a neutralizing human antibody. *Nature*, *393*(6686), 648–659. http://dx.
doi.org/10.1038/31405. PubMed PMID: 9641677, Epub 1998/06/26.

LaBonte, J., Lebbos, J., & Kirkpatrick, P. (2003). Enfuvirtide. *Nature Reviews. Drug Discovery*,
2(5), 345–346. http://dx.doi.org/10.1038/nrd1091. PubMed PMID: 12755128.

Lambotte, O., Taoufik, Y., de Goer, M. G., Wallon, C., Goujard, C., & Delfraissy, J. F.
(2000). Detection of infectious HIV in circulating monocytes from patients on prolonged highly active antiretroviral therapy. *Journal of Acquired Immune Deficiency Syndromes*, *23*(2), 114–119. PubMed PMID: 10737425.

Lapham, C. K., Ouyang, J., Chandrasekhar, B., Nguyen, N. Y., Dimitrov, D. S., &
Golding, H. (1996). Evidence for cell-surface association between fusin and the
CD4-gp120 complex in human cell lines. *Science*, *274*(5287), 602–605. PubMed PMID:
8849450, Epub 1996/10/25.

LaRosa, G. J., Davide, J. P., Weinhold, K., Waterbury, J. A., Profy, A. T., Lewis, J. A.,
et al. (1990). Conserved sequence and structural elements in the HIV-1 principal neutralizing determinant. *Science*, *249*(4971), 932–935. PubMed PMID: 2392685, Epub
1990/08/24.

Lasky, L. A., Groopman, J. E., Fennie, C. W., Benz, P. M., Capon, D. J., Dowbenko, D. J., et al.
(1986). Neutralization of the AIDS retrovirus by antibodies to a recombinant envelope glycoprotein. *Science*, *233*(4760), 209–212. PubMed PMID: 3014647, Epub 1986/07/11.

Laude, A. J., & Prior, I. A. (2004). Plasma membrane microdomains: Organization, function
and trafficking. *Molecular Membrane Biology*, *21*(3), 193–205. http://dx.doi.org/
10.1080/09687680410001700517. PubMed PMID: 15204627, Epub 2004/06/19;
UTKEY6LWED0W1J21 [pii].

Lawn, S. D., Roberts, B. D., Griffin, G. E., Folks, T. M., & Butera, S. T. (2000). Cellular
compartments of human immunodeficiency virus type 1 replication in vivo: Determination by presence of virion-associated host proteins and impact of opportunistic infection. *Journal of Virology*, *74*(1), 139–145. PubMed PMID: 10590100. PubMed Central
PMCID: PMC111522, Epub 1999/12/10.

Le Douce, V., Herbein, G., Rohr, O., & Schwartz, C. (2010). Molecular mechanisms of
HIV-1 persistence in the monocyte-macrophage lineage. *Retrovirology*, *7*, 32. http://
dx.doi.org/10.1186/1742-4690-7-32. PubMed PMID: 20380694. PubMed Central
PMCID: PMC2873506.

Lee, Y. M., Liu, B., & Yu, X. F. (1999). Formation of virus assembly intermediate complexes in
the cytoplasm by wild-type and assembly-defective mutant human immunodeficiency
virus type 1 and their association with membranes. *Journal of Virology*, *73*(7),
5654–5662. PubMed PMID: 10364315. PubMed Central PMCID: PMC112624, Epub
1999/06/11.

Lee, Y. M., & Yu, X. F. (1998). Identification and characterization of virus assembly intermediate complexes in HIV-1-infected CD4 + T cells. *Virology, 243*(1), 78–93. http://dx.
doi.org/10.1006/viro.1998.9064. PubMed PMID: 9527917, Epub 1998/04/07;
S0042-6822(98)99064-X [pii].

Leonard, C. K., Spellman, M. W., Riddle, L., Harris, R. J., Thomas, J. N., & Gregory, T. J.
(1990). Assignment of intrachain disulfide bonds and characterization of potential glycosylation sites of the type 1 recombinant human immunodeficiency virus envelope glycoprotein (gp120) expressed in Chinese hamster ovary cells. *The Journal of Biological
Chemistry, 265*(18), 10373–10382. PubMed PMID: 2355006, Epub 1990/06/25.

Levy, S., & Shoham, T. (2005). The tetraspanin web modulates immune-signalling complexes. *Nature Reviews. Immunology, 5*(2), 136–148. http://dx.doi.org/10.1038/
nri1548. PubMed PMID: 15688041, Epub 2005/02/03; nri1548 [pii].

Liao, F., Alkhatib, G., Peden, K. W., Sharma, G., Berger, E. A., & Farber, J. M. (1997).
STRL33, a novel chemokine receptor-like protein, functions as a fusion cofactor for
both macrophage-tropic and T cell line-tropic HIV-1. *The Journal of Experimental Medicine, 185*(11), 2015–2023. PubMed PMID: 9166430. PubMed Central PMCID:
PMC2196334, Epub 1997/06/02.

Lieberman-Blum, S. S., Fung, H. B., & Bandres, J. C. (2008). Maraviroc: A CCR5-receptor
antagonist for the treatment of HIV-1 infection. *Clinical Therapeutics, 30*(7), 1228–1250.
PubMed PMID: 18691983, Epub 2008/08/12; S0149-2918(08)80048-3 [pii].

Lifson, J. D., & Engleman, E. G. (1989). Role of CD4 in normal immunity and HIV infection. *Immunological Reviews, 109*, 93–117. PubMed PMID: 2475427.

Lindwasser, O. W., & Resh, M. D. (2002). Myristoylation as a target for inhibiting HIV
assembly: Unsaturated fatty acids block viral budding. *Proceedings of the National Academy
of Sciences of the United States of America, 99*(20), 13037–13042. http://dx.doi.org/
10.1073/pnas.212409999. PubMed PMID: 12244217. PubMed Central PMCID:
PMC130582, Epub 2002/09/24; 212409999 [pii].

Liu, L., Cimbro, R., Lusso, P., & Berger, E. A. (2011). Intraprotomer masking of third var-
iable loop (V3) epitopes by the first and second variable loops (V1V2) within the native
HIV-1 envelope glycoprotein trimer. *Proceedings of the National Academy of Sciences of the
United States of America, 108*(50), 20148–20153. http://dx.doi.org/10.1073/
pnas.1104840108. PubMed PMID: 22128330. PubMed Central PMCID:
PMC3250183.

MacArthur, R. D., & Novak, R. M. (2008). Reviews of anti-infective agents: Maraviroc:
The first of a new class of antiretroviral agents. *Clinical Infectious Diseases, 47*(2),
236–241. http://dx.doi.org/10.1086/589289. PubMed PMID: 18532888.

Maddon, P. J., Dalgleish, A. G., McDougal, J. S., Clapham, P. R., Weiss, R. A., & Axel, R.
(1986). The T4 gene encodes the AIDS virus receptor and is expressed in the immune
system and the brain. *Cell, 47*(3), 333–348. PubMed PMID: 3094962 0092-8674(86)
90590-8 [pii]; Epub 1986/11/07.

Marconi, V. C., Grandits, G. A., Weintrob, A. C., Chun, H., Landrum, M. L., Ganesan, A.,
et al. (2010). Outcomes of highly active antiretroviral therapy in the context of universal
access to healthcare: The U.S. Military HIV Natural History Study. *AIDS Research and
Therapy, 7*, 14. http://dx.doi.org/10.1186/1742-6405-7-14. PubMed PMID:
20507622. PubMed Central PMCID: PMC2894737.

Martin-Serrano, J., Eastman, S. W., Chung, W., & Bieniasz, P. D. (2005). HECT ubiquitin
ligases link viral and cellular PPXY motifs to the vacuolar protein-sorting pathway. *The
Journal of Cell Biology, 168*(1), 89–101. http://dx.doi.org/10.1083/jcb.200408155.
PubMed PMID: 15623582. PubMed Central PMCID: PMC2171676, Epub
2004/12/30; jcb.200408155 [pii].

Martin-Serrano, J., & Neil, S. J. (2011). Host factors involved in retroviral budding and
release. *Nature Reviews. Microbiology, 9*(7), 519–531. http://dx.doi.org/10.1038/
nrmicro2596. PubMed PMID: 21677686, Epub 2011/06/17; nrmicro2596 [pii].

Martin-Serrano, J., Yarovoy, A., Perez-Caballero, D., & Bieniasz, P. D. (2003). Divergent retroviral late-budding domains recruit vacuolar protein sorting factors by using alternative adaptor proteins. *Proceedings of the National Academy of Sciences of the United States of America, 100*(21), 12414–12419. http://dx.doi.org/10.1073/pnas.2133846100. PubMed PMID: 14519844. PubMed Central PMCID: PMC218772, Epub 2003/10/02; 2133846100 [pii].

Maxfield, F. R. (2002). Plasma membrane microdomains. *Current Opinion in Cell Biology, 14*(4), 483–487. PubMed PMID: 12383800, Epub 2002/10/18; S0955067402003514 [pii].

McLellan, J. S., Pancera, M., Carrico, C., Gorman, J., Julien, J. P., Khayat, R., et al. (2011). Structure of HIV-1 gp120 V1/V2 domain with broadly neutralizing antibody PG9. *Nature, 480*(7377), 336–343. http://dx.doi.org/10.1038/nature10696. PubMed PMID: 22113616. PubMed Central PMCID: PMC3406929.

Mink, M., Mosier, S. M., Janumpalli, S., Davison, D., Jin, L., Melby, T., et al. (2005). Impact of human immunodeficiency virus type 1 gp41 amino acid substitutions selected during enfuvirtide treatment on gp41 binding and antiviral potency of enfuvirtide in vitro. *Journal of Virology, 79*(19), 12447–12454. http://dx.doi.org/10.1128/JVI.79.19.12447-12454.2005. PubMed PMID: 16160172. PubMed Central PMCID: PMC1211558.

Moore, J. P. (1997). Coreceptors: Implications for HIV pathogenesis and therapy. *Science, 276*(5309), 51–52. PubMed PMID: 9122710, Epub 1997/04/04.

Morita, E., Sandrin, V., McCullough, J., Katsuyama, A., Baci Hamilton, I., & Sundquist, W. I. (2011). ESCRT-III protein requirements for HIV-1 budding. *Cell Host & Microbe, 9*(3), 235–242. http://dx.doi.org/10.1016/j.chom.2011.02.004. PubMed PMID: 21396898. PubMed Central PMCID: PMC3070458, Epub 2011/03/15; S1931-3128(11)00038-2 [pii].

Morita, E., & Sundquist, W. I. (2004). Retrovirus budding. *Annual Review of Cell and Developmental Biology, 20,* 395–425. http://dx.doi.org/10.1146/annurev.cellbio.20.010403.102350. PubMed PMID: 15473846, Epub 2004/10/12.

Moyle, G. J., Wildfire, A., Mandalia, S., Mayer, H., Goodrich, J., Whitcomb, J., et al. (2005). Epidemiology and predictive factors for chemokine receptor use in HIV-1 infection. *The Journal of Infectious Diseases, 191*(6), 866–872. http://dx.doi.org/10.1086/428096. PubMed PMID: 15717260, Epub 2005/02/18; JID33384 [pii].

Murakami, T., & Freed, E. O. (2000). The long cytoplasmic tail of gp41 is required in a cell type-dependent manner for HIV-1 envelope glycoprotein incorporation into virions. *Proceedings of the National Academy of Sciences of the United States of America, 97*(1), 343–348. PubMed PMID: 10618420. PubMed Central PMCID: PMC26665, Epub 2000/01/05.

Myers, G., Korber, B., Wain-Hobson, S., Smith, R. F., & Pavlakis, G. N. (1993). *Human retroviruses and AIDS 1993: A compilation and analysis of nucleic acid and amino acid sequences.* Los Alamos, NM: Theoretical Biology and Biophysics, Los Alamos National Laboratory.

Nabatov, A. A., Pollakis, G., Linnemann, T., Kliphius, A., Chalaby, M. I., & Paxton, W. A. (2004). Intrapatient alterations in the human immunodeficiency virus type 1 gp120 V1V2 and V3 regions differentially modulate coreceptor usage, virus inhibition by CC/CXC chemokines, soluble CD4, and the b12 and 2G12 monoclonal antibodies. *Journal of Virology, 78*(1), 524–530. PubMed PMID: 14671134. PubMed Central PMCID: PMC303404, Epub 2003/12/13.

Nara, P. L., Hwang, K. M., Rausch, D. M., Lifson, J. D., & Eiden, L. E. (1989). CD4 antigen-based antireceptor peptides inhibit infectivity of human immunodeficiency virus in vitro at multiple stages of the viral life cycle. *Proceedings of the National Academy of Sciences of the United States of America, 86*(18), 7139–7143. PubMed PMID: 2789382. PubMed Central PMCID: PMC298011.

Narayan, K. M., Agrawal, N., Du, S. X., Muranaka, J. E., Bauer, K., Leaman, D. P., et al. (2013). Prime-boost immunization of rabbits with HIV-1 gp120 elicits potent neutralization activity against a primary viral isolate. *PLoS One*, *8*(1), e52732. http://dx. doi.org/10.1371/journal.pone.0052732. PubMed PMID: 23326351. PubMed Central PMCID: PMC3541383.

Neil, S. J., Zang, T., & Bieniasz, P. D. (2008). Tetherin inhibits retrovirus release and is antagonized by HIV-1 Vpu. *Nature*, *451*(7177), 425–430. http://dx.doi.org/10.1038/nature06553. PubMed PMID: 18200009.

Nguyen, D. H., & Hildreth, J. E. (2000). Evidence for budding of human immunodeficiency virus type 1 selectively from glycolipid-enriched membrane lipid rafts. *Journal of Virology*, *74*(7), 3264–3272. PubMed PMID: 10708443.

Nydegger, S., Foti, M., Derdowski, A., Spearman, P., & Thali, M. (2003). HIV-1 egress is gated through late endosomal membranes. *Traffic*, *4*(12), 902–910. PubMed PMID: 14617353, Epub 2003/11/18; 145 [pii].

Nydegger, S., Khurana, S., Krementsov, D. N., Foti, M., & Thali, M. (2006). Mapping of tetraspanin-enriched microdomains that can function as gateways for HIV-1. *The Journal of Cell Biology*, *173*(5), 795–807. http://dx.doi.org/10.1083/jcb.200508165. PubMed PMID: 16735575. PubMed Central PMCID: PMC2063894, Epub 2006/06/01; jcb.200508165 [pii].

Ochsenbauer, C., Edmonds, T. G., Ding, H., Keele, B. F., Decker, J., Salazar, M. G., et al. (2012). Generation of transmitted/founder HIV-1 infectious molecular clones and characterization of their replication capacity in CD4 T lymphocytes and monocyte-derived macrophages. *Journal of Virology*, *86*(5), 2715–2728. http://dx.doi.org/10.1128/JVI.06157-11. PubMed PMID: 22190722. PubMed Central PMCID: PMC3302286.

Odorizzi, G., Cowles, C. R., & Emr, S. D. (1998). The AP-3 complex: A coat of many colours. *Trends in Cell Biology*, *8*(7), 282–288. PubMed PMID: 9714600, Epub 1998/08/26; S0962-8924(98)01295-1 [pii].

Ono, A. (2010). Relationships between plasma membrane microdomains and HIV-1 assembly. *Biology of the Cell*, *102*(6), 335–350. http://dx.doi.org/10.1042/BC20090165. PubMed PMID: 20356318. PubMed Central PMCID: PMC3056405, Epub 2010/04/02; BC20090165 [pii].

Ono, A., Huang, M., & Freed, E. O. (1997). Characterization of human immunodeficiency virus type 1 matrix revertants: Effects on virus assembly, Gag processing, and Env incorporation into virions. *Journal of Virology*, *71*(6), 4409–4418. PubMed PMID: 9151831. PubMed Central PMCID: PMC191659, Epub 1997/06/01.

Overbaugh, J., & Rudensey, L. M. (1992). Alterations in potential sites for glycosylation predominate during evolution of the simian immunodeficiency virus envelope gene in macaques. *Journal of Virology*, *66*(10), 5937–5948. PubMed PMID: 1527847. PubMed Central PMCID: PMC241471, Epub 1992/10/01.

Pan, X., Baldauf, H. M., Keppler, O. T., & Fackler, O. T. (2013). Restrictions to HIV-1 replication in resting CD4 + T lymphocytes. *Cell Research*, *23*(7), 876–885. http://dx.doi.org/10.1038/cr.2013.74. PubMed PMID: 23732522. PubMed Central PMCID: PMC3698640.

Panos, G., & Watson, D. C. (2014). Effect of HIV-1 subtype and tropism on treatment with chemokine coreceptor entry inhibitors; overview of viral entry inhibition. *Critical Reviews in Microbiology*. http://dx.doi.org/10.3109/1040841X.2013.867829 PubMed PMID: 24635642.

Parent, L. J., Bennett, R. P., Craven, R. C., Nelle, T. D., Krishna, N. K., Bowzard, J. B., et al. (1995). Positionally independent and exchangeable late budding functions of the Rous sarcoma virus and human immunodeficiency virus Gag proteins. *Journal of Virology*, *69*(9), 5455–5460. PubMed PMID: 7636991. PubMed Central PMCID: PMC189393, Epub 1995/09/01.

Park, J., & Morrow, C. D. (1992). The nonmyristylated Pr160gag-pol polyprotein of human immunodeficiency virus type 1 interacts with Pr55gag and is incorporated into viruslike particles. *Journal of Virology, 66*(11), 6304–6313. PubMed PMID: 1383561. PubMed Central PMCID: PMC240122, Epub 1992/11/01.

Parrish, N. F., Gao, F., Li, H., Giorgi, E. E., Barbian, H. J., Parrish, E. H., et al. (2013). Phenotypic properties of transmitted founder HIV-1. *Proceedings of the National Academy of Sciences of the United States of America, 110*(17), 6626–6633. http://dx.doi.org/10.1073/pnas.1304288110. PubMed PMID: 23542380. PubMed Central PMCID: PMC3637789.

Peitzsch, R. M., & McLaughlin, S. (1993). Binding of acylated peptides and fatty acids to phospholipid vesicles: Pertinence to myristoylated proteins. *Biochemistry, 32*(39), 10436–10443. PubMed PMID: 8399188, Epub 1993/10/05.

Pelchen-Matthews, A., Kramer, B., & Marsh, M. (2003). Infectious HIV-1 assembles in late endosomes in primary macrophages. *The Journal of Cell Biology, 162*(3), 443–455. http://dx.doi.org/10.1083/jcb.200304008. PubMed PMID: 12885763. PubMed Central PMCID: PMC2172706, Epub 2003/07/30; jcb.200304008 [pii].

Perelson, A. S., Neumann, A. U., Markowitz, M., Leonard, J. M., & Ho, D. D. (1996). HIV-1 dynamics in vivo: Virion clearance rate, infected cell life-span, and viral generation time. *Science, 271*(5255), 1582–1586. PubMed PMID: 8599114.

Peters, P. J., Bhattacharya, J., Hibbitts, S., Dittmar, M. T., Simmons, G., Bell, J., et al. (2004). Biological analysis of human immunodeficiency virus type 1 R5 envelopes amplified from brain and lymph node tissues of AIDS patients with neuropathology reveals two distinct tropism phenotypes and identifies envelopes in the brain that confer an enhanced tropism and fusigenicity for macrophages. *Journal of Virology, 78*(13), 6915–6926. PubMed PMID: 15194768.

Peters, P. J., Duenas-Decamp, M. J., Sullivan, W. M., & Clapham, P. R. (2007). Variation of macrophage tropism among HIV-1 R5 envelopes in brain and other tissues. *Journal of Neuroimmune Pharmacology, 2*(1), 32–41. http://dx.doi.org/10.1007/s11481-006-9042-2. PubMed PMID: 18040824.

Peters, P. J., Sullivan, W. M., Duenas-Decamp, M. J., Bhattacharya, J., Ankghuambom, C., Brown, R., et al. (2006). Non-macrophage-tropic human immunodeficiency virus type 1 R5 envelopes predominate in blood, lymph nodes, and semen: Implications for transmission and pathogenesis. *Journal of Virology, 80*(13), 6324–6332. http://dx.doi.org/10.1128/JVI.02328-05. PubMed PMID: 16775320. PubMed Central PMCID: PMC1488974.

Pfizer. Maraviroc Tablets NDA 22-128 2007 (April 24, 2007). *Antiviral drugs advisory committee (AVDAC) briefing document.* Available from http://www.fda.gov/ohrms/dockets/ac/07/briefing/2007-4283b1-01-pfizer.pdf.

Pham, Q. T., Bouchard, A., Grutter, M. G., & Berthoux, L. (2010). Generation of human TRIM5alpha mutants with high HIV-1 restriction activity. *Gene Therapy, 17*(7), 859–871. http://dx.doi.org/10.1038/gt.2010.40. PubMed PMID: 20357830.

Pickl, W. F., Pimentel-Muinos, F. X., & Seed, B. (2001). Lipid rafts and pseudotyping. *Journal of Virology, 75*(15), 7175–7183. http://dx.doi.org/10.1128/JVI.75.15.7175-7183.2001. PubMed PMID: 11435598. PubMed Central PMCID: PMC114446, Epub 2001/07/04.

Ping, L. H., Joseph, S. B., Anderson, J. A., Abrahams, M. R., Salazar-Gonzalez, J. F., Kincer, L. P., et al. (2013). Comparison of viral Env proteins from acute and chronic infections with subtype C human immunodeficiency virus type 1 identifies differences in glycosylation and CCR5 utilization and suggests a new strategy for immunogen design. *Journal of Virology, 87*(13), 7218–7233. http://dx.doi.org/10.1128/JVI.03577-12. PubMed PMID: 23616655. PubMed Central PMCID: PMC3700278.

Popov, S., Popova, E., Inoue, M., & Gottlinger, H. G. (2008). Human immunodeficiency virus type 1 Gag engages the Bro1 domain of ALIX/AIP1 through the nucleocapsid. *Journal of Virology, 82*(3), 1389–1398. http://dx.doi.org/10.1128/JVI.01912-07. PubMed PMID: 18032513. PubMed Central PMCID: PMC2224418.

Porcellini, S., Alberici, L., Gubinelli, F., Lupo, R., Olgiati, C., Rizzardi, G. P., et al. (2009). The F12-Vif derivative Chim3 inhibits HIV-1 replication in CD4 + T lymphocytes and CD34+-derived macrophages by blocking HIV-1 DNA integration. *Blood, 113*(15), 3443–3452. http://dx.doi.org/10.1182/blood-2008-06-158790. PubMed PMID: 19211937.

Puffer, B. A., Parent, L. J., Wills, J. W., & Montelaro, R. C. (1997). Equine infectious anemia virus utilizes a YXXL motif within the late assembly domain of the Gag p9 protein. *Journal of Virology, 71*(9), 6541–6546. PubMed PMID: 9261374. PubMed Central PMCID: PMC191930, Epub 1997/09/01.

Reis, M. N., de Alcantara, K. C., Cardoso, L. P., & Stefani, M. M. (2014). Polymorphisms in the HIV-1 gp41 env gene, natural resistance to enfuvirtide (T-20) and pol resistance among pregnant Brazilian women. *Journal of Medical Virology, 86*(1), 8–17. http://dx. doi.org/10.1002/jmv.23738. PubMed PMID: 24037943.

Resh, M. D. (2005). Intracellular trafficking of HIV-1 Gag: How Gag interacts with cell membranes and makes viral particles. *AIDS Reviews, 7*(2), 84–91. PubMed PMID: 16092502, Epub 2005/08/12.

Rolland, M., Edlefsen, P. T., Larsen, B. B., Tovanabutra, S., Sanders-Buell, E., Hertz, T., et al. (2012). Increased HIV-1 vaccine efficacy against viruses with genetic signatures in Env V2. *Nature, 490*(7420), 417–420. http://dx.doi.org/10.1038/nature11519. PubMed PMID: 22960785. PubMed Central PMCID: PMC3551291.

Rozera, G., Abbate, I., Bruselles, A., Vlassi, C., D'Offizi, G., Narciso, P., et al. (2009). Massively parallel pyrosequencing highlights minority variants in the HIV-1 env quasispecies deriving from lymphomonocyte sub-populations. *Retrovirology, 6*, 15. http://dx.doi.org/ 10.1186/1742-4690-6-15. PubMed PMID: 19216757. PubMed Central PMCID: PMC2660291, Epub 2009/02/17; 1742-4690-6-15 [pii].

Rubbert, A., Combadiere, C., Ostrowski, M., Arthos, J., Dybul, M., Machado, E., et al. (1998). Dendritic cells express multiple chemokine receptors used as coreceptors for HIV entry. *Journal of Immunology, 160*(8), 3933–3941. PubMed PMID: 9558100.

Rucker, J., Edinger, A. L., Sharron, M., Samson, M., Lee, B., Berson, J. F., et al. (1997). Utilization of chemokine receptors, orphan receptors, and herpesvirus-encoded receptors by diverse human and simian immunodeficiency viruses. *Journal of Virology, 71*(12), 8999–9007. PubMed PMID: 9371556. PubMed Central PMCID: PMC230200, Epub 1997/11/26.

Saarloos, M. N., Sullivan, B. L., Czerniewski, M. A., Parameswar, K. D., & Spear, G. T. (1997). Detection of HLA-DR associated with monocytotropic, primary, and plasma isolates of human immunodeficiency virus type 1. *Journal of Virology, 71*(2), 1640–1643. PubMed PMID: 8995692. PubMed Central PMCID: PMC191223, Epub 1997/02/01.

Sacktor, N., Lyles, R. H., Skolasky, R., Kleeberger, C., Selnes, O. A., Miller, E. N., et al. (2001). HIV-associated neurologic disease incidence changes: Multicenter AIDS Cohort Study, 1990–1998. *Neurology, 56*(2), 257–260. PubMed PMID: 11160967, Epub 2001/02/13.

Sagar, M., Wu, X., Lee, S., & Overbaugh, J. (2006). Human immunodeficiency virus type 1 V1-V2 envelope loop sequences expand and add glycosylation sites over the course of infection, and these modifications affect antibody neutralization sensitivity. *Journal of Virology, 80*(19), 9586–9598. http://dx.doi.org/10.1128/JVI.00141-06. PubMed PMID: 16973562. PubMed Central PMCID: PMC1617272.

Samson, M., Edinger, A. L., Stordeur, P., Rucker, J., Verhasselt, V., Sharron, M., et al. (1998). ChemR23, a putative chemoattractant receptor, is expressed in monocyte-derived dendritic cells and macrophages and is a coreceptor for SIV and some primary HIV-1 strains. *European Journal of Immunology, 28*(5), 1689–1700. http://dx.doi.org/10.1002/(SICI)1521-4141(199805)28:05<1689::AID-IMMU1689>3.0.CO;2-I. PubMed PMID: 9603476.

Sanders, R. W., van Anken, E., Nabatov, A. A., Liscaljet, I. M., Bontjer, I., Eggink, D., et al. (2008). The carbohydrate at asparagine 386 on HIV-1 gp120 is not essential for protein folding and function but is involved in immune evasion. *Retrovirology*, *5*, 10. http://dx.doi.org/10.1186/1742-4690-5-10. PubMed PMID: 18237398. PubMed Central PMCID: PMC2262092, Epub 2008/02/02; 1742-4690-5-10 [pii].

Sato, K., Aoki, J., Misawa, N., Daikoku, E., Sano, K., Tanaka, Y., et al. (2008). Modulation of human immunodeficiency virus type 1 infectivity through incorporation of tetraspanin proteins. *Journal of Virology*, *82*(2), 1021–1033. http://dx.doi.org/10.1128/JVI.01044-07. PubMed PMID: 17989173. PubMed Central PMCID: PMC2224585, Epub 2007/11/09; JVI.01044-07 [pii].

Sattentau, Q. J., & Moore, J. P. (1991). Conformational changes induced in the human immunodeficiency virus envelope glycoprotein by soluble CD4 binding. *The Journal of Experimental Medicine*, *174*(2), 407–415. PubMed PMID: 1713252. PubMed Central PMCID: PMC2118908, Epub 1991/08/01.

Sattentau, Q. J., & Weiss, R. A. (1988). The CD4 antigen: Physiological ligand and HIV receptor. *Cell*, *52*(5), 631–633. PubMed PMID: 2830988, Epub 1988/03/11; 0092-8674(88)90397-2 [pii].

Saunders, C. J., McCaffrey, R. A., Zharkikh, I., Kraft, Z., Malenbaum, S. E., Burke, B., et al. (2005). The V1, V2, and V3 regions of the human immunodeficiency virus type 1 envelope differentially affect the viral phenotype in an isolate-dependent manner. *Journal of Virology*, *79*(14), 9069–9080. http://dx.doi.org/10.1128/JVI.79.14.9069-9080.2005. PubMed PMID: 15994801. PubMed Central PMCID: PMC1168758, Epub 2005/07/05; 79/14/9069 [pii].

Schindler, M., Rajan, D., Banning, C., Wimmer, P., Koppensteiner, H., Iwanski, A., et al. (2010). Vpu serine 52 dependent counteraction of tetherin is required for HIV-1 replication in macrophages, but not in ex vivo human lymphoid tissue. *Retrovirology*, *7*, 1. http://dx.doi.org/10.1186/1742-4690-7-1. PubMed PMID: 20078884. PubMed Central PMCID: PMC2823648.

Schrier, R. D., McCutchan, J. A., Venable, J. C., Nelson, J. A., & Wiley, C. A. (1990). T-cell-induced expression of human immunodeficiency virus in macrophages. *Journal of Virology*, *64*(7), 3280–3288. PubMed PMID: 2112615. PubMed Central PMCID: PMC249555, Epub 1990/07/01.

Schwartz, M. D., Geraghty, R. J., & Panganiban, A. T. (1996). HIV-1 particle release mediated by Vpu is distinct from that mediated by p6. *Virology*, *224*(1), 302–309. http://dx.doi.org/10.1006/viro.1996.0532. PubMed PMID: 8862425.

Shankarappa, R., Margolick, J. B., Gange, S. J., Rodrigo, A. G., Upchurch, D., Farzadegan, H., et al. (1999). Consistent viral evolutionary changes associated with the progression of human immunodeficiency virus type 1 infection. *Journal of Virology*, *73*(12), 10489–10502. PubMed PMID: 10559367. PubMed Central PMCID: PMC113104.

Sharon, M., Kessler, N., Levy, R., Zolla-Pazner, S., Gorlach, M., & Anglister, J. (2003). Alternative conformations of HIV-1 V3 loops mimic beta hairpins in chemokines, suggesting a mechanism for coreceptor selectivity. *Structure*, *11*(2), 225–236. PubMed PMID: 12575942.

Simon-Loriere, E., Rossolillo, P., & Negroni, M. (2011). RNA structures, genomic organization and selection of recombinant HIV. *RNA Biology*, *8*(2), 280–286. PubMed PMID: 21422815, 15193 [pii]; Epub 2011/03/23.

Sirois, S., Sing, T., & Chou, K. C. (2005). HIV-1 gp120 V3 loop for structure-based drug design. *Current Protein & Peptide Science*, *6*(5), 413–422. PubMed PMID: 16248793, Epub 2005/10/27.

Solbak, S. M., Reksten, T. R., Hahn, F., Wray, V., Henklein, P., Henklein, P., et al. (2013). HIV-1 p6—A structured to flexible multifunctional membrane-interacting protein. *Biochimica et Biophysica Acta*, *1828*(2), 816–823. http://dx.doi.org/10.1016/j.bbamem.2012.11.010. PubMed PMID: 23174350.

Stamatatos, L., & Cheng-Mayer, C. (1998). An envelope modification that renders a primary, neutralization-resistant clade B human immunodeficiency virus type 1 isolate highly susceptible to neutralization by sera from other clades. *Journal of Virology, 72*(10), 7840–7845. PubMed PMID: 9733820. PubMed Central PMCID: PMC110102, Epub 1998/09/12.

Stamatatos, L., Wiskerchen, M., & Cheng-Mayer, C. (1998). Effect of major deletions in the V1 and V2 loops of a macrophage-tropic HIV type 1 isolate on viral envelope structure, cell entry, and replication. *AIDS Research and Human Retroviruses, 14*(13), 1129–1139. PubMed PMID: 9737584, Epub 1998/09/16.

Stein, B. S., Gowda, S. D., Lifson, J. D., Penhallow, R. C., Bensch, K. G., & Engleman, E. G. (1987). pH-Independent HIV entry into CD4-positive T cells via virus envelope fusion to the plasma membrane. *Cell, 49*(5), 659–668. PubMed PMID: 3107838, Epub 1987/06/05; 0092-8674(87)90542-3 [pii].

Stevenson, M. (2003). HIV-1 pathogenesis. *Nature Medicine, 9*(7), 853–860. http://dx.doi.org/10.1038/nm0703-853. PubMed PMID: 12835705.

Strack, B., Calistri, A., Craig, S., Popova, E., & Gottlinger, H. G. (2003). AIP1/ALIX is a binding partner for HIV-1 p6 and EIAV p9 functioning in virus budding. *Cell, 114*(6), 689–699. PubMed PMID: 14505569.

Strazza, M., Pirrone, V., Wigdahl, B., & Nonnemacher, M. R. (2011). Breaking down the barrier: The effects of HIV-1 on the blood-brain barrier. *Brain Research, 1399*, 96–115. http://dx.doi.org/10.1016/j.brainres.2011.05.015. PubMed PMID: 21641584. PubMed Central PMCID: PMC3139430.

Strizki, J. M., Albright, A. V., Sheng, H., O'Connor, M., Perrin, L., & Gonzalez-Scarano, F. (1996). Infection of primary human microglia and monocyte-derived macrophages with human immunodeficiency virus type 1 isolates: Evidence of differential tropism. *Journal of Virology, 70*(11), 7654–7662. PubMed PMID: 8892885. PubMed Central PMCID: PMC190834.

Swingler, S., Mann, A. M., Zhou, J., Swingler, C., & Stevenson, M. (2007). Apoptotic killing of HIV-1-infected macrophages is subverted by the viral envelope glycoprotein. *PLoS Pathogens, 3*(9), 1281–1290. http://dx.doi.org/10.1371/journal.ppat.0030134. PubMed PMID: 17907802. PubMed Central PMCID: PMC2323301.

Tachibana, I., & Hemler, M. E. (1999). Role of transmembrane 4 superfamily (TM4SF) proteins CD9 and CD81 in muscle cell fusion and myotube maintenance. *The Journal of Cell Biology, 146*(4), 893–904. PubMed PMID: 10459022. PubMed Central PMCID: PMC2156130, Epub 1999/08/25.

Thielen, A., Sichtig, N., Kaiser, R., Lam, J., Harrigan, P. R., & Lengauer, T. (2010). Improved prediction of HIV-1 coreceptor usage with sequence information from the second hypervariable loop of gp120. *The Journal of Infectious Diseases, 202*(9), 1435–1443. http://dx.doi.org/10.1086/656600. PubMed PMID: 20874088, Epub 2010/09/30.

Torres, J. V., Malley, A., Banapour, B., Anderson, D. E., Axthelm, M. K., Gardner, M. B., et al. (1993). An epitope on the surface envelope glycoprotein (gp130) of simian immunodeficiency virus (SIVmac) involved in viral neutralization and T cell activation. *AIDS Research and Human Retroviruses, 9*(5), 423–430. PubMed PMID: 7686386, Epub 1993/05/01.

Tremblay, M. J., Fortin, J. F., & Cantin, R. (1998). The acquisition of host-encoded proteins by nascent HIV-1. *Immunology Today, 19*(8), 346–351. PubMed PMID: 9709501, Epub 1998/08/26; S0167569998012869 [pii].

Usami, Y., Popov, S., & Gottlinger, H. G. (2007). Potent rescue of human immunodeficiency virus type 1 late domain mutants by ALIX/AIP1 depends on its CHMP4 binding site. *Journal of Virology, 81*(12), 6614–6622. http://dx.doi.org/10.1128/JVI.00314-07.

PubMed PMID: 17428861. PubMed Central PMCID: PMC1900090, Epub 2007/04/13; JVI.00314-07 [pii].

van't Wout, A. B., de Jong, M. D., Kootstra, N. A., Veenstra, J., Lange, J. M., Boucher, C. A., et al. (1996). Changes in cellular virus load and zidovudine resistance of syncytium-inducing and non-syncytium-inducing human immunodeficiency virus populations under zidovudine pressure: A clonal analysis. *The Journal of Infectious Diseases*, *174*(4), 845–849. PubMed PMID: 8843227, Epub 1996/10/01.

von Schwedler, U. K., Stuchell, M., Muller, B., Ward, D. M., Chung, H. Y., Morita, E., et al. (2003). The protein network of HIV budding. *Cell*, *114*(6), 701–713. PubMed PMID: 14505570, Epub 2003/09/25; S0092867403007141 [pii].

Wan, Y., Liu, L., Wu, L., Huang, X., Ma, L., & Xu, J. (2009). Deglycosylation or partial removal of HIV-1 CN54 gp140 V1/V2 domain enhances env-specific T cells. *AIDS Research and Human Retroviruses*, *25*(6), 607–617. http://dx.doi.org/10.1089/aid.2008.0289. PubMed PMID: 19500018.

Wang, W. K., Essex, M., & Lee, T. H. (1995). The highly conserved aspartic acid residue between hypervariable regions 1 and 2 of human immunodeficiency virus type 1 gp120 is important for early stages of virus replication. *Journal of Virology*, *69*(1), 538–542. PubMed PMID: 7983752. PubMed Central PMCID: PMC188606, Epub 1995/01/01.

Wang, W., Nie, J., Prochnow, C., Truong, C., Jia, Z., Wang, S., et al. (2013). A systematic study of the N-glycosylation sites of HIV-1 envelope protein on infectivity and antibody-mediated neutralization. *Retrovirology*, *10*(1), 14. http://dx.doi.org/10.1186/1742-4690-10-14. PubMed PMID: 23384254.

Wang, S. F., Tsao, C. H., Lin, Y. T., Hsu, D. K., Chiang, M. L., Lo, C. H., et al. (2014). Galectin-3 promotes HIV-1 budding via association with Alix and Gag p6. *Glycobiology*, *24*(11), 1022–1035. http://dx.doi.org/10.1093/glycob/cwu064. PubMed PMID: 24996823. PubMed Central PMCID: PMC4181451.

Wang, T., Xu, Y., Zhu, H., Andrus, T., Ivanov, S. B., Pan, C., et al. (2013). Successful isolation of infectious and high titer human monocyte-derived HIV-1 from two subjects with discontinued therapy. *PLoS One*, *8*(5), e65071. http://dx.doi.org/10.1371/journal.pone.0065071. PubMed PMID: 23741458. PubMed Central PMCID: PMC3669022.

Wei, X., Decker, J. M., Wang, S., Hui, H., Kappes, J. C., Wu, X., et al. (2003). Antibody neutralization and escape by HIV-1. *Nature*, *422*(6929), 307–312. http://dx.doi.org/10.1038/nature01470. PubMed PMID: 12646921.

Weiss, E. R., & Gottlinger, H. (2011). The role of cellular factors in promoting HIV budding. *Journal of Molecular Biology*, *410*(4), 525–533. http://dx.doi.org/10.1016/j.jmb.2011.04.055. PubMed PMID: 21762798. PubMed Central PMCID: PMC3139153.

Weng, J., Krementsov, D. N., Khurana, S., Roy, N. H., & Thali, M. (2009). Formation of syncytia is repressed by tetraspanins in human immunodeficiency virus type 1-producing cells. *Journal of Virology*, *83*(15), 7467–7474. http://dx.doi.org/10.1128/JVI.00163-09. PubMed PMID: 19458002. PubMed Central PMCID: PMC2708618, Epub 2009/05/22; JVI.00163-09 [pii].

Westby, M., Lewis, M., Whitcomb, J., Youle, M., Pozniak, A. L., James, I. T., et al. (2006). Emergence of CXCR4-using human immunodeficiency virus type 1 (HIV-1) variants in a minority of HIV-1-infected patients following treatment with the CCR5 antagonist maraviroc is from a pretreatment CXCR4-using virus reservoir. *Journal of Virology*, *80*(10), 4909–4920. http://dx.doi.org/10.1128/JVI.80.10.4909-4920.2006. PubMed PMID: 16641282. PubMed Central PMCID: PMC1472081.

Wightman, F., Solomon, A., Khoury, G., Green, J. A., Gray, L., Gorry, P. R., et al. (2010). Both CD31(+) and CD31(-) naive CD4(+) T cells are persistent HIV type 1-infected reservoirs in individuals receiving antiretroviral therapy. *The Journal of Infectious Diseases*, *202*(11), 1738–1748. http://dx.doi.org/10.1086/656721. PubMed PMID: 20979453.

Williams, B. G., Lima, V., & Gouws, E. (2011). Modelling the impact of antiretroviral therapy on the epidemic of HIV. *Current HIV Research*, *9*(6), 367–382. PubMed PMID: 21999772. PubMed Central PMCID: PMC3529404.

Wyatt, R., Moore, J., Accola, M., Desjardin, E., Robinson, J., & Sodroski, J. (1995). Involvement of the V1/V2 variable loop structure in the exposure of human immunodeficiency virus type 1 gp120 epitopes induced by receptor binding. *Journal of Virology*, *69*(9), 5723–5733. PubMed PMID: 7543586. PubMed Central PMCID: PMC189432, Epub 1995/09/01.

Wyatt, R., Sullivan, N., Thali, M., Repke, H., Ho, D., Robinson, J., et al. (1993). Functional and immunologic characterization of human immunodeficiency virus type 1 envelope glycoproteins containing deletions of the major variable regions. *Journal of Virology*, *67*(8), 4557–4565. PubMed PMID: 8331723. PubMed Central PMCID: PMC237840, Epub 1993/08/01.

Xiao, L., Owen, S. M., Goldman, I., Lal, A. A., deJong, J. J., Goudsmit, J., et al. (1998). CCR5 coreceptor usage of non-syncytium-inducing primary HIV-1 is independent of phylogenetically distinct global HIV-1 isolates: Delineation of consensus motif in the V3 domain that predicts CCR-5 usage. *Virology*, *240*(1), 83–92. http://dx.doi.org/10.1006/viro.1997.8924. PubMed PMID: 9448692, Epub 1998/02/04; S0042-6822(97)98924-8 [pii].

Xu, L., Pozniak, A., Wildfire, A., Stanfield-Oakley, S. A., Mosier, S. M., Ratcliffe, D., et al. (2005). Emergence and evolution of enfuvirtide resistance following long-term therapy involves heptad repeat 2 mutations within gp41. *Antimicrobial Agents and Chemotherapy*, *49*(3), 1113–1119. http://dx.doi.org/10.1128/AAC.49.3.1113-1119.2005. PubMed PMID: 15728911. PubMed Central PMCID: PMC549241.

Yuan, X., Yu, X., Lee, T. H., & Essex, M. (1993). Mutations in the N-terminal region of human immunodeficiency virus type 1 matrix protein block intracellular transport of the Gag precursor. *Journal of Virology*, *67*(11), 6387–6394. PubMed PMID: 8411340. PubMed Central PMCID: PMC238073, Epub 1993/11/01.

Zamarchi, R., Allavena, P., Borsetti, A., Stievano, L., Tosello, V., Marcato, N., et al. (2002). Expression and functional activity of CXCR-4 and CCR-5 chemokine receptors in human thymocytes. *Clinical and Experimental Immunology*, *127*(2), 321–330. PubMed PMID: 11876757. PubMed Central PMCID: PMC1906330.

Zerhouni, B., Nelson, J. A., & Saha, K. (2004). Isolation of CD4-independent primary human immunodeficiency virus type 1 isolates that are syncytium inducing and acutely cytopathic for CD8+ lymphocytes. *Journal of Virology*, *78*(3), 1243–1255. PubMed PMID: 14722279. PubMed Central PMCID: PMC321385.

Zhang, L., Carruthers, C. D., He, T., Huang, Y., Cao, Y., Wang, G., et al. (1997). HIV type 1 subtypes, coreceptor usage, and CCR5 polymorphism. *AIDS Research and Human Retroviruses*, *13*(16), 1357–1366. PubMed PMID: 9359654.

Zhang, L., He, T., Huang, Y., Chen, Z., Guo, Y., Wu, S., et al. (1998). Chemokine coreceptor usage by diverse primary isolates of human immunodeficiency virus type 1. *Journal of Virology*, *72*(11), 9307–9312. PubMed PMID: 9765480. PubMed Central PMCID: PMC110352, Epub 1998/10/10.

Zhou, W., Parent, L. J., Wills, J. W., & Resh, M. D. (1994). Identification of a membrane-binding domain within the amino-terminal region of human immunodeficiency virus type 1 Gag protein which interacts with acidic phospholipids. *Journal of Virology*, *68*(4), 2556–2569. PubMed PMID: 8139035. PubMed Central PMCID: PMC236733, Epub 1994/04/01.

Zhu, T. (2000). HIV-1 genotypes in peripheral blood monocytes. *Journal of Leukocyte Biology*, *68*(3), 338–344. PubMed PMID: 10985249.

Zhu, T. (2002). HIV-1 in peripheral blood monocytes: An underrated viral source. *The Journal of Antimicrobial Chemotherapy*, *50*(3), 309–311. PubMed PMID: 12205054.

Zimmerman, P. A., Buckler-White, A., Alkhatib, G., Spalding, T., Kubofcik, J., Combadiere, C., et al. (1997). Inherited resistance to HIV-1 conferred by an inactivating mutation in CC chemokine receptor 5: Studies in populations with contrasting clinical phenotypes, defined racial background, and quantified risk. *Molecular Medicine*, *3*(1), 23–36. PubMed PMID: 9132277.

Zouali, M. (1995). B-cell superantigens: Implications for selection of the human antibody repertoire. *Immunology Today*, *16*(8), 399–405. PubMed PMID: 7546197, Epub 1995/08/01; 0167569995800093 [pii].

INDEX

Note: Page numbers followed by "*f*" indicate figures and "*t*" indicate tables.

P

Pathogen-associated molecular patterns (PAMPs), 74–75

Pestivirus
- adaptive immune response, elimination of, 116–117
- approved/tentative species, 51–54, 52*f*
- basic features of, 48–49, 49*f*
- in bats, 53
- border disease, 48, 59
- BVDV infection (*see* Bovine viral diarrhea virus (BVDV))
- clovenhoofed animals, 53
- CSFV (*see* Classical swine fever virus (CSFV))
- cytopathic and noncytopathic viruses, 53
- detergent treatment, 53
- diaplacental infection, 54–55
- E proteins, 53
- fetal infection, 54–55
- *Flaviviridae* family, 48–49
- genome
  - CP1 and NCP1 strains, 102–104, 103*f*
  - cp CSFV mutants, attenuated phenotype, 113–114
  - CP7, NS2 protease, 108–109
  - DIs, autonomous RNA replication, 111–113
  - human hepatitis C virus, 60–61, 61*f*
  - Jiv, NS2 protease, 110–111
  - noncp BVDV-2, genome duplication in, 114
  - 3′-nontranslated regions, 66–68, 67*f*
  - 5′-nontranslated regions, 61–66, 62*f*
  - NS2, cINS/Jiv/DNAJ-C14 insertions, 106–108
  - NS3, insertions upstream of, 104–106
  - NS3, viral sequences upstream of, 106
  - open reading frame (ORF) coding, 60–61
  - RNA recombination, 114–116
  - single-stranded RNA, 53, 60
  - strain-specific insertions in, 101–102
- innate immune responses, inhibition of
  - IFN-1 response, 118–121
  - viral RNA replication, control of, 117–118
- intrauterine infection, 54–55
- livestock farming, losses in, 50
- NFAR protein, 121–123
- nonstructural proteins
  - NS2, 68–69, 91–92
  - NS2–3, 68–69, 92–94
  - NS3, 68–69, 94–96
  - NS4A, 68–69, 96–97
  - NS5A, 68–69, 98–99
  - NS4B, 68–69, 97–98
  - NS5B, 68–69, 99–100
  - p7, 68–69, 90–91
- N^{pro}, 68–69
  - 23 and 14 kDa product, 69–70
  - C53, 69–70
  - C-terminal substrate site, 71
  - Cys69, 69–71, 73
  - Cys168/Ser169 site, 70
  - exaltation of ND virus, 76–77
  - Glu22 and His49, 69–71, 73
  - IFN-1 response, 74–77
  - IRF-3 degradation, 74–76
  - ND virus replication, 76–77
  - PAMP receptors, 74–75
  - structure of, 70–71, 72*f*
  - virus/host interactions, 74
- persistently infected (PI) animals, 54–55
- in rats, 53
- reintroduction, risk of, 50
- in soybean cyst nematode, 53
- structural proteins
  - core protein C, 68–69, 78
  - E1, 68–69, 83–85
  - E2, 68–69, 85–90
  - E^{rns}, 68–69, 78–83, 79*f*
- survival strategies, 50
- transmission, 54
- vaccination
  - DIVA vaccines, 128–129
  - killed vaccines, 123–124, 128
  - live attenuated vaccines, 123–128
  - replicons, 128

Phosphonoacetic acid (PAA), 184–185

Pilchard herpesvirus 1, 172

Plasmacytoid dendritic cells (PDCs), 77, 83

Poplar mosaic virus (PopMV), 27–28

Potato virus X (PVX), 27–28

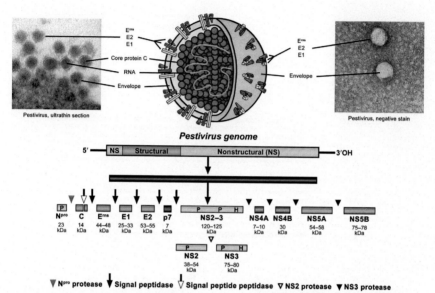

Norbert Tautz *et al.*, Figure 1 Basic features of pestiviruses. The figure shows a diagram of a pestiviral particle flanked by electron microscopic pictures (BVDV in ultrathin section on the left; CSFV in negative stain on the right). Below, the viral genome is shown in a schematic representation with the single long ORF encoding a polyprotein indicated below. Processing of the polyprotein by the proteases specified at the bottom leads to the shown viral proteins. P, protease domain; H, helicase. Electron microscopy: Harald Granzow and Frank Weiland, Friedrich-Loeffler-Institut; graphic design: Mandy Jörn, Friedrich-Loeffler-Institut. Details shown in the figure are addressed in the text.

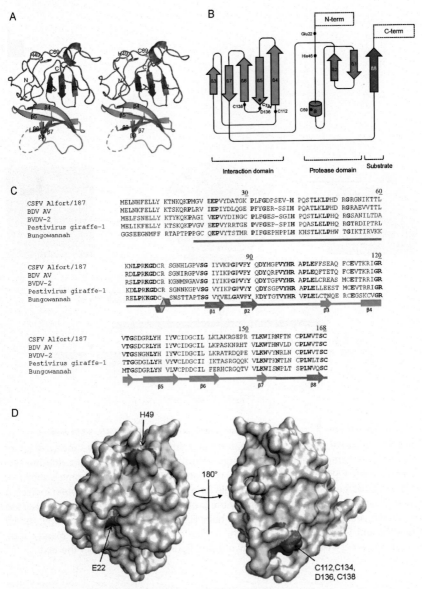

Norbert Tautz et al., Figure 6 Structure of N^pro. (A) Ribbon representation as stereoview of a C168A N^pro mutant. The protease domain is shown in the upper part and the zinc-binding domain in the lower part. Disordered residues (residues 145–149) are indicated by a dashed line. Secondary structural elements, as well as the N- and C-termini, are labeled. (B) Schematic presentation of the arrangement of secondary structure elements of N^pro that contribute to the zinc-binding interaction domain (left), and the protease domain with the C-terminal substrate region (right). The β-sheets are shown as arrows, the α-helix with the active site residue as cylinder. The location

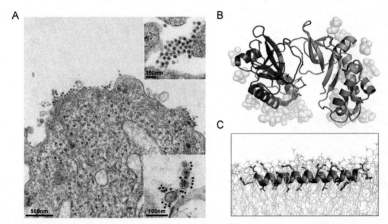

Norbert Tautz *et al.*, Figure 7 E^{rns} association with viral particles and structure. (A) Electron micrograph of a cell infected with BVDV. The viral particles are clearly visible on the cell surface. The inset in the upper right corner shows the viral particles at a higher magnification. The inset in the lower right corner shows CSFV particles labeled with anti-E^{rns} antibodies and gold particles which clearly demonstrates the presence of E^{rns} on the viral particle (electron microscopy: Harald Granzow and Frank Weiland, Friedrich-Loeffler Institut). (B) Structure of the E^{rns} ectodomain as a dimer as originally published (Krey et al., 2012), with each monomer in different shades. The transparent spheres show the glycosylation of E^{rns}, and disulfide bonds are shown as thin sticks. Image was rendered with PyMOL using the structure deposited in the PDB database (4DVK) (PyMOL: The PyMOL Molecular Graphics System, Version 1.7.4 Schrödinger, LLC). (C) An MD simulation of the E^{rns} membrane anchor taken from Aberle et al. (2014).

and number of important amino acids are given. (C) Alignment of pestivirus N^{pro} sequences. The alignment includes the CSFV strain Alfort 187, Border Disease virus (BDV) strain AV, BVDV-2, Pestivirus giraffe-1, and the Bungowannah virus (GenBank Accession Numbers X87939.1, ABV54604.1, AAV69983.1, NP_777520.1, and DQ901403.1, respectively). Amino-acid numbering corresponds to the CSFV Alfort 187 sequence. The secondary structure elements are indicated below the sequence. α-Helix and β-strands are shown as coil and arrows, respectively. (D) The spatial distribution of the residues involved in N^{pro}-mediated proteasomal degradation of IRF3. The residues that are essential for the N^{pro}-mediated IRF3 degradation (dark gray) are mapped onto the surface of N^{pro}. The residues localize to two protein surfaces, one in each domain. The protease domain surface cluster (left) includes residues Glu22 and His49, and the zinc-binding domain cluster (right) is formed by the residues in the TRASH motif. *Panels (A), (C), and (D) are from Gottipati et al. (2013) with kind permission of Kyung H. Choi, Department of Biochemistry and Molecular Biology, Sealy Center for Structural Biology and Molecular Biophysics, The University of Texas Medical Branch at Galveston, Galveston, Texas, USA.*

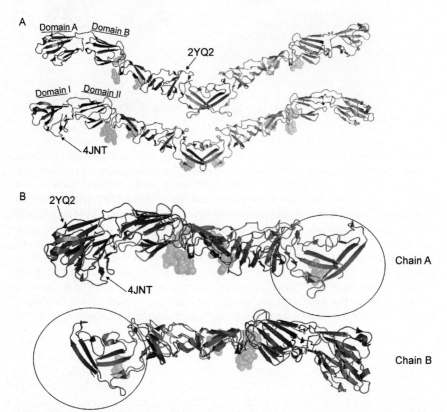

Norbert Tautz _et al.,_ Figure 8 Structure of E2. (A) Structure of the BVDV E2 dimer at pH 8 (2YQ2 (El Omari et al., 2013)) and at pH 5.5 (4JNT (Li et al., 2013)), with both monomers in different colors. Glycans in the structures are shown as transparent spheres (the glycosylation at the base of the dimer is not present in the 2YQ2 structure as it was mutated in the crystallized protein). Indicated are domain A or I and domain B or II. (B) Monomers of the above structures superimposed. Note the domain swap in structure 2YQ2, not present in 4JNT (enclosed in the circle). Images were rendered with PyMOL using the structures deposited in the PDB database (PyMOL: The PyMOL Molecular Graphics System, Version 1.7.4 Schrödinger, LLC).

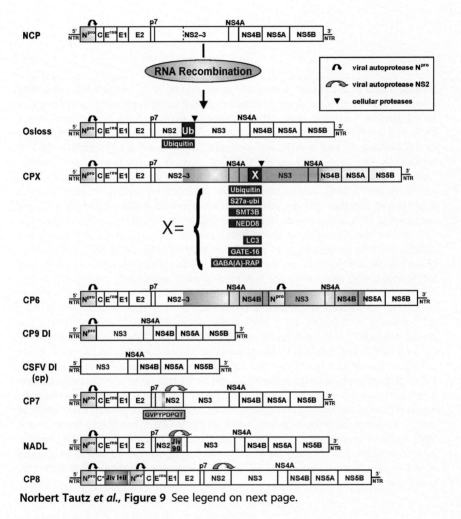

Norbert Tautz *et al.,* **Figure 9** See legend on next page.

Norbert Tautz *et al.*, **Figure 9** Genome organizations of cp pestiviruses. Schematic representation of genome organizations of autonomously replicating cp pestiviruses or subgenomic pestiviral RNAs. Those genomes evolved from noncp pestiviruses (NCP; top) by RNA recombination. The ORF including the encoded proteins as well as the flanking 5'- and 3'-NTRs are indicated. The N-terminal autoprotease N^{pro} is highlighted. The genome of cp BVDV Osloss contains an insertion of an ubiquitin (Ub)-coding sequence at the NS2/NS3 gene border. CPX represents cp BVDV genomes which contain a large duplication of viral sequences encoding at least NS2–4A (highlighted) together with an insertion of host-derived mRNA sequences (X) right upstream of NS3. X represents one of the sequences listed in the bars below the CPX genome: polyubiquitin, ubiquitin fusion protein S27a-ubi, ubiquitin-like proteins (SMT3B, NEDD8), or proteins with a ubiquitin-like fold (LC3, GATE-16, GABA(A)-RAP). These cell-derived proteins are substrates to cellular proteases which generate the N-terminus of NS3 (processing site indicated by black arrowhead). Analogous to CPX, the genome of CP6 encompasses a large duplication of viral sequences (highlighted). Instead of cell-derived sequences, a duplication of N^{pro} resides upstream of NS3; N^{pro} cleavage is indicated by a black arrow. In analogy, N^{pro} also generates the N-terminus of NS3 in the polyprotein of DI9. CSFV DI, a cp subgenome, encodes only the start methionine followed by the genomic sequence encoding NS3 to NS5B. BVDV CP7 encompasses a 27-base insertion (highlighted) central in NS2. This insertion is critical for high efficient cleavage at the NS2/3 site by the NS2 cysteine protease (indicated by a large gray arrow). Alternatively, the NS2 protease can be activated by various fragments of the cellular Jiv-mRNA (highlighted) located in the NS2 gene of BVDV NADL or within the structural protein-coding region of BVDV CP8. While the insertion in the NADL genome represents a minimal Jiv fragment, the insertion in the CP8 genome is more complex and contains besides two Jiv fragments also additional sequences as well as a duplications of core and N^{pro} which facilitate the generation of an authentic core protein.

Printed in the United States
By Bookmasters